Contractor's Pricing Guide: Residential Repair & Remodeling Costs

with RSMeans data

2022

Thomas Lane, Senior Editor

Vice President, Data
Chris Gaudreau

Vice President, Product
Ted Kail

Principal Engineer
Bob Mewis

Engineers: Architectural Divisions
1, 3, 4, 5, 6, 7, 8, 9, 10, 11, 12, 13, 41
Matthew Doheny, Manager
Sam Babbitt
Sergeleen Edouard
Richard Goldin
Scott Keller
Thomas Lane
Giles Munyard

Engineers: Civil Divisions and Wages
2, 31, 32, 33, 34, 35, 44, 46
Derrick Hale, PE, Manager
Christopher Babbitt
Stephen Bell
Michael Lynch
Elisa Mello
David Yazbek

Engineers: Mechanical, Electrical, Plumbing & Conveying Divisions
14, 21, 22, 23, 25, 26, 27, 28, 48
Joseph Kelble, Manager
Brian Adams
Michelle Curran
Antonio D'Aulerio

Thomas Lyons
Jake MacDonald

Contributing Engineers
John Melin, PE, Manager
Paul Cowan
Barry Hutchinson
Gerard Lafond, PE
Matthew Sorrentino

Technical Project Lead
Kevin Souza

Production
Debra Panarelli, Manager
Jonathan Forgit
Edward Green
Sharon Larsen
Sheryl Rose
Janice Thalin

Data Quality
Euricka Thomas, Manager
David Byars
Audrey Considine
Ellen D'Amico

Cover Design
Blaire Collins

Innovation
Chris Gaudreau, Vice President
Kedar Gaikwad
Todd Glowac
Srini Narla
Joseph Woughter

D1614165

GƆRDIAN®

Contractor's Pricing Guide:
Residential Repair & Remodeling Costs

with RSMeans data

CARLSBAD CITY LIBRARY
CARLSBAD CA 92011
DISCARD

2022

692
CoN
2022

RSMeans data from Gordian
Construction Publishers & Consultants
30 Patewood Dr., Suite 350
Greenville, SC 29615
United States of America
1.800.448.8182
RSMeans.com

Copyright 2021 by The Gordian Group Inc.
All rights reserved.
Cover photo © iStock.com/quackersnaps

Printed in the United States of America
ISSN 1542-9156
ISBN 978-1-955341-19-6

Gordian's authors, editors, and engineers apply diligence and judgment in locating and using reliable sources for the information published. However, Gordian makes no express or implied warranty or guarantee in connection with the content of the information contained herein, including the accuracy, correctness, value, sufficiency, or completeness of the data, methods, and other information contained herein. Gordian makes no express or implied warranty of merchantability or fitness for a particular purpose. Gordian shall have no liability to any customer or third party for any loss, expense, or damage, including consequential, incidental, special, or punitive damage, including lost profits or lost revenue, caused directly or indirectly by any error or omission, or arising out of, or in connection with, the information contained herein. For the purposes of this paragraph, "Gordian" shall include The Gordian Group, Inc., and its divisions, subsidiaries, successors, parent companies, and their employees, partners, principals, agents and representatives, and any third-party providers or sources of information or data. Gordian grants the purchaser of this publication limited license to use the cost data contained herein for purchaser's internal business purposes in connection with construction estimating and related work. The publication, and all cost data contained herein, may not be reproduced, integrated into any software or computer program, developed into a database or other electronic compilation, stored in an information storage or retrieval system, or transmitted or distributed to anyone in any form or by any means, electronic or mechanical, including photocopying or scanning, without prior written permission of Gordian. This publication is subject to protection under copyright law, trade secret law and other intellectual property laws of the United States, Canada and other jurisdictions. Gordian and its affiliates exclusively own and retain all rights, title and interest in and to this publication and the cost data contained herein including, without limitation, all copyright, patent, trademark and trade secret rights. Except for the limited license contained herein, your purchase of this publication does not grant you any intellectual property rights in the publication or the cost data.

AUG 1 9 2022

0342 $77.00 per copy (in United States)
Price is subject to change without prior notice.

Table of Contents

Foreword

The Value of RSMeans data from Gordian

Since 1942, RSMeans data has been the industry-standard materials, labor, and equipment cost information database for contractors, facility owners and managers, architects, engineers, and anyone else that requires the latest localized construction cost information. More than 75 years later, the objective remains the same: to provide facility and construction professionals with the most current and comprehensive construction cost database possible.

With the constant influx of new construction methods and materials, in addition to ever-changing labor and material costs, last year's cost data is not reliable for today's designs, estimates, or budgets. Gordian's cost engineers apply real-world construction experience to identify and quantify new building products and methodologies, adjust productivity rates, and adjust costs to local market conditions across the nation. This adds up to more than 22,000 hours in cost research annually. This unparalleled construction cost expertise is why so many facility and construction professionals rely on RSMeans data year over year.

About Gordian

Gordian is the leading provider of facility and construction cost data, software, and services for all phases of the building lifecycle. From planning to design, procurement, construction, and operations, Gordian delivers groundbreaking solutions to contractors, architects, engineers, business and financial officers, and facility owners in local, state, and federal governments, education, healthcare, and other industries. A pioneer of Job Order Contracting (JOC), Gordian's offerings also include our proprietary RSMeans data and Facility Intelligence solutions. We develop and maintain the largest collection of labor, material, and equipment data and associated costs for all areas of construction. Gordian's solutions are accessed through our innovative software platforms and supported by a team of industry experts proven to help clients maximize efficiency, optimize cost savings, and increase building quality.

Our Commitment

At Gordian, we do more than talk about the quality of our data and the usefulness of its application. We stand behind all of our RSMeans data—from historical cost indexes to construction materials and techniques—to craft current costs and predict future trends. If you have any questions about our products or services, please call us toll-free at 800.448.8182 or visit our website at gordian.com.

How the Cost Data Is Built: An Overview

A Powerful Construction Tool

You have in your hands one of the most powerful construction tools available today. A successful project is built on the foundation of an accurate and dependable estimate. This data will enable you to construct such an estimate.

For the casual user the data is designed to be:

- quickly and easily understood so you can get right to your estimate.
- filled with valuable information so you can understand the necessary factors that go into the cost estimate.

For the regular user, the data is designed to be:

- a handy reference that can be quickly referred to for key costs.
- a comprehensive, fully reliable source of current construction costs, so you'll be prepared to estimate any project.
- a source for project costs, product selections, and alternate materials and methods.

To meet all of these requirements we have organized the data into the following clearly defined sections.

Unit Price Section

The cost data has been divided into 23 sections representing the order of construction. Within each section is a listing of components and variations of those components that are applicable to that section. Costs are shown for the various actions that must be done to that component.

Reference Section

This section includes information on Location Factors and a listing of abbreviations.

Location Factors

You can adjust total project costs to over 900 locations throughout the U.S. and Canada by using the data in this section.

Abbreviations

A listing of the abbreviations used throughout, along with the terms they represent, is included.

Index

A comprehensive listing of all terms and subjects will help you quickly find what you need when you are not sure where it falls in the order of construction.

The Scope of this Data

This data is designed to be as comprehensive and easy to use as possible. To that end we have made certain assumptions and limited its scope in two key ways:

1. We have established material prices based on a national average.
2. We have computed labor costs based on residential wage rates.

Project Size

This data is intended for use by those involved primarily in residential repair and remodeling construction costing between $10,000 and $100,000.

With reasonable exercise of judgment, the figures can be used for any building work. For other types of projects, such as new home construction or commercial buildings, consult the appropriate RSMeans data set for more information.

How to Use the Cost Data: The Details

What's Behind the Numbers? The Development of Cost Data

RSMeans data engineers continually monitor developments in the construction industry in order to ensure reliable, thorough, and up-to-date cost information. While overall construction costs may vary relative to general economic conditions, price fluctuations within the industry are dependent upon many factors. Individual price variations may, in fact, be opposite to overall economic trends. Therefore, costs are constantly tracked and complete updates are performed yearly. Also, new items are frequently added in response to changes in materials and methods.

Costs in U.S. Dollars

All costs represent U.S. national averages and are given in U.S. dollars. The City Cost Index (CCI) with RSMeans data or Location Factors can be used to adjust costs to a particular location. The CCI or Location Factors for Canada can be used to adjust U.S. national averages to local costs in Canadian dollars. No exchange rate conversion is necessary because it has already been factored in.

Material Costs

RSMeans data engineers contact manufacturers, dealers, distributors, and contractors all across the U.S. and Canada to determine national average material costs. If you have access to current material costs for your specific location, you may wish to make adjustments to reflect differences from the national average. Included within material costs are fasteners for a normal installation. RSMeans data engineers use manufacturers' recommendations, written specifications, and/or standard construction practices for the sizing and spacing of fasteners. Adjustments to material costs may be required for your specific application or location. The manufacturer's warranty is assumed. Extended warranties are not included in the material costs. **Material costs do not include sales tax.**

Labor Costs

Labor costs are based upon a mathematical average of trade-specific wages in 30 major U.S. cities. The type of wage (union, open shop, or residential) is identified on the inside back cover of printed publications or selected by the estimator when using the electronic products. Markups for the wages can also be found on the inside back cover of printed publications and/or under the labor references found in the electronic products.

- If wage rates in your area vary from those used, or if rate increases are expected within a given year, labor costs should be adjusted accordingly.

Labor costs reflect productivity based on actual working conditions. In addition to actual installation, these figures include time spent during a normal weekday on tasks, such as material receiving and handling, mobilization at the site, site movement, breaks, and cleanup.

Productivity data is developed over an extended period so as not to be influenced by abnormal variations and reflects a typical average.

Equipment Costs

Equipment costs include not only rental but also operating costs for equipment under normal use. The operating costs include parts and labor for routine servicing, such as the repair and replacement of pumps, filters, and worn lines. Normal operating expendables, such as fuel, lubricants, tires, and electricity (where applicable), are also included. Extraordinary operating expendables with highly variable wear patterns, such as diamond bits and blades, are excluded. These costs are included under materials. Equipment rental rates are obtained from industry sources throughout North America—contractors, suppliers, dealers, manufacturers, and distributors.

Rental rates can also be treated as reimbursement costs for contractor-owned equipment. Owned equipment costs include depreciation, loan payments, interest, taxes, insurance, storage, and major repairs.

Equipment costs do not include operators' wages.

Factors Affecting Costs

Costs can vary depending upon a number of variables. Here's a listing of some factors that affect costs and points to consider.

Quality—The prices for materials and the workmanship upon which productivity is based represent sound construction work. They are also in line with industry standards and manufacturer specifications and are frequently used by federal, state, and local governments.

Overtime—We have made no allowance for overtime. If you anticipate premium time or work beyond normal working hours, be sure to make an appropriate adjustment to your labor costs.

Productivity—The productivity, daily output, and labor-hour figures for each line item are based on an eight-hour work day in daylight hours in moderate temperatures and up to a 14' working height unless otherwise indicated. For work that extends beyond normal work hours or is performed under adverse conditions, productivity may decrease.

Size of Project—The size, scope of work, and type of construction project will have a significant impact on cost. Economies of scale can reduce costs for large projects. Unit costs can often run higher for small projects.

Location—Material prices are for metropolitan areas. However, in dense urban areas, traffic and site storage limitations may increase costs. Beyond a 20-mile radius of metropolitan areas, extra trucking or transportation charges may also increase the material costs slightly. On the other hand, lower wage rates may be in effect. Be sure to consider both of these factors when preparing an estimate, particularly if the job site is located in a central city or remote rural location. In addition, highly specialized subcontract items may require travel and per-diem expenses for mechanics.

Other Factors—

- season of year
- contractor management
- weather conditions
- local union restrictions
- building code requirements
- availability of:
  - adequate energy
  - skilled labor
  - building materials
- owner's special requirements/restrictions
- safety requirements
- environmental considerations
- access

Unpredictable Factors—General business conditions influence "in-place" costs of all items. Substitute materials and construction methods may have to be employed. These may affect the installed cost and/or life cycle costs. Such factors may be difficult to evaluate and cannot necessarily be predicted on the basis of the job's location in a particular section of the country. Thus, where these factors apply, you may find significant but unavoidable cost variations for which you will have to apply a measure of judgment to your estimate.

Final Checklist

Estimating can be a straightforward process provided you remember the basics. Here's a checklist of some of the steps you should remember to complete before finalizing your estimate.

Did you remember to:

- factor in the City Cost Index or Location Factors for your locale?
- take into consideration which items have been marked up and by how much?

- mark up the entire estimate sufficiently for your purposes?
- read the background information on techniques and technical matters that could impact your project time span and cost?
- include all components of your project in the final estimate?
- double check your figures for accuracy?
- call RSMeans data engineers if you have any questions about your estimate or the data you've used? Remember, Gordian stands behind all of our products, including our extensive RSMeans data solutions. If you have any questions about your estimate, about the costs you've used from our data, or even about the technical aspects of the job that may affect your estimate, feel free to call the Gordian RSMeans editors at 1.800.448.8182.

Unit Price Section

Table of Contents

Table of Contents (cont.)

RSMeans data: Unit Prices—How They Work

The prices in this publication include all overhead and profit mark-ups for the installing contractor. If a general contractor is involved, an additional 10% should be added to the cost shown in the "Total" column.

The costs for material, labor, and any equipment are shown here. Material costs include delivery and a 10% mark-up. No sales taxes are included. Labor costs include all mark-ups as shown on the inside back cover of this data set. Equipment costs also include a 10% mark-up.

The cost in the Total column should be multiplied by the quantity of each item to determine the total cost for that item. The sum of all items' total costs becomes the project's total cost.

The data set is arranged into 23 categories that follow the order of construction for residential projects. The category is shown as white letters on a black background at the top of each page.

A brief specification of the cost item is shown here.

Graphics representing some of the cost items are found at the beginning of each category.

The cost data is arranged in a tree structure format. Major cost sections are left justified and shown as bold headings within a grey background. Cost items and sub-cost items are found under the bold headings. The complete description should read as "bold heading, cost item, sub-cost item."

For example, the identified item's description is "Beam/Girder, Solid Wood Beam, 2" x 8"."

Each action performed to the cost item is conveniently listed to the right of the description. See the explanation below for the scope of work included as part of the action.

The unit of measure upon which all the costs are based is shown here. An abbreviations list is included within the Reference Section found at the back of this data set.

Rough Frame / Structure

Rough Framing

Beam / Girder	Unit	Material	Labor	Equip.	Total	Specification
Solid Wood Beam						
2" x 8"						
Demolish	L.F.		1.01		1.01	Cost includes material and labor to install 2" x 8" beam.
Install	L.F.	1.57	1.49		3.06	
Demolish and Install	L.F.	1.57	2.50		4.07	
Clean	L.F.	.06	.33		.39	
Paint	L.F.	.33	2.18		2.51	
Minimum Charge	Job		242		242	
2" x 10"						
Demolish	L.F.		1.25		1.25	Cost includes material and labor to install 2" x 10" beam.
Install	L.F.	2.19	1.61		3.80	
Demolish and Install	L.F.	2.19	2.86		5.05	
Clean	L.F.	.07	.40		.47	
Paint	L.F.	.21	1.35		1.56	
Minimum Charge	Job		242		242	
2" x 12"						
Demolish	L.F.		1.51		1.51	Cost includes material and labor to install 2" x 12" beam.
Install	L.F.	3.05	1.76		4.81	
Demolish and Install	L.F.	3.05	3.27		6.32	
Clean	L.F.	.07	.47		.54	
Paint	L.F.	.50	3.23		3.73	
Minimum Charge	Job		242		242	
4" x 8"						
Demolish	L.F.		3.09		3.09	Cost includes material and labor to install 4" x 8" beam.
Install	L.F.	6.15	2.25	2.18	10.58	
Demolish and Install	L.F.	6.15	5.34	2.18	13.67	
Clean	L.F.	.07	.40		.47	
Paint	L.F.	.21	1.35		1.56	
Minimum Charge	Job		242		242	
4" x 10"						
Demolish	L.F.		3.87		3.87	Cost includes material and labor to install 4" x 10" beam.
Install	L.F.	7.75	2.37	2.30	12.42	
Demolish and Install	L.F.	7.75	6.24	2.30	16.29	
Clean	L.F.	.07	.47		.54	
Paint	L.F.	.23	1.57		1.80	
Minimum Charge	Job		242		242	
4" x 12"						
Demolish	L.F.		4.64		4.64	Cost includes material and labor to install 4" x 12" beam.
Install	L.F.	9.50	2.50	2.42	14.42	
Demolish and Install	L.F.	9.50	7.14	2.42	19.06	
Clean	L.F.	.08	.53		.61	
Paint	L.F.	.50	3.23		3.73	
Minimum Charge	Job		242		242	

Action	Scope
Demolish	Includes removing the item and hauling the debris to a truck, dumpster, or storage area.
Install	Includes the installation of the item with normal placement and spacing of fasteners.
Demolish and Install	Includes the demolish action and the install action.
Reinstall	Includes the reinstallation of a removed item.
Clean	Includes cleaning of the item.
Paint	Includes up to two coats of finish on the item (unless otherwise indicated).
Minimum Charge	Includes the minimum labor or the minimum labor and equipment cost. Use this charge for small quantities of material.

Job Costs

| Scaffolding | Temporary Electrical Service | Air Compressor |

Construction Fees		Unit	Material	Labor	Equip.	Total	Specification
Asbestos Test	Minimum Charge	Job		290		290	Minimum charge for testing.
Architectural	Minimum Charge	Day	2500			2500	Minimum labor charge for architect.
Engineering	Minimum Charge	Day	2150			2150	Minimum labor charge for field engineer.

Construction Permits		Unit	Material	Labor	Equip.	Total	Specification
Job (less than 5K)	Minimum Charge	Job	65			65	Permit fees, job less than 5K.
Job (5K to 10K)	Minimum Charge	Job	158			158	Permit fees, job more than 5K, but less than 10K.
Job (10K to 25K)	Minimum Charge	Job	305			305	Permit fees, job more than 10K, but less than 25K.
Job (25K to 50K)	Minimum Charge	Job	590			590	Permit fees, job more than 25K, but less than 50K.
Job (50K to 100K)	Minimum Charge	Job	1100			1100	Permit fees, job more than 50K, but less than 100K.
Job (100K to 250K)	Minimum Charge	Job	2275			2275	Permit fees, job more than 100K, but less than 250K.
Job (250K to 500K)	Minimum Charge	Job	4400			4400	Permit fees, job more than 250K, but less than 500K.
Job (500K to 1M)	Minimum Charge	Job	7975			7975	Permit fees, job more than 500K, but less than 1 million.

1

Trade Labor		Unit	Material	Labor	Equip.	Total	Specification
Carpenter							
Daily (8 hours)							
	Install	Day		485		485	Daily labor rate for one carpenter.
	Minimum Charge	Job		242		242	
Weekly (40 hours)							
	Install	Week		2425		2425	Weekly labor rate for one carpenter.
	Minimum Charge	Job		242		242	
Drywaller							
Daily (8 hours)							
	Install	Day		485		485	Daily labor rate for one carpenter.
	Minimum Charge	Job		242		242	
Weekly (40 hours)							
	Install	Week		2425		2425	Weekly labor rate for one carpenter.
	Minimum Charge	Job		242		242	
Roofer							
Daily (8 hours)							
	Install	Day		455		455	Daily labor rate for one skilled roofer.
	Minimum Charge	Job		227		227	
Weekly (40 hours)							
	Install	Week		2275		2275	Weekly labor rate for one skilled
	Minimum Charge	Job		227		227	roofer.
Painter							
Daily (8 hours)							
	Install	Day		405		405	Daily labor rate for one painter.
	Minimum Charge	Job		202		202	
Weekly (40 hours)							
	Install	Week		2025		2025	Weekly labor rate for one painter.
	Minimum Charge	Job		202		202	
Mason							
Daily (8 hours)							
	Install	Day		480		480	Daily labor rate for one skilled mason.
	Minimum Charge	Job		239		239	
Weekly (40 hours)							
	Install	Week		2400		2400	Weekly labor rate for one skilled
	Minimum Charge	Job		239		239	mason.
Electrician							
Daily (8 hours)							
	Install	Day		570		570	Daily labor rate for an electrician.
	Minimum Charge	Job		284		284	
Weekly (40 hours)							
	Install	Week		2850		2850	Weekly labor rate for an electrician.
	Minimum Charge	Job		284		284	
Plumber							
Daily (8 hours)							
	Install	Day		545		545	Daily labor rate for a plumber.
	Minimum Charge	Job		273		273	
Weekly (40 hours)							
	Install	Week		2725		2725	Weekly labor rate for a plumber.
	Minimum Charge	Job		273		273	
Common Laborer							
Daily (8 hours)							
	Install	Day		375		375	Daily labor rate for a common
	Minimum Charge	Job		187		187	laborer.

Job Costs

Trade Labor		Unit	Material	Labor	Equip.	Total	Specification
Weekly (40 hours)		Week		1875		1875	Weekly labor rate for a common
	Install						laborer.
	Minimum Charge	Job		187		187	

Temporary Utilities		Unit	Material	Labor	Equip.	Total	Specification
Power Pole							
Job Site Electricity							
	Install	Month	65.50			65.50	Includes average cost of electricity per
	Minimum Charge	Job		284		284	month.
Water Service							
Job Site Water							
	Install	Month	87			87	Includes average costs for water for
	Minimum Charge	Month	87			87	one month.
Telephone							
	Install	Month	97.50			97.50	Includes average cost for local and
	Minimum Charge	Month	97.50			97.50	long distance telephone service for
							one month.
Chemical Toilet							
	Install	Month			345	345	Monthly rental of portable toilet.
	Minimum Charge	Day			17.20	17.20	
Office Trailer							
Place and Remove							
	Install	Job	415			415	Includes cost to a contractor to deliver,
	Minimum Charge	Job		229		229	place and level an office trailer and
							remove at end of job.
Storage Container							
Place and Remove							
	Install	Job	238			238	Includes cost to a contractor to deliver
	Minimum Charge	Job		229		229	and place a storage container and
							remove at end of job.
Fencing							
	Install	L.F.	3.44			3.44	Includes material costs for temporary
	Minimum Charge	Job		187		187	fencing.
Railing							
	Install	L.F.	20.50	12.80	.92	34.22	Includes material and labor to install
	Minimum Charge	Job		187		187	primed steel pipe.

Equipment Rental		Unit	Material	Labor	Equip.	Total	Specification
Flatbed Truck							
Daily Rental							
	Install	Day			375	375	Daily rental of flatbed truck.
	Minimum Charge	Job			375	375	
Weekly Rental							
	Install	Week			1875	1875	Weekly rental of flatbed truck.
	Minimum Charge	Job			375	375	
Monthly Rental							
	Install	Month			7475	7475	Monthly rental of flatbed truck.
	Minimum Charge	Job			375	375	

Equipment Rental		Unit	Material	Labor	Equip.	Total	Specification
Dump Truck							
Daily Rental							
	Install	Day			430	430	Daily rental of dump truck.
	Minimum Charge	Day			430	430	
Weekly Rental							
	Install	Week			2150	2150	Weekly rental of dump truck.
	Minimum Charge	Day			430	430	
Monthly Rental							
	Install	Month			8625	8625	Monthly rental of dump truck.
	Minimum Charge	Day			430	430	
Forklift							
Daily Rental							
	Install	Day			445	445	Daily rental of forklift.
	Minimum Charge	Day			445	445	
Weekly Rental							
	Install	Week			2225	2225	Weekly rental of forklift.
	Minimum Charge	Day			445	445	
Monthly Rental							
	Install	Month			8875	8875	Monthly rental of forklift.
	Minimum Charge	Day			445	445	
Bobcat							
Daily Rental							
	Install	Day			289	289	Daily rental of wheeled, skid steer
	Minimum Charge	Day			289	289	loader.
Weekly Rental							
	Install	Week			1450	1450	Weekly rental of wheeled, skid steer
	Minimum Charge	Day			289	289	loader.
Monthly Rental							
	Install	Month			5800	5800	Monthly rental of wheeled, skid steer
	Minimum Charge	Day			289	289	loader.
Air Compressor							
Daily Rental							
	Install	Day			174	174	Daily rental of air compressor.
	Minimum Charge	Day			174	174	
Weekly Rental							
	Install	Week			870	870	Weekly rental of air compressor.
	Minimum Charge	Day			174	174	
Monthly Rental							
	Install	Month			3475	3475	Monthly rental of air compressor.
	Minimum Charge	Day			174	174	
Jackhammer							
Daily Rental							
	Install	Day			41.50	41.50	Daily rental of air tool, jackhammer.
	Minimum Charge	Day			41.50	41.50	
Weekly Rental							
	Install	Week			207	207	Weekly rental of air tool, jackhammer.
	Minimum Charge	Day			41.50	41.50	
Monthly Rental							
	Install	Month			825	825	Monthly rental of air tool,
	Minimum Charge	Day			41.50	41.50	jackhammer.
Concrete Mixer							
Daily Rental							
	Install	Day			125	125	Daily rental of concrete mixer.
	Minimum Charge	Day			125	125	

Equipment Rental		Unit	Material	Labor	Equip.	Total	Specification
Weekly Rental							
	Install	Week			625	625	Weekly rental of concrete mixer.
	Minimum Charge	Day			125	125	
Monthly Rental							
	Install	Month			2500	2500	Monthly rental of concrete mixer.
	Minimum Charge	Day			125	125	
Concrete Bucket							
Daily Rental							
	Install	Day			43	43	Daily rental of concrete bucket.
	Minimum Charge	Day			43	43	
Weekly Rental							
	Install	Week			216	216	Weekly rental of concrete bucket.
	Minimum Charge	Day			43	43	
Monthly Rental							
	Install	Month			860	860	Monthly rental of concrete bucket.
	Minimum Charge	Day			43	43	
Concrete Pump							
Daily Rental							
	Install	Day			830	830	Daily rental of concrete pump.
	Minimum Charge	Day			830	830	
Weekly Rental							
	Install	Week			4150	4150	Weekly rental of concrete pump.
	Minimum Charge	Day			830	830	
Monthly Rental							
	Install	Month			16600	16600	Monthly rental of concrete pump.
	Minimum Charge	Day			830	830	
Generator							
Daily Rental							
	Install	Day			60.50	60.50	Daily rental of generator.
	Minimum Charge	Day			60.50	60.50	
Weekly Rental							
	Install	Week			305	305	Weekly rental of generator.
	Minimum Charge	Day			60.50	60.50	
Monthly Rental							
	Install	Month			1225	1225	Monthly rental of generator.
	Minimum Charge	Day			60.50	60.50	
Welder							
Daily Rental							
	Install	Day			52.50	52.50	Daily rental of welder.
	Minimum Charge	Day			52.50	52.50	
Weekly Rental							
	Install	Week			262	262	Weekly rental of welder.
	Minimum Charge	Day			52.50	52.50	
Monthly Rental							
	Install	Month			1050	1050	Monthly rental of welder.
	Minimum Charge	Day			52.50	52.50	
Sandblaster							
Daily Rental							
	Install	Day			92.50	92.50	Daily rental of sandblaster.
	Minimum Charge	Day			92.50	92.50	
Weekly Rental							
	Install	Week			460	460	Weekly rental of sandblaster.
	Minimum Charge	Day			92.50	92.50	

Job Costs

Equipment Rental

Equipment Rental		Unit	Material	Labor	Equip.	Total	Specification
Monthly Rental							
	Install	Month			1850	1850	Monthly rental of sandblaster.
	Minimum Charge	Day			92.50	92.50	
Space Heater							
Daily Rental							
	Install	Day			68.50	68.50	Daily rental of space heater.
	Minimum Charge	Day			45	45	
Weekly Rental							
	Install	Day			340	340	Weekly rental of space heater.
	Minimum Charge	Day			45	45	
Monthly Rental							
	Install	Day			1375	1375	Monthly rental of space heater.
	Minimum Charge	Day			45	45	

Move / Reset Contents

Move / Reset Contents		Unit	Material	Labor	Equip.	Total	Specification
Small Room							
	Install	Room		45.50		45.50	Includes cost to remove and reset the
	Minimum Charge	Job		187		187	contents of a small room.
Average Room							
	Install	Room		55		55	Includes cost to remove and reset the
	Minimum Charge	Job		187		187	contents of an average room.
Large Room							
	Install	Room		75.50		75.50	Includes cost to remove and reset the
	Minimum Charge	Job		187		187	contents of a large room.
Extra Large Room							
	Install	Room		120		120	Includes cost to remove and reset the
	Minimum Charge	Job		187		187	contents of an extra large room.
Store Contents							
Weekly							
	Install	Week	310			310	Storage fee, per week.
Monthly							
	Install	Month	935			935	Storage fee, per month.

Cover and Protect

Cover and Protect		Unit	Material	Labor	Equip.	Total	Specification
Cover / Protect Walls							
	Install	SF Wall		.14		.14	Includes labor costs to cover and
	Minimum Charge	Job		187		187	protect walls.
Cover / Protect Floors							
	Install	SF Flr.		.14		.14	Includes labor costs to cover and
	Minimum Charge	Job		187		187	protect floors.

Temporary Bracing

Temporary Bracing		Unit	Material	Labor	Equip.	Total	Specification
Shim Support Piers							
	Install	Ea.	5.15	144		149.15	Includes labor and material to install
	Minimum Charge	Job		430		430	shims support for piers.

For customer support on your Contractor's Pricing Guide: Residential Repair & Remodeling, call 800.448.8182.

Job Costs

Jack and Relevel

	Unit	Material	Labor	Equip.	Total	Specification
Single Story						
Install	S.F.	1.79	1.73		3.52	Includes labor and equipment to jack
Minimum Charge	Job		430		430	and relevel single-story building.
Two-story						
Install	S.F.	1.79	2.88		4.67	Includes labor and equipment to jack
Minimum Charge	Job		430		430	and relevel two-story building.

Dumpster Rental

	Unit	Material	Labor	Equip.	Total	Specification
Small Load						
Install	Week			645	645	Dumpster, weekly rental, 2 dumps, 5
Minimum Charge	Week			645	645	yard container.
Medium Load						
Install	Week			735	735	Dumpster, weekly rental, 2 dumps, 10
Minimum Charge	Week			735	735	yard container.
Large Load						
Install	Week			1125	1125	Dumpster, weekly rental, 2 dumps, 30
Minimum Charge	Week			1125	1125	yard container.

Debris Hauling

	Unit	Material	Labor	Equip.	Total	Specification
Per Ton						
Install	Ton		119	62.50	181.50	Includes hauling of debris by trailer or
Minimum Charge	Ton		890	470	1360	dump truck.
Per Cubic Yard						
Install	C.Y.		39.50	21	60.50	Includes hauling of debris by trailer or
Minimum Charge	Job		890	470	1360	dump truck.
Per Pick-up Truck Load						
Install	Ea.		208		208	Includes hauling of debris by pick-up
Minimum Charge	Ea.		208		208	truck.

Dump Fees

	Unit	Material	Labor	Equip.	Total	Specification
Per Cubic Yard						
Install	C.Y.	59.50			59.50	Dump charges, building materials.
Minimum Charge	C.Y.	59.50			59.50	
Per Ton						
Install	Ton	142			142	Dump charges, building materials.
Minimum Charge	Ton	142			142	

Scaffolding

	Unit	Material	Labor	Equip.	Total	Specification
4' to 6' High						
Install	Month	153			153	Includes monthly rental of scaffolding,
Minimum Charge	Month	153			153	steel tubular, 30" wide, 7' long, 5' high.
7' to 11' High						
Install	Month	191			191	Includes monthly rental of scaffolding,
Minimum Charge	Month	191			191	steel tubular, 30" wide, 7' long, 10' high.

For customer support on your Contractor's Pricing Guide: Residential Repair & Remodeling, call 800.448.8182.

Job Costs

Scaffolding

Scaffolding		Unit	Material	Labor	Equip.	Total	Specification
12' to 16' High							
	Install	Month	229			229	Includes monthly rental of scaffolding, steel tubular, 30" wide, 7' long, 15' high.
	Minimum Charge	Month	229			229	
17' to 21' High							
	Install	Month	325			325	Includes monthly rental of scaffolding, steel tubular, 30" wide, 7' long, 20' high.
	Minimum Charge	Month	325			325	
22' to 26' High							
	Install	Month	365			365	Includes monthly rental of scaffolding, steel tubular, 30" wide, 7' long, 25' high.
	Minimum Charge	Month	365			365	
27' to 30' High							
	Install	Month	400			400	Includes monthly rental of scaffolding, steel tubular, 30" wide, 7' long, 30' high.
	Minimum Charge	Month	400			400	

Construction Clean-up

Construction Clean-up		Unit	Material	Labor	Equip.	Total	Specification
Initial							
	Clean	S.F.		.47		.47	Includes labor for general clean-up.
	Minimum Charge	Job		187		187	
Progressive							
	Clean	S.F.		.23		.23	Includes labor for progressive site/job clean-up on a weekly basis.
	Minimum Charge	Job		187		187	

8

Building Demolition		Unit	Material	Labor	Equip.	Total	Specification
Wood-framed Building							
1 Story Demolition							
	Demolish	S.F.		1.81	2.10	3.91	Includes demolition of 1 floor only light wood-framed building, including hauling of debris. Dump fees not included.
	Minimum Charge	Job		2850	3325	6175	
2nd Story Demolition							
	Demolish	S.F.		1.84	2.13	3.97	Includes demolition of 2nd floor only light wood-framed building, including hauling of debris. Dump fees not included.
	Minimum Charge	Job		2850	3325	6175	
3rd Story Demolition							
	Demolish	S.F.		2.39	2.78	5.17	Includes demolition of 3rd floor only light wood-framed building, including hauling of debris. Dump fees not included.
	Minimum Charge	Job		2850	3325	6175	
Masonry Building							
	Demolish	S.F.		2.57	2.99	5.56	Includes demolition of one story masonry building, including hauling of debris. Dump fees not included.
	Minimum Charge	Job		2850	3325	6175	
Concrete Building							
Unreinforced							
	Demolish	S.F.		5.20	1.45	6.65	Includes demolition of a one story non-reinforced concrete building, including hauling of debris. Dump fees not included.
	Minimum Charge	Job		2850	3325	6175	
Reinforced							
	Demolish	S.F.		5.95	1.66	7.61	Includes demolition of a one story reinforced concrete building, including hauling of debris. Dump fees not included.
	Minimum Charge	Job		2850	3325	6175	
Steel Building							
	Demolish	S.F.		3.43	3.98	7.41	Includes demolition of a metal building, including hauling of debris. Dump fees not included. No salvage value assumed for scrap metal.
	Minimum Charge	Job		2850	3325	6175	

Job Preparation

General Clean-up		Unit	Material	Labor	Equip.	Total	Specification
Remove Debris							
	Clean	S.F.		1.20	.37	1.57	Includes removal of debris caused by the loss including contents items and loose debris.
	Minimum Charge	Job		600	187	787	
Muck-out							
	Clean	S.F.		.73		.73	Includes labor to remove 2" of muck and mud after a flood loss.
	Minimum Charge	Job		187		187	
Deodorize / Disinfect							
	Clean	SF Flr.		.25		.25	Includes deodorizing building with application of mildicide and disinfectant to floors and walls - will vary depending on the extent of deodorization required.
	Minimum Charge	Job		187		187	
Clean Walls							
Heavy							
	Clean	S.F.		.16		.16	Includes labor and material for heavy cleaning (multiple applications) with detergent and solvent.
	Minimum Charge	Job		187		187	
Clean Ceiling							
Light							
	Clean	S.F.		.27		.27	Includes labor and material to clean light to moderate smoke and soot.
	Minimum Charge	Job		187		187	
Heavy							
	Clean	S.F.		.40		.40	Includes labor and material to clean heavy smoke and soot.
	Minimum Charge	Job		187		187	

Flood Clean-up		Unit	Material	Labor	Equip.	Total	Specification
Emergency Service Call							
After Hours / Weekend							
	Minimum Charge	Job		249		249	Includes minimum charges for deflooding clean-up after hours or on weekends.
Grey Water / Sewage							
	Minimum Charge	Job		187		187	Includes minimum charges for deflooding clean-up.
Water Extraction							
Clean (non-grey)							
	Clean	S.F.	.03	.75		.78	Includes labor to remove clean uncontaminated water from the loss.
	Minimum Charge	Job		187		187	
Grey (non-solids)							
	Clean	S.F.	.07	1.50		1.57	Includes labor to remove non-solid grey water from the loss.
	Minimum Charge	Job		187		187	
Sewage / Solids							
	Clean	SF Flr.	.10	1.87		1.97	Includes labor to remove water from the loss that may contain sewage or solids.
	Minimum Charge	Job		187		187	

For customer support on your Contractor's Pricing Guide: Residential Repair & Remodeling, call 800.448.8182.

Flood Clean-up		Unit	Material	Labor	Equip.	Total	Specification
Mildicide Walls							
Topographical							
	Clean	SF Wall		.34		.34	Clean-up of deflooding including
	Minimum Charge	Job		187		187	topographical mildicide application.
Injection Treatments							
	Clean	SF Flr.	47.50	5.75		53.25	Clean-up of deflooding including
	Minimum Charge	Job		187		187	mildicide injection.
Carpet Cleaning							
Uncontaminated							
	Clean	SF Flr.		.23		.23	Includes labor to clean carpet.
	Minimum Charge	Job		187		187	
Contaminated							
	Clean	SF Flr.		.37		.37	Includes labor to clean contaminated
	Minimum Charge	Job		187		187	carpet.
Stairway							
	Clean	Ea.	.10	4.40		4.50	Includes labor to clean contaminated
	Minimum Charge	Job		187		187	carpet on a stairway.
Carpet Treatment							
Disinfect / Deodorize							
	Clean	SF Flr.		.23		.23	Includes disinfecting and deodorizing
	Minimum Charge	Job		187		187	of carpet.
Disinfect For Sewage							
	Clean	SF Flr.		.60		.60	Includes labor to disinfect and
	Minimum Charge	Job		187		187	deodorize carpet exposed to sewage.
Mildicide							
	Clean	SF Flr.	.02	.23		.25	Includes labor and materials to apply
	Minimum Charge	Job		187		187	mildicide to carpet.
Thermal Fog Area							
	Clean	SF Flr.	.01	.58		.59	Includes labor and materials to
	Minimum Charge	Job		187		187	perform thermal fogging during
							deflooding clean-up.
Wet Fog / ULV Area							
	Clean	SF Flr.	.03	.33		.36	Includes labor and materials to
	Minimum Charge	Job		187		187	perform wet fogging / ULV during
							deflooding clean-up.
Steam Clean Fixtures							
	Clean	Ea.	.01	37.50		37.51	Includes labor and materials to steam
	Minimum Charge	Job		187		187	clean fixtures during deflooding
							clean-up.
Pressure Wash Sewage							
	Clean	SF Flr.		3.40		3.40	Includes labor to pressure wash
	Minimum Charge	Job		187		187	sewage contamination.
Air Mover							
	Minimum Charge	Day			90.50	90.50	Daily rental of air mover.
Dehumidifier							
Medium Size							
	Minimum Charge	Day			48	48	Daily rental of dehumidifier, medium.
Large Size							
	Minimum Charge	Day			405	405	Daily rental of dehumidifier, large.

11

Job Preparation

Flood Clean-up

	Unit	Material	Labor	Equip.	Total	Specification
Air Purifier						
HEPA Filter-Bacteria						
Minimum Charge	Ea.	390			390	HEPA filter for air purifier.

Asbestos Removal

	Unit	Material	Labor	Equip.	Total	Specification
Asbestos Analysis						
Minimum Charge	Job		290		290	Minimum charge for testing.
Asbestos Pipe Insulation						
1/2" to 3/4" Diameter						
Demolish	L.F.	.36	7.50		7.86	Minimum charge to remove asbestos. Includes full Tyvek suits for workers changed four times per eight-hour shift with respirators, air monitoring and supervision by qualified professionals.
Minimum Charge	Job		4650		4650	
1" to 3" Diameter						
Demolish	L.F.	1.11	23.50		24.61	Minimum charge to remove asbestos. Includes full Tyvek suits for workers changed four times per eight-hour shift with respirators, air monitoring and supervision by qualified professionals.
Minimum Charge	Job		4650		4650	
Asbestos-based Siding						
Demolish	S.F.	2.48	51.50		53.98	Minimum charge to remove asbestos. Includes full Tyvek suits for workers changed four times per eight-hour shift with respirators, air monitoring and supervision by qualified professionals.
Minimum Charge	Job		4650		4650	
Asbestos-based Plaster						
Demolish	S.F.	1.90	40		41.90	Minimum charge to remove asbestos. Includes full Tyvek suits for workers changed four times per eight-hour shift with respirators, air monitoring and supervision by qualified professionals.
Minimum Charge	Job		4650		4650	
Encapsulate Asbestos Ceiling						
Install	S.F.	.59	.22		.81	Includes encapsulation with penetrating sealant sprayed onto ceiling and walls with airless sprayer.
Minimum Charge	Job		4650		4650	
Scrape Acoustical Ceiling						
Demolish	S.F.	.07	1.31		1.38	Minimum charge to remove asbestos. Includes full Tyvek suits for workers changed four times per eight-hour shift with respirators, air monitoring and supervision by qualified professionals.
Minimum Charge	Job		4650		4650	

For customer support on your Contractor's Pricing Guide: Residential Repair & Remodeling, call 800.448.8182.

Job Preparation

Asbestos Removal

		Unit	Material	Labor	Equip.	Total	Specification
Seal Ceiling Asbestos							
	Install	S.F.	2.78	4.37		7.15	Includes attaching plastic cover to ceiling area and sealing where asbestos exists, taping all seams, caulk or foam insulation in cracks and negative air system.
	Minimum Charge	Job		4650		4650	
Seal Wall Asbestos							
	Install	S.F.	2.78	3.89		6.67	Includes attaching plastic cover to wall area and sealing where asbestos exists, taping all seams, caulk or foam insulation in cracks and negative air system.
	Minimum Charge	Job		4650		4650	
Seal Floor Asbestos							
	Install	S.F.	4.46	7.75		12.21	Includes attaching plastic cover to floor area and sealing where asbestos exists, taping all seams, caulk or foam insulation in cracks and negative air system.
	Minimum Charge	Job		4650		4650	
Negative Air Vent System							
	Install	Day			80.50	80.50	Daily rental of negative air vent system.
	Minimum Charge	Week			405	405	
HEPA Vacuum Cleaner							
	Install	Day			92.50	92.50	Daily rental of HEPA vacuum, 16 gallon.
	Minimum Charge	Week			465	465	
Airless Sprayer							
	Install	Day			85.50	85.50	Daily rental of airless sprayer.
	Minimum Charge	Week			430	430	
Decontamination Unit							
	Install	Day			149	149	Daily rental of decontamination unit.
	Minimum Charge	Week			745	745	
Light Stand							
	Install	Day			28.50	28.50	Daily rental of floodlight w/tripod.
	Minimum Charge	Week			142	142	

Room Demolition

		Unit	Material	Labor	Equip.	Total	Specification
Strip Typical Room							
	Demolish	S.F.		3.74		3.74	Includes 4 labor hours for general demolition.
	Minimum Charge	Job		208		208	
Strip Typical Bathroom							
	Demolish	S.F.		4.99		4.99	Includes 4 labor hours for general demolition.
	Minimum Charge	Job		208		208	
Strip Typical Kitchen							
	Demolish	S.F.		4.79		4.79	Includes 4 labor hours for general demolition.
	Minimum Charge	Job		208		208	
Strip Typical Laundry							
	Demolish	S.F.		4.27		4.27	Includes 4 labor hours for general demolition.
	Minimum Charge	Job		208		208	

For customer support on your Contractor's Pricing Guide: Residential Repair & Remodeling, call 800.448.8182.

Selective Demolition		Unit	Material	Labor	Equip.	Total	Specification
Concrete Slab							
4″ Unreinforced							
	Demolish	S.F.		.26	.70	.96	Includes labor and equipment to
	Minimum Charge	Job		580	1575	2155	remove unreinforced concrete slab and haul debris to truck or dumpster.
6″ Unreinforced							
	Demolish	S.F.		.29	.79	1.08	Includes labor and equipment to
	Minimum Charge	Job		580	1575	2155	remove unreinforced concrete slab and haul debris to truck or dumpster.
8″ Unreinforced							
	Demolish	S.F.		.33	.91	1.24	Includes labor and equipment to
	Minimum Charge	Job		580	1575	2155	remove unreinforced concrete slab and haul debris to truck or dumpster.
4″ Reinforced							
	Demolish	S.F.		.26	.70	.96	Includes labor and equipment to
	Minimum Charge	Job		580	1575	2155	remove reinforced concrete slab and haul debris to truck or dumpster.
6″ Reinforced							
	Demolish	S.F.		.29	.79	1.08	Includes labor and equipment to
	Minimum Charge	Job		580	1575	2155	remove reinforced concrete slab and haul debris to truck or dumpster.
8″ Reinforced							
	Demolish	S.F.		.33	.91	1.24	Includes labor and equipment to
	Minimum Charge	Job		580	1575	2155	remove reinforced concrete slab and haul debris to truck or dumpster.
Brick Wall							
4″ Thick							
	Demolish	S.F.		3.45	.66	4.11	Includes labor and equipment to
	Minimum Charge	Job		1050	203	1253	remove brick wall and haul debris to a central location for disposal.
8″ Thick							
	Demolish	S.F.		4.74	.91	5.65	Includes labor and equipment to
	Minimum Charge	Job		1050	203	1253	remove brick wall and haul debris to a central location for disposal.
12″ Thick							
	Demolish	S.F.		6.30	1.22	7.52	Includes labor and equipment to
	Minimum Charge	Job		1050	203	1253	remove brick wall and haul debris to a central location for disposal.
Concrete Masonry Wall							
4″ Thick							
	Demolish	S.F.		3.83	.74	4.57	Includes minimum labor and
	Minimum Charge	Job		1050	203	1253	equipment to remove 4″ block wall and haul debris to a central location for disposal.
6″ Thick							
	Demolish	S.F.		4.74	.91	5.65	Includes minimum labor and
	Minimum Charge	Job		1050	203	1253	equipment to remove 6″ block wall and haul debris to a central location for disposal.

For customer support on your Contractor's Pricing Guide: Residential Repair & Remodeling, call 800.448.8182.

Job Preparation

Selective Demolition		Unit	Material	Labor	Equip.	Total	Specification
8″ Thick							
	Demolish	S.F.		5.85	1.12	6.97	Includes minimum labor and
	Minimum Charge	Job		1050	203	1253	equipment to remove 8″ block wall and haul debris to a central location for disposal.
12″ Thick							
	Demolish	S.F.		8.45	1.62	10.07	Includes minimum labor and
	Minimum Charge	Job		1050	203	1253	equipment to remove 12″ block wall and haul debris to a central location for disposal.
Tree Removal							
Small, 4″ - 8″							
	Demolish	Ea.		134	109	243	Minimum charge, common labor
	Minimum Charge	Job		187		187	trade.
Medium, 9″ - 12″							
	Demolish	Ea.		201	164	365	Minimum charge, common labor
	Minimum Charge	Job		187		187	trade.
Large, 13″ - 18″							
	Demolish	Ea.		241	197	438	Minimum charge, tree removal.
	Minimum Charge	Job		615	135	750	
Very Large, 19″ - 24″							
	Demolish	Ea.		300	246	546	Minimum charge, tree removal.
	Minimum Charge	Job		615	135	750	
Stump Removal							
	Demolish	Ea.		89	173	262	Includes minimum charges for a site
	Minimum Charge	Job		340	100	440	demolition crew and backhoe / loader.

Backhoe/Loader - Wheel Type	Tractor Loader - Wheel Type	Backhoe - Crawler Type

Excavation		Unit	Material	Labor	Equip.	Total	Specification
Footings							
Hand Trenching							
	Install	L.F.		2.34		2.34	Includes hand excavation for footing
	Minimum Charge	Job		93.50		93.50	or trench up to 10" deep and 12" wide.
Machine Trenching							
	Install	L.F.		.89	.36	1.25	Includes machine excavation for
	Minimum Charge	Job		445	181	626	footing or trench up to 10" deep and 12" wide.
Trenching w / Backfill							
Hand							
	Install	L.F.		3.74		3.74	Includes hand excavation, backfill and
	Minimum Charge	Job		93.50		93.50	compaction.
Machine							
	Install	L.F.		4.43	10.75	15.18	Includes machine excavation, backfill
	Minimum Charge	Job		187	105	292	and compaction.

Backfill		Unit	Material	Labor	Equip.	Total	Specification
Hand (no compaction)							
	Install	L.C.Y.		31		31	Includes backfill by hand in average
	Minimum Charge	Job		187		187	soil without compaction.
Hand (w / compaction)							
	Install	E.C.Y.		44		44	Includes backfill by hand in average
	Minimum Charge	Job		187		187	soil with compaction.
Machine (no compaction)							
	Install	L.C.Y.		4.43	1.81	6.24	Includes backfill, by 55 HP wheel
	Minimum Charge	Job		445	181	626	loader, of trenches from loose material piled adjacent without compaction.
Machine (w / compaction)							
	Install	L.C.Y.		11.80	4.82	16.62	Includes backfill, by 55 HP wheel
	Minimum Charge	Job		445	181	626	loader, of trenches from loose material piled adjacent with compaction by vibrating tampers.

Concrete Footing and Formwork

Wood Pile

Piles	Unit	Material	Labor	Equip.	Total	Specification
Treated Wood, 12″ Butt						
To 30′						
Demolish	L.F.		5.95	1.75	7.70	Includes material, labor and
Install	V.L.F.	18.90	5.60	3.70	28.20	equipment to install up to 30′ long
Demolish and Install	V.L.F.	18.90	11.55	5.45	35.90	treated wood piling.
Paint	L.F.	.53	2.31		2.84	
Minimum Charge	Job		3525	2325	5850	
30′ to 40′						
Demolish	L.F.		5.95	1.75	7.70	Includes material, labor and
Install	V.L.F.	19.60	5	3.30	27.90	equipment to install 30′ to 40′ long
Demolish and Install	V.L.F.	19.60	10.95	5.05	35.60	treated wood piling.
Paint	L.F.	.53	2.31		2.84	
Minimum Charge	Job		3525	2325	5850	
40′ to 50′						
Demolish	L.F.		5.95	1.75	7.70	Includes material, labor and
Install	V.L.F.	27	4.88	3.21	35.09	equipment to install 40′ to 50′ long
Demolish and Install	V.L.F.	27	10.83	4.96	42.79	treated wood piling.
Paint	L.F.	.53	2.31		2.84	
Minimum Charge	Job		3525	2325	5850	
50′ to 60′						
Demolish	L.F.		5.95	1.75	7.70	Includes material, labor and
Install	V.L.F.	29	4.39	2.89	36.28	equipment to install 50′ to 60′ long
Demolish and Install	V.L.F.	29	10.34	4.64	43.98	treated wood piling.
Paint	L.F.	.53	2.31		2.84	
Minimum Charge	Job		3525	2325	5850	
60′ to 80′						
Demolish	L.F.		5.95	1.75	7.70	Includes material, labor and
Install	V.L.F.	27.50	4.18	3.94	35.62	equipment to install 60′ to 80′ long
Demolish and Install	V.L.F.	27.50	10.13	5.69	43.32	treated wood piling.
Paint	L.F.	.53	2.31		2.84	
Minimum Charge	Job		3525	2325	5850	
Untreated Wood, 12″ Butt						
To 30′						
Demolish	L.F.		5.95	1.75	7.70	Includes material, labor and
Install	V.L.F.	15.95	5.60	3.70	25.25	equipment to install up to 30′ long
Demolish and Install	V.L.F.	15.95	11.55	5.45	32.95	untreated wood piling.
Paint	L.F.	.53	2.31		2.84	
Minimum Charge	Job		3525	2325	5850	

Foundation

Piles

	Unit	Material	Labor	Equip.	Total	Specification
30' to 40'						
Demolish	L.F.		5.95	1.75	7.70	Includes material, labor and
Install	V.L.F.	16.85	5	3.30	25.15	equipment to install 30' to 40' long
Demolish and Install	V.L.F.	16.85	10.95	5.05	32.85	untreated wood piling.
Paint	L.F.	.53	2.31		2.84	
Minimum Charge	Job		3525	2325	5850	
40' to 50'						
Demolish	L.F.		5.95	1.75	7.70	Includes material, labor and
Install	V.L.F.	17.70	4.88	3.21	25.79	equipment to install 40' to 50' long
Demolish and Install	V.L.F.	17.70	10.83	4.96	33.49	untreated wood piling.
Paint	L.F.	.53	2.31		2.84	
Minimum Charge	Job		3525	2325	5850	
50' to 60'						
Demolish	L.F.		5.95	1.75	7.70	Includes material, labor and
Install	V.L.F.	25	4.39	2.89	32.28	equipment to install 50' to 60' long
Demolish and Install	V.L.F.	25	10.34	4.64	39.98	untreated wood piling.
Paint	L.F.	.53	2.31		2.84	
Minimum Charge	Job		3525	2325	5850	
60' to 80'						
Demolish	L.F.		5.95	1.75	7.70	Includes material, labor and
Install	V.L.F.	28.50	4.18	2.75	35.43	equipment to install 60' to 80' long
Demolish and Install	V.L.F.	28.50	10.13	4.50	43.13	untreated wood piling.
Paint	L.F.	.53	2.31		2.84	
Minimum Charge	Job		3525	2325	5850	

Precast Concrete

	Unit	Material	Labor	Equip.	Total	Specification
10" Square						
Demolish	L.F.		6.10	1.79	7.89	Includes material, labor and
Install	V.L.F.	15.35	5	3.30	23.65	equipment to install a precast,
Demolish and Install	V.L.F.	15.35	11.10	5.09	31.54	prestressed, 40' long, 10" square
Paint	L.F.	.53	2.31		2.84	concrete pile.
Minimum Charge	Job		3525	2325	5850	
12" Square						
Demolish	L.F.		6.50	1.91	8.41	Includes material, labor and
Install	V.L.F.	27.50	5.15	3.40	36.05	equipment to install a precast,
Demolish and Install	V.L.F.	27.50	11.65	5.31	44.46	prestressed, 40' long, 12" square
Paint	L.F.	.53	2.31		2.84	concrete pile.
Minimum Charge	Job		3525	2325	5850	
14" Square						
Demolish	L.F.		6.65	1.96	8.61	Includes material, labor and
Install	V.L.F.	31	5.85	3.86	40.71	equipment to install a precast,
Demolish and Install	V.L.F.	31	12.50	5.82	49.32	prestressed, 40' long, 14" square
Paint	L.F.	.53	2.31		2.84	concrete pile.
Minimum Charge	Job		3525	2325	5850	
16" Square						
Demolish	L.F.		7	2.06	9.06	Includes material, labor and
Install	V.L.F.	32.50	6.25	4.13	42.88	equipment to install a precast,
Demolish and Install	V.L.F.	32.50	13.25	6.19	51.94	prestressed, 40' long, 16" square
Paint	L.F.	.53	2.31		2.84	concrete pile.
Minimum Charge	Job		3525	2325	5850	
18" Square						
Demolish	L.F.		7.60	2.23	9.83	Includes material, labor and
Install	V.L.F.	56	6.75	6.35	69.10	equipment to install a precast,
Demolish and Install	V.L.F.	56	14.35	8.58	78.93	prestressed, 40' long, 18" square
Paint	L.F.	.53	2.31		2.84	concrete pile.
Minimum Charge	Job		3525	2325	5850	

Piles	Unit	Material	Labor	Equip.	Total	Specification
Cross Bracing						
Treated 2″ x 6″						
Demolish	L.F.		1.22		1.22	Cost includes material and labor to
Install	L.F.	1.33	6.95		8.28	install 2″ x 6″ lumber for cross-bracing
Demolish and Install	L.F.	1.33	8.17		9.50	of foundation piling or posts.
Reinstall	L.F.		5.58		5.58	
Clean	S.F.		.47		.47	
Paint	L.F.	.48	1.55		2.03	
Minimum Charge	Job		865		865	
Treated 2″ x 8″						
Demolish	L.F.		1.22		1.22	Cost includes material and labor to
Install	L.F.	1.77	7.20		8.97	install 2″ x 8″ lumber for cross-bracing
Demolish and Install	L.F.	1.77	8.42		10.19	of foundation piling or posts.
Reinstall	L.F.		5.76		5.76	
Clean	S.F.		.47		.47	
Paint	L.F.	.64	1.61		2.25	
Minimum Charge	Job		865		865	
Treated 2″ x 10″						
Demolish	L.F.		1.26		1.26	Cost includes material and labor to
Install	L.F.	2.05	7.45		9.50	install 2″ x 10″ lumber for
Demolish and Install	L.F.	2.05	8.71		10.76	cross-bracing of foundation piling or
Reinstall	L.F.		5.96		5.96	posts.
Clean	S.F.		.47		.47	
Paint	L.F.	.77	1.65		2.42	
Minimum Charge	Job		865		865	
Treated 2″ x 12″						
Demolish	L.F.		1.51		1.51	Cost includes material and labor to
Install	L.F.	3.25	7.70		10.95	install 2″ x 12″ lumber for
Demolish and Install	L.F.	3.25	9.21		12.46	cross-bracing of foundation piling or
Reinstall	L.F.		6.17		6.17	posts.
Clean	S.F.		.47		.47	
Paint	L.F.	.90	1.68		2.58	
Minimum Charge	Job		865		865	
Metal Pipe						
6″ Concrete Filled						
Demolish	L.F.		8.45	2.48	10.93	Includes material, labor and
Install	V.L.F.	23	7.30	4.82	35.12	equipment to install concrete filled
Demolish and Install	V.L.F.	23	15.75	7.30	46.05	steel pipe pile.
Paint	L.F.	.25	2.31		2.56	
Minimum Charge	Job		3525	2325	5850	
6″ Unfilled						
Demolish	L.F.		8.45	2.48	10.93	Includes material, labor and
Install	V.L.F.	19.60	6.70	4.41	30.71	equipment to install unfilled steel pipe
Demolish and Install	V.L.F.	19.60	15.15	6.89	41.64	pile.
Paint	L.F.	.25	2.31		2.56	
Minimum Charge	Job		3525	2325	5850	
8″ Concrete Filled						
Demolish	L.F.		8.45	2.48	10.93	Includes material, labor and
Install	V.L.F.	26	7.65	5.05	38.70	equipment to install concrete filled
Demolish and Install	V.L.F.	26	16.10	7.53	49.63	steel pipe pile.
Paint	L.F.	.25	2.31		2.56	
Minimum Charge	Job		3525	2325	5850	
8″ Unfilled						
Demolish	L.F.		8.45	2.48	10.93	Includes material, labor and
Install	V.L.F.	20.50	7.05	4.63	32.18	equipment to install unfilled steel pipe
Demolish and Install	V.L.F.	20.50	15.50	7.11	43.11	pile.
Paint	L.F.	.25	2.31		2.56	
Minimum Charge	Job		3525	2325	5850	

Foundation

Piles

Piles	Unit	Material	Labor	Equip.	Total	Specification
10″ Concrete Filled						
Demolish	L.F.		8.70	2.56	11.26	Includes material, labor and
Install	V.L.F.	31.50	7.80	5.15	44.45	equipment to install concrete filled
Demolish and Install	V.L.F.	31.50	16.50	7.71	55.71	steel pipe pile.
Paint	L.F.	.25	2.31		2.56	
Minimum Charge	Job		3525	2325	5850	
10″ Unfilled						
Demolish	L.F.		8.70	2.56	11.26	Includes material, labor and
Install	V.L.F.	25	7.05	4.63	36.68	equipment to install unfilled steel pipe
Demolish and Install	V.L.F.	25	15.75	7.19	47.94	pile.
Paint	L.F.	.25	2.31		2.56	
Minimum Charge	Job		3525	2325	5850	
12″ Concrete Filled						
Demolish	L.F.		8.70	2.56	11.26	Includes material, labor and
Install	V.L.F.	38	8.45	5.55	52	equipment to install concrete filled
Demolish and Install	V.L.F.	38	17.15	8.11	63.26	steel pipe pile.
Paint	L.F.	.25	2.31		2.56	
Minimum Charge	Job		3525	2325	5850	
12″ Unfilled						
Demolish	L.F.		8.70	2.56	11.26	Includes material, labor and
Install	V.L.F.	38.50	7.40	4.87	50.77	equipment to install unfilled steel pipe
Demolish and Install	V.L.F.	38.50	16.10	7.43	62.03	pile.
Paint	L.F.	.25	2.31		2.56	
Minimum Charge	Job		3525	2325	5850	

Steel H Section

	Unit	Material	Labor	Equip.	Total	Specification
H Section 8 x 8 x 36#						
Demolish	L.F.		5.90	1.73	7.63	Includes material, labor and
Install	V.L.F.	18.75	5.50	3.61	27.86	equipment to install steel H section
Demolish and Install	V.L.F.	18.75	11.40	5.34	35.49	pile.
Paint	S.F.	.22	.34		.56	
Minimum Charge	Job		3525	2325	5850	
H Section 10 x 10 x 57#						
Demolish	L.F.		6.45	1.90	8.35	Includes material, labor and
Install	V.L.F.	28.50	5.75	3.79	38.04	equipment to install steel H section
Demolish and Install	V.L.F.	28.50	12.20	5.69	46.39	pile.
Paint	S.F.	.22	.34		.56	
Minimum Charge	Job		3525	2325	5850	
H Section 12 x 12 x 74#						
Demolish	L.F.		6.65	1.96	8.61	Includes material, labor and
Install	V.L.F.	36.50	5.95	5.60	48.05	equipment to install steel H section
Demolish and Install	V.L.F.	36.50	12.60	7.56	56.66	pile.
Paint	S.F.	.22	.34		.56	
Minimum Charge	Job		3525	2325	5850	
H Section 14 x 14 x 89#						
Demolish	L.F.		7	2.06	9.06	Includes material, labor and
Install	V.L.F.	49.50	6.50	6.15	62.15	equipment to install steel H section
Demolish and Install	V.L.F.	49.50	13.50	8.21	71.21	pile.
Paint	S.F.	.22	.34		.56	
Minimum Charge	Job		3525	2325	5850	

For customer support on your Contractor's Pricing Guide: Residential Repair & Remodeling, call 800.448.8182.

Foundation

Foundation Post

	Unit	Material	Labor	Equip.	Total	Specification
Wood Foundation Post						
4″ x 4″						
Demolish	L.F.		.96	.28	1.24	Cost includes material and labor to
Install	L.F.	1.38	6.25		7.63	install 4″ x 4″ treated wood post by
Demolish and Install	L.F.	1.38	7.21	.28	8.87	hand in normal soil conditions.
Reinstall	L.F.		4.99		4.99	
Minimum Charge	Job		865		865	
4″ x 6″						
Demolish	L.F.		.96	.28	1.24	Cost includes material and labor to
Install	L.F.	2.08	6.25		8.33	install 4″ x 6″ treated wood post by
Demolish and Install	L.F.	2.08	7.21	.28	9.57	hand in normal soil conditions.
Reinstall	L.F.		4.99		4.99	
Minimum Charge	Job		865		865	
6″ x 6″						
Demolish	L.F.		.96	.28	1.24	Cost includes material and labor to
Install	L.F.	3.29	6.25		9.54	install 6″ x 6″ treated wood post by
Demolish and Install	L.F.	3.29	7.21	.28	10.78	hand in normal soil conditions.
Reinstall	L.F.		4.99		4.99	
Minimum Charge	Job		865		865	
Water Jet						
Demolish	Ea.		111		111	Includes material, labor and
Install	Ea.	69	855		924	equipment to water jet a 6″ x 6″ wood
Demolish and Install	Ea.	69	966		1035	post under an existing building. Cost
Reinstall	Ea.		854.86		854.86	may vary depending upon location
Minimum Charge	Job		750		750	and access.
Shim						
Install	Ea.	1.30	86.50		87.80	Includes labor and material to trim or
Minimum Charge	Job		865		865	readjust existing foundation posts under an existing building. Access of 2 - 4 foot assumed.
Cross-bracing						
Treated 2″ x 6″						
Demolish	L.F.		1.22		1.22	Cost includes material and labor to
Install	L.F.	1.33	6.95		8.28	install 2″ x 6″ lumber for cross-bracing
Demolish and Install	L.F.	1.33	8.17		9.50	of foundation piling or posts.
Reinstall	L.F.		5.58		5.58	
Clean	S.F.		.47		.47	
Paint	L.F.	.48	1.55		2.03	
Minimum Charge	Job		865		865	
Treated 2″ x 8″						
Demolish	L.F.		1.22		1.22	Cost includes material and labor to
Install	L.F.	1.77	7.20		8.97	install 2″ x 8″ lumber for cross-bracing
Demolish and Install	L.F.	1.77	8.42		10.19	of foundation piling or posts.
Reinstall	L.F.		5.76		5.76	
Clean	S.F.		.47		.47	
Paint	L.F.	.64	1.61		2.25	
Minimum Charge	Job		865		865	
Treated 2″ x 10″						
Demolish	L.F.		1.26		1.26	Cost includes material and labor to
Install	L.F.	2.05	7.45		9.50	install 2″ x 10″ lumber for
Demolish and Install	L.F.	2.05	8.71		10.76	cross-bracing of foundation piling or
Reinstall	L.F.		5.96		5.96	posts.
Clean	S.F.		.47		.47	
Paint	L.F.	.77	1.65		2.42	
Minimum Charge	Job		865		865	

For customer support on your Contractor's Pricing Guide: Residential Repair & Remodeling, call 800.448.8182.

Foundation

Foundation Post

Treated 2" x 12"	Unit	Material	Labor	Equip.	Total	Specification
Demolish	L.F.		1.51		1.51	Cost includes material and labor to
Install	L.F.	3.25	7.70		10.95	install 2" x 12" lumber for
Demolish and Install	L.F.	3.25	9.21		12.46	cross-bracing of foundation piling or
Reinstall	L.F.		6.17		6.17	posts.
Clean	S.F.		.47		.47	
Paint	L.F.	.90	1.68		2.58	
Minimum Charge	Job		865		865	

Concrete Footing

	Unit	Material	Labor	Equip.	Total	Specification
Continuous						
12" w x 6" d						
Demolish	L.F.		.64	1.76	2.40	Includes material and labor to install
Install	L.F.	6.70	3.72	.04	10.46	continuous reinforced concrete footing
Demolish and Install	L.F.	6.70	4.36	1.80	12.86	poured by chute cast against earth
Minimum Charge	Job		970		970	including reinforcing, forming and
						finishing.
12" w x 12" d						
Demolish	L.F.		.77	2.11	2.88	Includes material and labor to install
Install	L.F.	15.20	7.55	.08	22.83	continuous reinforced concrete footing
Demolish and Install	L.F.	15.20	8.32	2.19	25.71	poured by chute cast against earth
Minimum Charge	Job		970		970	including reinforcing, forming and
						finishing.
18" w x 10" d						
Demolish	L.F.		.77	2.11	2.88	Includes material and labor to install
Install	L.F.	14.15	9.55	.10	23.80	continuous reinforced concrete footing
Demolish and Install	L.F.	14.15	10.32	2.21	26.68	poured by chute cast against earth
Minimum Charge	Job		970		970	including reinforcing, forming and
						finishing.
24" w x 24" d						
Demolish	L.F.		1.16	3.17	4.33	Includes material and labor to install
Install	L.F.	42	28.50	.31	70.81	continuous reinforced concrete footing
Demolish and Install	L.F.	42	29.66	3.48	75.14	poured by chute cast against earth
Minimum Charge	Job		970		970	including reinforcing, forming and
						finishing.
Stem Wall						
Single Story						
Demolish	L.F.		16.75	13.70	30.45	Includes material and labor to install
Install	L.F.	34.50	125	1.34	160.84	stem wall 24" above and 18" below
Demolish and Install	L.F.	34.50	141.75	15.04	191.29	grade for a single story structure.
Minimum Charge	Job		970		970	Includes 12" wide and 8" deep
						footing.
Two Story						
Demolish	L.F.		16.75	13.70	30.45	Includes material and labor to install
Install	L.F.	42	179	1.92	222.92	stem wall 24" above and 18" below
Demolish and Install	L.F.	42	195.75	15.62	253.37	grade typical for a two story structure,
Minimum Charge	Job		970		970	includes 18" wide and 10" deep
						footing.

For customer support on your Contractor's Pricing Guide: Residential Repair & Remodeling, call 800.448.8182.

Concrete Slab	Unit	Material	Labor	Equip.	Total	Specification
Reinforced						
4″						
Demolish	S.F.		.26	.70	.96	Cost includes material and labor to
Install	S.F.	3.62	.48		4.10	install 4″ reinforced concrete
Demolish and Install	S.F.	3.62	.74	.70	5.06	slab-on-grade poured by chute
Clean	S.F.	.12	.42		.54	including forms, vapor barrier, wire
Paint	S.F.	.20	.91		1.11	mesh, 3000 PSI concrete, float finish,
Minimum Charge	Job		238		238	and curing.
6″						
Demolish	S.F.		.29	.79	1.08	Cost includes material and labor to
Install	S.F.	4.92	.73	.01	5.66	install 6″ reinforced concrete
Demolish and Install	S.F.	4.92	1.02	.80	6.74	slab-on-grade poured by chute
Clean	S.F.	.12	.42		.54	including forms, vapor barrier, wire
Paint	S.F.	.20	.91		1.11	mesh, 3000 PSI concrete, float finish,
Minimum Charge	Job		238		238	and curing.
8″						
Demolish	S.F.		.33	.91	1.24	Cost includes material and labor to
Install	S.F.	6.25	.95	.01	7.21	install 8″ reinforced concrete
Demolish and Install	S.F.	6.25	1.28	.92	8.45	slab-on-grade poured by chute
Clean	S.F.	.12	.42		.54	including forms, vapor barrier, wire
Paint	S.F.	.20	.91		1.11	mesh, 3000 PSI concrete, float finish,
Minimum Charge	Job		238		238	and curing.
Unreinforced						
4″						
Demolish	S.F.		.26	.70	.96	Cost includes material and labor to
Install	S.F.	2.30	1.17	.01	3.48	install 4″ concrete slab-on-grade
Demolish and Install	S.F.	2.30	1.43	.71	4.44	poured by chute, including forms,
Clean	S.F.	.12	.42		.54	vapor barrier, 3000 PSI concrete, float
Paint	S.F.	.20	.91		1.11	finish, and curing.
Minimum Charge	Job		238		238	
6″						
Demolish	S.F.		.29	.79	1.08	Cost includes material and labor to
Install	S.F.	3.36	1.20	.01	4.57	install 6″ concrete slab-on-grade
Demolish and Install	S.F.	3.36	1.49	.80	5.65	poured by chute, including forms,
Clean	S.F.	.12	.42		.54	vapor barrier, 3000 PSI concrete, float
Paint	S.F.	.20	.91		1.11	finish, and curing.
Minimum Charge	Job		238		238	
8″						
Demolish	S.F.		.33	.91	1.24	Cost includes material and labor to
Install	S.F.	4.59	1.26	.01	5.86	install 8″ concrete slab-on-grade
Demolish and Install	S.F.	4.59	1.59	.92	7.10	poured by chute, including forms,
Clean	S.F.	.12	.42		.54	vapor barrier, 3000 PSI concrete, float
Paint	S.F.	.20	.91		1.11	finish, and curing.
Minimum Charge	Job		238		238	
Lightweight						
Demolish	S.F.		1.11	.36	1.47	Includes material and labor to install
Install	S.F.	1.54	3.54		5.08	lightweight concrete, 3″ - 4″ thick.
Demolish and Install	S.F.	1.54	4.65	.36	6.55	
Clean	S.F.	.12	.42		.54	
Paint	S.F.	.20	.91		1.11	
Minimum Charge	Job		238		238	

Concrete Slab	Unit	Material	Labor	Equip.	Total	Specification
Gunite						
Demolish	S.F.		1.37	.40	1.77	Includes material and labor to install
Install	S.F.	.57	1.49	.23	2.29	gunite on flat plane per inch of
Demolish and Install	S.F.	.57	2.86	.63	4.06	thickness using 1" as base price. No
Clean	S.F.	.12	.42		.54	forms or reinforcing are included.
Paint	S.F.	.20	.91		1.11	
Minimum Charge	Job		1325		1325	
Epoxy Inject						
Install	L.F.	3.54	7.95		11.49	Includes labor and material to inject
Minimum Charge	Job		238		238	epoxy into cracks.
Slabjacking						
Install	S.F.	.57	1.39	.13	2.09	Includes labor, material and
Minimum Charge	Job		238		238	equipment to install slabjacking under
						an existing building.
Scrape and Paint						
Paint	S.F.	.23	1.01		1.24	Includes labor and materials to scrape
Minimum Charge	Job		202		202	and paint a concrete floor.
Vapor Barrier						
Install	S.F.	.04	.13		.17	Cost includes material and labor to
Minimum Charge	Job		242		242	install 4 mil polyethylene vapor barrier
						10' wide sheets with 6" overlaps.

Rough Framing

Beam / Girder	Unit	Material	Labor	Equip.	Total	Specification
Solid Wood Beam						
2" x 8"						
Demolish	L.F.		1.01		1.01	Cost includes material and labor to
Install	L.F.	1.57	1.49		3.06	install 2" x 8" beam.
Demolish and Install	L.F.	1.57	2.50		4.07	
Clean	L.F.	.06	.33		.39	
Paint	L.F.	.33	2.18		2.51	
Minimum Charge	Job		242		242	
2" x 10"						
Demolish	L.F.		1.25		1.25	Cost includes material and labor to
Install	L.F.	2.19	1.61		3.80	install 2" x 10" beam.
Demolish and Install	L.F.	2.19	2.86		5.05	
Clean	L.F.	.07	.40		.47	
Paint	L.F.	.21	1.35		1.56	
Minimum Charge	Job		242		242	
2" x 12"						
Demolish	L.F.		1.51		1.51	Cost includes material and labor to
Install	L.F.	3.05	1.76		4.81	install 2" x 12" beam.
Demolish and Install	L.F.	3.05	3.27		6.32	
Clean	L.F.	.07	.47		.54	
Paint	L.F.	.50	3.23		3.73	
Minimum Charge	Job		242		242	
4" x 8"						
Demolish	L.F.		3.09		3.09	Cost includes material and labor to
Install	L.F.	6.15	2.25	2.18	10.58	install 4" x 8" beam.
Demolish and Install	L.F.	6.15	5.34	2.18	13.67	
Clean	L.F.	.07	.40		.47	
Paint	L.F.	.21	1.35		1.56	
Minimum Charge	Job		242		242	
4" x 10"						
Demolish	L.F.		3.87		3.87	Cost includes material and labor to
Install	L.F.	7.75	2.37	2.30	12.42	install 4" x 10" beam.
Demolish and Install	L.F.	7.75	6.24	2.30	16.29	
Clean	L.F.	.07	.47		.54	
Paint	L.F.	.23	1.57		1.80	
Minimum Charge	Job		242		242	
4" x 12"						
Demolish	L.F.		4.64		4.64	Cost includes material and labor to
Install	L.F.	9.50	2.50	2.42	14.42	install 4" x 12" beam.
Demolish and Install	L.F.	9.50	7.14	2.42	19.06	
Clean	L.F.	.08	.53		.61	
Paint	L.F.	.50	3.23		3.73	
Minimum Charge	Job		242		242	

Rough Frame / Structure

Beam / Girder		Unit	Material	Labor	Equip.	Total	Specification
6″ x 8″							
	Demolish	L.F.		4.64		4.64	Cost includes material and labor to
	Install	L.F.	12.60	4.28	4.15	21.03	install 6″ x 8″ beam.
	Demolish and Install	L.F.	12.60	8.92	4.15	25.67	
	Clean	L.F.	.07	.47		.54	
	Paint	L.F.	.23	1.57		1.80	
	Minimum Charge	Job		242		242	
6″ x 10″							
	Demolish	L.F.		5.80		5.80	Cost includes material and labor to
	Install	L.F.	6.55	1.94		8.49	install 6″ x 10″ triple beam.
	Demolish and Install	L.F.	6.55	7.74		14.29	
	Clean	L.F.	.08	.53		.61	
	Paint	L.F.	.28	1.79		2.07	
	Minimum Charge	Job		242		242	
6″ x 12″							
	Demolish	L.F.		6.90		6.90	Cost includes material and labor to
	Install	L.F.	9.15	2.04		11.19	install 6″ x 12″ triple beam.
	Demolish and Install	L.F.	9.15	8.94		18.09	
	Clean	L.F.	.09	.60		.69	
	Paint	L.F.	.31	2.02		2.33	
	Minimum Charge	Job		242		242	
Steel I-type							
W8 x 31							
	Demolish	L.F.		10.75	10.25	21	Includes material and labor to install
	Install	L.F.	61.50	3.09	2.53	67.12	W shaped steel beam / girder.
	Demolish and Install	L.F.	61.50	13.84	12.78	88.12	
	Minimum Charge	Job		555		555	
W8 x 48							
	Demolish	L.F.		10.75	10.25	21	Includes material and labor to install
	Install	L.F.	95	3.23	2.64	100.87	W shaped steel beam / girder.
	Demolish and Install	L.F.	95	13.98	12.89	121.87	
	Minimum Charge	Job		555		555	
W8 x 67							
	Demolish	L.F.		10.75	10.25	21	Includes material and labor to install
	Install	L.F.	133	3.39	2.77	139.16	W shaped steel beam / girder.
	Demolish and Install	L.F.	133	14.14	13.02	160.16	
	Minimum Charge	Job		555		555	
W10 x 45							
	Demolish	L.F.		10.75	10.25	21	Includes material and labor to install
	Install	L.F.	89	3.23	2.64	94.87	W shaped steel beam / girder.
	Demolish and Install	L.F.	89	13.98	12.89	115.87	
	Minimum Charge	Job		555		555	
W10 x 68							
	Demolish	L.F.		10.75	10.25	21	Includes material and labor to install
	Install	L.F.	134	3.39	2.77	140.16	W shaped steel beam / girder.
	Demolish and Install	L.F.	134	14.14	13.02	161.16	
	Minimum Charge	Job		555		555	
W12 x 50							
	Demolish	L.F.		10.75	10.25	21	Includes material and labor to install
	Install	L.F.	99	3.23	2.64	104.87	W shaped steel beam / girder.
	Demolish and Install	L.F.	99	13.98	12.89	125.87	
	Minimum Charge	Job		555		555	
W12 x 87							
	Demolish	L.F.		15.80	15.05	30.85	Includes material and labor to install
	Install	L.F.	173	3.39	2.77	179.16	W shaped steel beam / girder.
	Demolish and Install	L.F.	173	19.19	17.82	210.01	
	Minimum Charge	Job		555		555	

For customer support on your Contractor's Pricing Guide: Residential Repair & Remodeling, call 800.448.8182.

Rough Frame / Structure

Beam / Girder

Beam / Girder		Unit	Material	Labor	Equip.	Total	Specification
W12 x 120							
	Demolish	L.F.		15.80	15.05	30.85	Includes material and labor to install
	Install	L.F.	238	3.47	2.84	244.31	W shaped steel beam / girder.
	Demolish and Install	L.F.	238	19.27	17.89	275.16	
	Minimum Charge	Job		555		555	
W14 x 74							
	Demolish	L.F.		15.80	15.05	30.85	Includes material and labor to install
	Install	L.F.	146	3.39	2.77	152.16	W shaped steel beam / girder.
	Demolish and Install	L.F.	146	19.19	17.82	183.01	
	Minimum Charge	Job		555		555	
W14 x 120							
	Demolish	L.F.		15.80	15.05	30.85	Includes material and labor to install
	Install	L.F.	238	3.47	2.84	244.31	W shaped steel beam / girder.
	Demolish and Install	L.F.	238	19.27	17.89	275.16	
	Minimum Charge	Job		555		555	

Column

Column		Unit	Material	Labor	Equip.	Total	Specification
Concrete							
Small Diameter							
	Demolish	L.F.		17.55	3.38	20.93	Includes material and labor to install
	Install	V.L.F.	55.50	37.50	8.60	101.60	small diameter concrete column.
	Demolish and Install	V.L.F.	55.50	55.05	11.98	122.53	
	Minimum Charge	Job		800	315	1115	
Large Diameter							
	Demolish	L.F.		52.50	10.15	62.65	Includes material and labor to install
	Install	V.L.F.	84.50	44	10.15	138.65	large diameter concrete column.
	Demolish and Install	V.L.F.	84.50	96.50	20.30	201.30	
	Minimum Charge	Job		800	315	1115	

Floor Framing System

Floor Framing System		Unit	Material	Labor	Equip.	Total	Specification
12″ O.C.							
2″ x 6″ Joists							
	Demolish	S.F.		.86		.86	Cost includes material and labor to
	Install	S.F.	1.24	.77		2.01	install 2″ x 6″ joists, 12″ O.C.
	Demolish and Install	S.F.	1.24	1.63		2.87	including box or band joist. Blocking
	Reinstall	S.F.		.62		.62	or bridging not included.
	Minimum Charge	Job		242		242	
2″ x 8″ Joists							
	Demolish	S.F.		.89		.89	Cost includes material and labor to
	Install	S.F.	1.84	.88		2.72	install 2″ x 8″ joists, 12″ O.C.
	Demolish and Install	S.F.	1.84	1.77		3.61	including box or band joist. Blocking
	Reinstall	S.F.		.71		.71	or bridging not included.
	Minimum Charge	Job		242		242	
2″ x 10″ Joists							
	Demolish	S.F.		.91		.91	Cost includes material and labor to
	Install	S.F.	2.55	1.08		3.63	install 2″ x 10″ joists, 12″ O.C.
	Demolish and Install	S.F.	2.55	1.99		4.54	including box or band joist. Blocking
	Reinstall	S.F.		.86		.86	or bridging not included.
	Minimum Charge	Job		242		242	

For customer support on your Contractor's Pricing Guide: Residential Repair & Remodeling, call 800.448.8182.

Floor Framing System

Floor Framing System		Unit	Material	Labor	Equip.	Total	Specification
2″ x 12″ Joists							
	Demolish	S.F.		.94		.94	Cost includes material and labor to
	Install	S.F.	3.56	1.11		4.67	install 2″ x 12″ joists, 12″ O.C.
	Demolish and Install	S.F.	3.56	2.05		5.61	including box or band joist. Blocking
	Reinstall	S.F.		.89		.89	or bridging not included.
	Minimum Charge	Job		242		242	
Block / Bridge							
	Demolish	Ea.		1.30		1.30	Includes material and labor to install
	Install	Ea.	2.92	2.72		5.64	set of cross bridging or per block of
	Demolish and Install	Ea.	2.92	4.02		6.94	solid bridging for 2″ x 10″ joists cut to
	Minimum Charge	Job		242		242	size on site.

16″ O.C.

		Unit	Material	Labor	Equip.	Total	Specification
2″ x 6″ Joists							
	Demolish	S.F.		.64		.64	Cost includes material and labor to
	Install	S.F.	.96	.58		1.54	install 2″ x 6″ joists including box or
	Demolish and Install	S.F.	.96	1.22		2.18	band joist installed 16″ O.C. Does not
	Reinstall	S.F.		.46		.46	include beams, blocking or bridging.
	Minimum Charge	Job		242		242	
2″ x 8″ Joists							
	Demolish	S.F.		.66		.66	Cost includes material and labor to
	Install	S.F.	1.42	.66		2.08	install 2″ x 8″ joists including box or
	Demolish and Install	S.F.	1.42	1.32		2.74	band joist installed 16″ O.C. Does not
	Reinstall	S.F.		.53		.53	include beams, blocking or bridging.
	Minimum Charge	Job		242		242	
2″ x 10″ Joists							
	Demolish	S.F.		.68		.68	Cost includes material and labor to
	Install	S.F.	1.96	.81		2.77	install 2″ x 10″ joists including box or
	Demolish and Install	S.F.	1.96	1.49		3.45	band joist 16″ O.C. Does not include
	Reinstall	S.F.		.65		.65	beams, blocking or bridging.
	Minimum Charge	Job		242		242	
2″ x 12″ Joists							
	Demolish	S.F.		.71		.71	Cost includes material and labor to
	Install	S.F.	2.74	.83		3.57	install 2″ x 12″ joists including box or
	Demolish and Install	S.F.	2.74	1.54		4.28	band joist installed 16″ O.C. Does not
	Reinstall	S.F.		.66		.66	include beams, blocking or bridging.
	Minimum Charge	Job		242		242	
Block / Bridge							
	Demolish	Ea.		1.30		1.30	Includes material and labor to install
	Install	Ea.	2.92	2.72		5.64	set of cross bridging or per block of
	Demolish and Install	Ea.	2.92	4.02		6.94	solid bridging for 2″ x 10″ joists cut to
	Minimum Charge	Job		242		242	size on site.

24″ O.C.

		Unit	Material	Labor	Equip.	Total	Specification
2″ x 6″ Joists							
	Demolish	S.F.		.43		.43	Cost includes material and labor to
	Install	S.F.	.68	.39		1.07	install 2″ x 6″ joists including box or
	Demolish and Install	S.F.	.68	.82		1.50	band joist installed 24″ O.C. Does not
	Reinstall	S.F.		.31		.31	include beams, blocking or bridging.
	Minimum Charge	Job		242		242	
2″ x 8″ Joists							
	Demolish	S.F.		.44		.44	Cost includes material and labor to
	Install	S.F.	1.01	.44		1.45	install 2″ x 8″ joists including box or
	Demolish and Install	S.F.	1.01	.88		1.89	band joist installed 24″ O.C. Does not
	Reinstall	S.F.		.35		.35	include beams, blocking or bridging.
	Minimum Charge	Job		242		242	

Floor Framing System

Floor Framing System	Unit	Material	Labor	Equip.	Total	Specification
2″ x 10″ Joists						
Demolish	S.F.		.46		.46	Cost includes material and labor to
Install	S.F.	1.40	.54		1.94	install 2″ x 10″ joists including box or
Demolish and Install	S.F.	1.40	1		2.40	band joist installed 24″ O.C. Does not
Reinstall	S.F.		.43		.43	include beams, blocking or bridging.
Minimum Charge	Job		242		242	
2″ x 12″ Joists						
Demolish	S.F.		.47		.47	Cost includes material and labor to
Install	S.F.	1.95	.55		2.50	install 2″ x 12″ joists including box or
Demolish and Install	S.F.	1.95	1.02		2.97	band joist installed 24″ O.C. Does not
Reinstall	S.F.		.44		.44	include beams, blocking or bridging.
Minimum Charge	Job		242		242	
Block / Bridge						
Demolish	Ea.		1.30		1.30	Includes material and labor to install
Install	Ea.	2.92	2.72		5.64	set of cross bridging or per block of
Demolish and Install	Ea.	2.92	4.02		6.94	solid bridging for 2″ x 10″ joists cut to
Minimum Charge	Job		242		242	size on site.
Engineered Lumber, Joist						
9-1/2″						
Demolish	S.F.		.68		.68	Includes material and labor to install
Install	S.F.	4.20	.48		4.68	engineered lumber truss / joists per
Demolish and Install	S.F.	4.20	1.16		5.36	S.F. of floor area based on joists 16″
Reinstall	S.F.		.39		.39	O.C. Beams, supports and bridging
Minimum Charge	Job		242		242	are not included.
11-7/8″						
Demolish	S.F.		.71		.71	Includes material and labor to install
Install	S.F.	4.42	.49		4.91	engineered lumber truss / joists per
Demolish and Install	S.F.	4.42	1.20		5.62	S.F. of floor area based on joists 16″
Reinstall	S.F.		.40		.40	O.C. Beams, supports and bridging
Minimum Charge	Job		242		242	are not included.
14″						
Demolish	S.F.		.73		.73	Includes material and labor to install
Install	S.F.	5.10	.53		5.63	engineered lumber truss / joists per
Demolish and Install	S.F.	5.10	1.26		6.36	S.F. of floor area based on joists 16″
Reinstall	S.F.		.43		.43	O.C. Beams, supports and bridging
Minimum Charge	Job		242		242	are not included.
16″						
Demolish	S.F.		.76		.76	Includes material and labor to install
Install	S.F.	7.45	.56		8.01	engineered lumber truss / joists per
Demolish and Install	S.F.	7.45	1.32		8.77	S.F. of floor area based on joists 16″
Reinstall	S.F.		.45		.45	O.C. Beams, supports and bridging
Minimum Charge	Job		242		242	are not included.
Block / Bridge						
Demolish	Pr.		1.30		1.30	Includes material and labor to install
Install	Pr.	.02	3.59		3.61	set of steel, one-nail type cross
Demolish and Install	Pr.	.02	4.89		4.91	bridging for trusses placed 16″ O.C.
Minimum Charge	Job		242		242	
Block / Bridge						
2″ x 6″						
Demolish	Ea.		1.30		1.30	Includes material and labor to install
Install	Ea.	1.42	2.18		3.60	set of cross bridging or per block of
Demolish and Install	Ea.	1.42	3.48		4.90	solid bridging for 2″ x 6″ joists cut to
Minimum Charge	Job		242		242	size on site.

For customer support on your Contractor's Pricing Guide: Residential Repair & Remodeling, call 800.448.8182.

Rough Frame / Structure

Floor Framing System	Unit	Material	Labor	Equip.	Total	Specification
2" x 8"						
Demolish	Ea.		1.30		1.30	Includes material and labor to install
Install	Ea.	2.10	2.42		4.52	set of cross bridging or per block of
Demolish and Install	Ea.	2.10	3.72		5.82	solid bridging for 2" x 8" joists cut to
Minimum Charge	Job		242		242	size on site.
2" x 10"						
Demolish	Ea.		1.30		1.30	Includes material and labor to install
Install	Ea.	2.92	2.72		5.64	set of cross bridging or per block of
Demolish and Install	Ea.	2.92	4.02		6.94	solid bridging for 2" x 10" joists cut to
Minimum Charge	Job		242		242	size on site.
2" x 12"						
Demolish	Ea.		1.30		1.30	Includes material and labor to install
Install	Ea.	4.08	3.21		7.29	set of cross bridging or per block of
Demolish and Install	Ea.	4.08	4.51		8.59	solid bridging for 2" x 12" joists cut to
Minimum Charge	Job		242		242	size on site.
Ledger Strips						
1" x 2"						
Demolish	L.F.		.35		.35	Cost includes material and labor to
Install	L.F.	.33	1.40		1.73	install 1" x 2" ledger strip nailed to the
Demolish and Install	L.F.	.33	1.75		2.08	face of studs, beams or joist.
Minimum Charge	Job		242		242	
1" x 3"						
Demolish	L.F.		.35		.35	Includes material and labor to install
Install	L.F.	.69	1.61		2.30	up to 1" x 4" ledger strip nailed to the
Demolish and Install	L.F.	.69	1.96		2.65	face of studs, beams or joist.
Minimum Charge	Job		242		242	
1" x 4"						
Demolish	L.F.		.35		.35	Includes material and labor to install
Install	L.F.	.69	1.61		2.30	up to 1" x 4" ledger strip nailed to the
Demolish and Install	L.F.	.69	1.96		2.65	face of studs, beams or joist.
Minimum Charge	Job		242		242	
2" x 2"						
Demolish	L.F.		.38		.38	Cost includes material and labor to
Install	L.F.	.51	1.47		1.98	install 2" x 2" ledger strip nailed to the
Demolish and Install	L.F.	.51	1.85		2.36	face of studs, beams or joist.
Minimum Charge	Job		242		242	
2" x 4"						
Demolish	L.F.		.42		.42	Cost includes material and labor to
Install	L.F.	.73	1.94		2.67	install 2" x 4" ledger strip nailed to the
Demolish and Install	L.F.	.73	2.36		3.09	face of studs, beams or joist.
Reinstall	L.F.		1.55		1.55	
Minimum Charge	Job		242		242	
Ledger Boards						
2" x 6"						
Demolish	L.F.		.42		.42	Cost includes material and labor to
Install	L.F.	4.31	3.03		7.34	install 2" x 6" ledger board fastened to
Demolish and Install	L.F.	4.31	3.45		7.76	a wall, joists or studs.
Reinstall	L.F.		2.42		2.42	
Minimum Charge	Job		242		242	
4" x 6"						
Demolish	L.F.		.69		.69	Cost includes material and labor to
Install	L.F.	8.70	3.46		12.16	install 4" x 6" ledger board fastened to
Demolish and Install	L.F.	8.70	4.15		12.85	a wall, joists or studs.
Reinstall	L.F.		2.77		2.77	
Minimum Charge	Job		242		242	

Floor Framing System

Floor Framing System		Unit	Material	Labor	Equip.	Total	Specification
4″ x 8″							
	Demolish	L.F.		.92		.92	Cost includes material and labor to
	Install	L.F.	10.30	4.03		14.33	install 4″ x 8″ ledger board fastened to
	Demolish and Install	L.F.	10.30	4.95		15.25	a wall, joists or studs.
	Reinstall	L.F.		3.23		3.23	
	Minimum Charge	Job		242		242	
Sill Plate Per L.F.							
2″ x 4″							
	Demolish	L.F.		.36		.36	Cost includes material and labor to
	Install	L.F.	.94	1.76		2.70	install 2″ x 4″ pressure treated lumber,
	Demolish and Install	L.F.	.94	2.12		3.06	drilled and installed on foundation
	Reinstall	L.F.		1.41		1.41	bolts 48″ O.C. Bolts, nuts and washers
	Minimum Charge	Job		242		242	are not included.
2″ x 6″							
	Demolish	L.F.		.53		.53	Cost includes material and labor to
	Install	L.F.	1.31	1.94		3.25	install 2″ x 6″ pressure treated lumber,
	Demolish and Install	L.F.	1.31	2.47		3.78	drilled and installed on foundation
	Reinstall	L.F.		1.55		1.55	bolts 48″ O.C. Bolts, nuts and washers
	Minimum Charge	Job		242		242	not included.
2″ x 8″							
	Demolish	L.F.		.71		.71	Cost includes material and labor to
	Install	L.F.	1.75	2.15		3.90	install 2″ x 8″ pressure treated lumber,
	Demolish and Install	L.F.	1.75	2.86		4.61	drilled and installed on foundation
	Reinstall	L.F.		1.72		1.72	bolts 48″ O.C. Bolts, nuts and washers
	Minimum Charge	Job		242		242	not included.
Earthquake Strapping							
	Install	Ea.	2.68	3.03		5.71	Includes labor and material to replace
	Minimum Charge	Job		242		242	an earthquake strap.

Subflooring

Subflooring		Unit	Material	Labor	Equip.	Total	Specification
Plywood							
1/2″							
	Demolish	S.F.		1.08		1.08	Cost includes material and labor to
	Install	SF Flr.	.64	.52		1.16	install 1/2″ plywood subfloor, CD
	Demolish and Install	SF Flr.	.64	1.60		2.24	standard interior grade, plugged and
	Minimum Charge	Job		242		242	touch sanded.
5/8″							
	Demolish	S.F.		1.09		1.09	Cost includes material and labor to
	Install	SF Flr.	.69	.58		1.27	install 5/8″ plywood subfloor, CD
	Demolish and Install	SF Flr.	.69	1.67		2.36	standard interior grade, plugged and
	Reinstall	SF Flr.		.46		.46	touch sanded.
	Minimum Charge	Job		242		242	
3/4″							
	Demolish	S.F.		1.11		1.11	Cost includes material and labor to
	Install	SF Flr.	.79	.62		1.41	install 3/4″ plywood subfloor, CD
	Demolish and Install	SF Flr.	.79	1.73		2.52	standard interior grade, plugged and
	Reinstall	SF Flr.		.50		.50	touch sanded.
	Minimum Charge	Job		242		242	

33

For customer support on your Contractor's Pricing Guide: Residential Repair & Remodeling, call 800.448.8182.

Rough Frame / Structure

Subflooring

Subflooring		Unit	Material	Labor	Equip.	Total	Specification
Particle Board							
1/2″							
	Demolish	S.F.		.54		.54	Cost includes material and labor to
	Install	S.F.	1.10	.67		1.77	install 1/2″ particle board subfloor.
	Demolish and Install	S.F.	1.10	1.21		2.31	
	Minimum Charge	Job		242		242	
5/8″							
	Demolish	S.F.		.55		.55	Cost includes material and labor to
	Install	SF Flr.	1.13	.70		1.83	install 5/8″ particle board subfloor.
	Demolish and Install	SF Flr.	1.13	1.25		2.38	
	Minimum Charge	Job		242		242	
3/4″							
	Demolish	S.F.		.55		.55	Cost includes material and labor to
	Install	SF Flr.	1.17	.77		1.94	install 3/4″ particle board subfloor.
	Demolish and Install	SF Flr.	1.17	1.32		2.49	
	Minimum Charge	Job		242		242	
Plank Board							
	Demolish	S.F.		1.65		1.65	Cost includes material and labor to
	Install	SF Flr.	2.32	1.08		3.40	install 1″ x 6″ standard grade plank
	Demolish and Install	SF Flr.	2.32	2.73		5.05	flooring.
	Reinstall	SF Flr.		1.08		1.08	
	Minimum Charge	Job		242		242	
Felt Underlay							
	Demolish	S.F.		.10		.10	Includes material and labor to install
	Install	S.F.	.04	.13		.17	#15 felt building paper.
	Demolish and Install	S.F.	.04	.23		.27	
	Minimum Charge	Job		242		242	
Prep (for flooring)							
	Install	S.F.		.45		.45	Preparation of subflooring for
	Minimum Charge	Job		235		235	installation of new finished flooring. Cost reflects average time to prep area.

Wall Framing System

Wall Framing System		Unit	Material	Labor	Equip.	Total	Specification
2″ x 4″							
8′ High							
	Demolish	L.F.		5.55		5.55	Cost includes material and labor to
	Install	L.F.	7.45	8.05		15.50	install 2″ x 4″ x 8′ high wall system,
	Demolish and Install	L.F.	7.45	13.60		21.05	including studs, treated bottom plate,
	Reinstall	L.F.		6.45		6.45	double top plate and one row of fire
	Minimum Charge	Job		242		242	blocking.
9′ High							
	Demolish	L.F.		6.20		6.20	Cost includes material and labor to
	Install	L.F.	8	8.05		16.05	install 2″ x 4″ x 9′ wall system
	Demolish and Install	L.F.	8	14.25		22.25	including studs, treated bottom plate,
	Reinstall	L.F.		6.45		6.45	double top plate and one row of fire
	Minimum Charge	Job		242		242	blocking.
10′ High							
	Demolish	L.F.		6.95		6.95	Cost includes material and labor to
	Install	L.F.	8.55	8.05		16.60	install 2″ x 4″ x 10′ wall system
	Demolish and Install	L.F.	8.55	15		23.55	including studs, treated bottom plate,
	Reinstall	L.F.		6.45		6.45	double top plate and one row of fire
	Minimum Charge	Job		242		242	blocking.

For customer support on your Contractor's Pricing Guide: Residential Repair & Remodeling, call 800.448.8182.

Wall Framing System	Unit	Material	Labor	Equip.	Total	Specification
12' High						
Demolish	L.F.		8.30		8.30	Cost includes material and labor to
Install	L.F.	9.65	10.10		19.75	install 2" x 4" x 12' wall system
Demolish and Install	L.F.	9.65	18.40		28.05	including studs, treated bottom plate,
Reinstall	L.F.		8.07		8.07	double top plate and one row of fire
Minimum Charge	Job		242		242	blocking.
2" x 6"						
8' High						
Demolish	L.F.		5.20		5.20	Cost includes material and labor to
Install	L.F.	10.90	8.95		19.85	install 2" x 6" x 8' wall system
Demolish and Install	L.F.	10.90	14.15		25.05	including studs, treated bottom plate,
Reinstall	L.F.		7.17		7.17	double top plate and one row of fire
Minimum Charge	Job		242		242	blocking.
9' High						
Demolish	L.F.		5.85		5.85	Cost includes material and labor to
Install	L.F.	11.70	8.95		20.65	install 2" x 6" x 9' wall system
Demolish and Install	L.F.	11.70	14.80		26.50	including studs, treated bottom plate,
Reinstall	L.F.		7.17		7.17	double top plate and one row of fire
Minimum Charge	Job		242		242	blocking.
10' High						
Demolish	L.F.		6.40		6.40	Cost includes material and labor to
Install	L.F.	12.50	8.95		21.45	install 2" x 6" x 10' wall system
Demolish and Install	L.F.	12.50	15.35		27.85	including studs, treated bottom plate,
Reinstall	L.F.		7.17		7.17	double top plate and one row of fire
Minimum Charge	Job		242		242	blocking.
12' High						
Demolish	L.F.		7.55		7.55	Cost includes material and labor to
Install	L.F.	14.15	11		25.15	install 2" x 6" x 12' wall system
Demolish and Install	L.F.	14.15	18.55		32.70	including studs, treated bottom plate,
Reinstall	L.F.		8.80		8.80	double top plate and one row of fire
Minimum Charge	Job		242		242	blocking.
Fireblock						
2" x 4", 16" O.C. System						
Demolish	L.F.		.69		.69	Cost includes material and labor to
Install	L.F.	.73	1.61		2.34	install 2" x 4" fireblocks in wood frame
Demolish and Install	L.F.	.73	2.30		3.03	walls per L.F. of wall to be blocked.
Reinstall	L.F.		1.29		1.29	
Minimum Charge	Job		242		242	
2" x 4", 24" O.C. System						
Demolish	L.F.		.69		.69	Cost includes material and labor to
Install	L.F.	.73	1.61		2.34	install 2" x 4" fireblocks in wood frame
Demolish and Install	L.F.	.73	2.30		3.03	walls per L.F. of wall to be blocked.
Reinstall	L.F.		1.29		1.29	
Minimum Charge	Job		242		242	
2" x 6", 16" O.C. System						
Demolish	L.F.		1.04		1.04	Cost includes material and labor to
Install	L.F.	1.07	1.61		2.68	install 2" x 6" fireblocks in wood frame
Demolish and Install	L.F.	1.07	2.65		3.72	walls per L.F. of wall to be blocked.
Reinstall	L.F.		1.29		1.29	
Minimum Charge	Job		242		242	
2" x 6", 24" O.C. System						
Demolish	L.F.		1.04		1.04	Cost includes material and labor to
Install	L.F.	1.07	1.61		2.68	install 2" x 6" fireblocks in wood frame
Demolish and Install	L.F.	1.07	2.65		3.72	walls per L.F. of wall to be blocked.
Reinstall	L.F.		1.29		1.29	
Minimum Charge	Job		242		242	

Rough Frame / Structure

Wall Framing System

Wall Framing System	Unit	Material	Labor	Equip.	Total	Specification
Bracing						
1" x 3"						
Demolish	L.F.		.40		.40	Cost includes material and labor to
Install	L.F.	.51	3.03		3.54	install 1" x 3" let-in bracing.
Demolish and Install	L.F.	.51	3.43		3.94	
Minimum Charge	Job		242		242	
1" x 4"						
Demolish	L.F.		.40		.40	Cost includes material and labor to
Install	L.F.	.69	3.03		3.72	install 1" x 4" let-in bracing.
Demolish and Install	L.F.	.69	3.43		4.12	
Minimum Charge	Job		242		242	
1" x 6"						
Demolish	L.F.		.40		.40	Cost includes material and labor to
Install	L.F.	1.08	3.23		4.31	install 1" x 6" let-in bracing.
Demolish and Install	L.F.	1.08	3.63		4.71	
Minimum Charge	Job		242		242	
2" x 3"						
Demolish	L.F.		.52		.52	Cost includes material and labor to
Install	L.F.	.51	3.23		3.74	install 2" x 3" let-in bracing.
Demolish and Install	L.F.	.51	3.75		4.26	
Reinstall	L.F.		2.58		2.58	
Minimum Charge	Job		242		242	
2" x 4"						
Demolish	L.F.		.52		.52	Cost includes material and labor to
Install	L.F.	.73	3.23		3.96	install 2" x 4" let-in bracing.
Demolish and Install	L.F.	.73	3.75		4.48	
Reinstall	L.F.		2.58		2.58	
Minimum Charge	Job		242		242	
2" x 6"						
Demolish	L.F.		.52		.52	Cost includes material and labor to
Install	L.F.	1.07	3.46		4.53	install 2" x 6" let-in bracing.
Demolish and Install	L.F.	1.07	3.98		5.05	
Reinstall	L.F.		2.77		2.77	
Minimum Charge	Job		242		242	
2" x 8"						
Demolish	L.F.		.52		.52	Cost includes material and labor to
Install	L.F.	1.57	3.46		5.03	install 2" x 8" let-in bracing.
Demolish and Install	L.F.	1.57	3.98		5.55	
Reinstall	L.F.		2.77		2.77	
Minimum Charge	Job		242		242	
Earthquake Strapping						
Install	Ea.	2.68	3.03		5.71	Includes labor and material to replace
Minimum Charge	Job		242		242	an earthquake strap.
Hurricane Clips						
Install	Ea.	1.42	3.34		4.76	Includes labor and material to install a
Minimum Charge	Job		242		242	hurricane clip.
Stud						
2" x 4"						
Demolish	L.F.		.42		.42	Cost includes material and labor to
Install	L.F.	.72	.88		1.60	install 2" x 4" wall stud per L.F. of stud.
Demolish and Install	L.F.	.72	1.30		2.02	
Reinstall	L.F.		.70		.70	
Minimum Charge	Job		242		242	

For customer support on your Contractor's Pricing Guide: Residential Repair & Remodeling, call 800.448.8182.

Wall Framing System		Unit	Material	Labor	Equip.	Total	Specification
2″ x 6″							
	Demolish	L.F.		.52		.52	Cost includes material and labor to
	Install	L.F.	1.07	.97		2.04	install 2″ x 6″ stud per L.F. of stud.
	Demolish and Install	L.F.	1.07	1.49		2.56	
	Reinstall	L.F.		.77		.77	
	Minimum Charge	Job		242		242	
Plates							
2″ x 4″							
	Demolish	L.F.		.38		.38	Cost includes material and labor to
	Install	L.F.	.73	1.21		1.94	install 2″ x 4″ plate per L.F. of plate.
	Demolish and Install	L.F.	.73	1.59		2.32	
	Reinstall	L.F.		.97		.97	
	Minimum Charge	Job		242		242	
2″ x 6″							
	Demolish	L.F.		.40		.40	Cost includes material and labor to
	Install	L.F.	1.07	1.29		2.36	install 2″ x 6″ plate per L.F. of plate.
	Demolish and Install	L.F.	1.07	1.69		2.76	
	Reinstall	L.F.		1.03		1.03	
	Minimum Charge	Job		242		242	
Headers							
2″ x 6″							
	Demolish	L.F.		.37		.37	Includes material and labor to install
	Install	L.F.	1.07	2.69		3.76	header over wall openings and
	Demolish and Install	L.F.	1.07	3.06		4.13	around floor, ceiling and roof
	Clean	L.F.	.01	.47		.48	openings or flush beam.
	Paint	L.F.	.24	1.61		1.85	
	Minimum Charge	Job		242		242	
2″ x 8″							
	Demolish	L.F.		.40		.40	Includes material and labor to install
	Install	L.F.	1.57	2.85		4.42	header over wall openings and
	Demolish and Install	L.F.	1.57	3.25		4.82	around floor, ceiling and roof
	Clean	L.F.	.01	.47		.48	openings or flush beam.
	Paint	L.F.	.33	2.18		2.51	
	Minimum Charge	Job		242		242	
2″ x 10″							
	Demolish	L.F.		.40		.40	Includes material and labor to install
	Install	L.F.	2.19	3.03		5.22	header over wall openings and
	Demolish and Install	L.F.	2.19	3.43		5.62	around floor, ceiling and roof
	Clean	L.F.	.01	.48		.49	openings or flush beam.
	Paint	L.F.	.41	2.69		3.10	
	Minimum Charge	Job		242		242	
2″ x 12″							
	Demolish	L.F.		.42		.42	Includes material and labor to install
	Install	L.F.	3.05	3.23		6.28	header over wall openings and
	Demolish and Install	L.F.	3.05	3.65		6.70	around floor, ceiling and roof
	Clean	L.F.	.01	.48		.49	openings or flush beam.
	Paint	L.F.	.50	3.23		3.73	
	Minimum Charge	Job		242		242	
4″ x 8″							
	Demolish	L.F.		.74		.74	Includes material and labor to install
	Install	L.F.	6.15	3.72		9.87	header over wall openings and
	Demolish and Install	L.F.	6.15	4.46		10.61	around floor, ceiling and roof
	Clean	L.F.	.01	.47		.48	openings or flush beam.
	Paint	L.F.	.33	2.18		2.51	
	Minimum Charge	Job		242		242	

Rough Frame / Structure

Wall Framing System

Wall Framing System	Unit	Material	Labor	Equip.	Total	Specification
4" x 10"						
Demolish	L.F.		.79		.79	Includes material and labor to install
Install	L.F.	7.75	4.03		11.78	header over wall openings and
Demolish and Install	L.F.	7.75	4.82		12.57	around floor, ceiling and roof
Clean	L.F.	.01	.48		.49	openings or flush beam.
Paint	L.F.	.41	2.69		3.10	
Minimum Charge	Job		242		242	
4" x 12"						
Demolish	L.F.		.83		.83	Includes material and labor to install
Install	L.F.	9.50	5.10		14.60	header over wall openings and
Demolish and Install	L.F.	9.50	5.93		15.43	around floor, ceiling and roof
Clean	L.F.	.01	.48		.49	openings or flush beam.
Paint	L.F.	.50	3.23		3.73	
Minimum Charge	Job		242		242	
6" x 8"						
Demolish	L.F.		.74		.74	Includes material and labor to install
Install	L.F.	12.60	2.69		15.29	header over wall openings and
Demolish and Install	L.F.	12.60	3.43		16.03	around floor, ceiling and roof
Clean	L.F.	.01	.47		.48	openings or flush beam.
Paint	L.F.	.33	2.18		2.51	
Minimum Charge	Job		242		242	
6" x 10"						
Demolish	L.F.		.79		.79	Includes material and labor to install
Install	L.F.	16.15	5.85		22	header over wall openings and
Demolish and Install	L.F.	16.15	6.64		22.79	around floor, ceiling and roof
Clean	L.F.	.01	.48		.49	openings or flush beam.
Paint	L.F.	.41	2.69		3.10	
Minimum Charge	Job		242		242	
6" x 12"						
Demolish	L.F.		.83		.83	Includes material and labor to install
Install	L.F.	18.55	6.90		25.45	header over wall openings and
Demolish and Install	L.F.	18.55	7.73		26.28	around floor, ceiling and roof
Clean	L.F.	.01	.48		.49	openings or flush beam.
Paint	L.F.	.50	1.12		1.62	
Minimum Charge	Job		242		242	

Rough-in Opening

Rough-in Opening	Unit	Material	Labor	Equip.	Total	Specification
Door w / 2" x 4" Lumber						
3' Wide						
Demolish	Ea.		13		13	Includes material and labor to install
Install	Ea.	30.50	15.15		45.65	header, double studs each side,
Demolish and Install	Ea.	30.50	28.15		58.65	cripples, blocking and nails, up to 3'
Reinstall	Ea.		12.10		12.10	opening in 2" x 4" stud wall 8' high.
Minimum Charge	Job		242		242	
4' Wide						
Demolish	Ea.		13		13	Includes material and labor to install
Install	Ea.	32	15.15		47.15	header, double studs each side,
Demolish and Install	Ea.	32	28.15		60.15	cripples, blocking and nails, up to 4'
Reinstall	Ea.		12.10		12.10	opening in 2" x 4" stud wall 8' high.
Minimum Charge	Job		242		242	
5' Wide						
Demolish	Ea.		13		13	Includes material and labor to install
Install	Ea.	39.50	15.15		54.65	header, double studs each side,
Demolish and Install	Ea.	39.50	28.15		67.65	cripples, blocking and nails, up to 5'
Reinstall	Ea.		12.10		12.10	opening in 2" x 4" stud wall 8' high.
Minimum Charge	Job		242		242	

For customer support on your Contractor's Pricing Guide: Residential Repair & Remodeling, call 800.448.8182.

Rough-in Opening		Unit	Material	Labor	Equip.	Total	Specification
6' Wide							
	Demolish	Ea.		13		13	Includes material and labor to install
	Install	Ea.	43	15.15		58.15	header, double studs each side,
	Demolish and Install	Ea.	43	28.15		71.15	cripples, blocking and nails, up to 6'
	Reinstall	Ea.		12.10		12.10	opening in 2" x 4" stud wall 8' high.
	Minimum Charge	Job		242		242	
8' Wide							
	Demolish	Ea.		13.85		13.85	Includes material and labor to install
	Install	Ea.	59	16.15		75.15	header, double studs each side,
	Demolish and Install	Ea.	59	30		89	cripples, blocking and nails, up to 8'
	Reinstall	Ea.		12.91		12.91	opening in 2" x 4" stud wall 8' high.
	Minimum Charge	Job		242		242	
10' Wide							
	Demolish	Ea.		13.85		13.85	Includes material and labor to install
	Install	Ea.	85.50	16.15		101.65	header, double studs each side,
	Demolish and Install	Ea.	85.50	30		115.50	cripples, blocking and nails, up to 10'
	Reinstall	Ea.		12.91		12.91	opening in 2" x 4" stud wall 8' high.
	Minimum Charge	Job		242		242	
12' Wide							
	Demolish	Ea.		13.85		13.85	Includes material and labor to install
	Install	Ea.	165	16.15		181.15	header, double studs each side,
	Demolish and Install	Ea.	165	30		195	cripples, blocking and nails, up to 12'
	Reinstall	Ea.		12.91		12.91	opening in 2" x 4" stud wall 8' high.
	Minimum Charge	Job		242		242	
Door w / 2" x 6" Lumber							
3' Wide							
	Demolish	Ea.		13		13	Includes material and labor to install
	Install	Ea.	41.50	15.15		56.65	header, double studs each side,
	Demolish and Install	Ea.	41.50	28.15		69.65	cripples, blocking and nails, up to 3'
	Reinstall	Ea.		12.10		12.10	opening in 2" x 6" stud wall 8' high.
	Minimum Charge	Job		242		242	
4' Wide							
	Demolish	Ea.		13		13	Includes material and labor to install
	Install	Ea.	43	15.15		58.15	header, double studs each side,
	Demolish and Install	Ea.	43	28.15		71.15	cripples, blocking and nails, up to 4'
	Reinstall	Ea.		12.10		12.10	opening in 2" x 6" stud wall 8' high.
	Minimum Charge	Job		242		242	
5' Wide							
	Demolish	Ea.		13		13	Includes material and labor to install
	Install	Ea.	50.50	15.15		65.65	header, double studs each side,
	Demolish and Install	Ea.	50.50	28.15		78.65	cripples, blocking and nails up to 5'
	Reinstall	Ea.		12.10		12.10	opening in 2" x 6" stud wall 8' high.
	Minimum Charge	Job		242		242	
6' Wide							
	Demolish	Ea.		13		13	Includes material and labor to install
	Install	Ea.	54	15.15		69.15	header, double studs each side,
	Demolish and Install	Ea.	54	28.15		82.15	cripples, blocking and nails up to 6'
	Reinstall	Ea.		12.10		12.10	opening in 2" x 6" stud wall 8' high.
	Minimum Charge	Job		242		242	
8' Wide							
	Demolish	Ea.		13.85		13.85	Includes material and labor to install
	Install	Ea.	70	16.15		86.15	header, double studs each side,
	Demolish and Install	Ea.	70	30		100	cripples, blocking and nails up to 8'
	Reinstall	Ea.		12.91		12.91	opening in 2" x 6" stud wall 8' high.
	Minimum Charge	Job		242		242	
10' Wide							
	Demolish	Ea.		13.85		13.85	Includes material and labor to install
	Install	Ea.	96.50	16.15		112.65	header, double studs each side,
	Demolish and Install	Ea.	96.50	30		126.50	cripples, blocking and nails up to 10'
	Reinstall	Ea.		12.91		12.91	opening in 2" x 6" stud wall 8' high.
	Minimum Charge	Job		242		242	

Rough Frame / Structure

Rough-in Opening

12' Wide

	Unit	Material	Labor	Equip.	Total	Specification
Demolish	Ea.		13.85		13.85	Includes material and labor to install
Install	Ea.	176	16.15		192.15	header, double studs each side,
Demolish and Install	Ea.	176	30		206	cripples, blocking and nails, up to 12'
Reinstall	Ea.		12.91		12.91	opening in 2" x 6" stud wall 8' high.
Minimum Charge	Job		242		242	

Rough-in Opening

Window w / 2" x 4" Lumber

2' Wide

	Unit	Material	Labor	Equip.	Total	Specification
Demolish	Ea.		17.30		17.30	Includes material and labor to install
Install	Ea.	32	20		52	header, double studs each side of
Demolish and Install	Ea.	32	37.30		69.30	opening, cripples, blocking, nails and
Reinstall	Ea.		16.13		16.13	sub-sills for opening up to 2' in a 2" x
Minimum Charge	Job		242		242	4" stud wall 8' high.

3' Wide

	Unit	Material	Labor	Equip.	Total	Specification
Demolish	Ea.		17.30		17.30	Includes material and labor to install
Install	Ea.	37.50	20		57.50	header, double studs each side of
Demolish and Install	Ea.	37.50	37.30		74.80	opening, cripples, blocking, nails and
Reinstall	Ea.		16.13		16.13	sub-sills for opening up to 3' in a 2" x
Minimum Charge	Job		242		242	4" stud wall 8' high.

4' Wide

	Unit	Material	Labor	Equip.	Total	Specification
Demolish	Ea.		17.30		17.30	Includes material and labor to install
Install	Ea.	40.50	20		60.50	header, double studs each side of
Demolish and Install	Ea.	40.50	37.30		77.80	opening, cripples, blocking, nails and
Reinstall	Ea.		16.13		16.13	sub-sills for opening up to 4' in a 2" x
Minimum Charge	Job		242		242	4" stud wall 8' high.

5' Wide

	Unit	Material	Labor	Equip.	Total	Specification
Demolish	Ea.		17.30		17.30	Includes material and labor to install
Install	Ea.	48.50	20		68.50	header, double studs each side of
Demolish and Install	Ea.	48.50	37.30		85.80	opening, cripples, blocking, nails and
Reinstall	Ea.		16.13		16.13	sub-sills for opening up to 5' in a 2" x
Minimum Charge	Job		242		242	4" stud wall 8' high.

6' Wide

	Unit	Material	Labor	Equip.	Total	Specification
Demolish	Ea.		17.30		17.30	Includes material and labor to install
Install	Ea.	53.50	20		73.50	header, double studs each side of
Demolish and Install	Ea.	53.50	37.30		90.80	opening, cripples, blocking, nails and
Reinstall	Ea.		16.13		16.13	sub-sills for opening up to 6' in a 2" x
Minimum Charge	Job		242		242	4" stud wall 8' high.

7' Wide

	Unit	Material	Labor	Equip.	Total	Specification
Demolish	Ea.		17.30		17.30	Includes material and labor to install
Install	Ea.	67	20		87	header, double studs each side of
Demolish and Install	Ea.	67	37.30		104.30	opening, cripples, blocking, nails and
Reinstall	Ea.		16.13		16.13	sub-sills for opening up to 7' in a 2" x
Minimum Charge	Job		242		242	4" stud wall 8' high.

8' Wide

	Unit	Material	Labor	Equip.	Total	Specification
Demolish	Ea.		18.90		18.90	Includes material and labor to install
Install	Ea.	74.50	22		96.50	header, double studs each side of
Demolish and Install	Ea.	74.50	40.90		115.40	opening, cripples, blocking, nails and
Reinstall	Ea.		17.60		17.60	sub-sills for opening up to 8' in a 2" x
Minimum Charge	Job		242		242	4" stud wall 8' high.

For customer support on your Contractor's Pricing Guide: Residential Repair & Remodeling, call 800.448.8182.

Rough-in Opening		Unit	Material	Labor	Equip.	Total	Specification
10' Wide							
	Demolish	Ea.		18.90		18.90	Includes material and labor to install
	Install	Ea.	103	22		125	header, double studs each side of
	Demolish and Install	Ea.	103	40.90		143.90	opening, cripples, blocking, nails and
	Reinstall	Ea.		17.60		17.60	sub-sills for opening up to 10' in a 2"
	Minimum Charge	Job		242		242	x 4" stud wall 8' high.
12' Wide							
	Demolish	Ea.		18.90		18.90	Includes material and labor to install
	Install	Ea.	187	22		209	header, double studs each side of
	Demolish and Install	Ea.	187	40.90		227.90	opening, cripples, blocking, nails and
	Reinstall	Ea.		17.60		17.60	sub-sills for opening up to 12' in a 2"
	Minimum Charge	Job		242		242	x 4" stud wall 8' high.

Window w / 2" x 6" Lumber

		Unit	Material	Labor	Equip.	Total	Specification
2' Wide							
	Demolish	Ea.		17.30		17.30	Includes material and labor to install
	Install	Ea.	45.50	20		65.50	header, double studs each side of
	Demolish and Install	Ea.	45.50	37.30		82.80	opening, cripples, blocking, nails and
	Reinstall	Ea.		16.13		16.13	sub-sills for opening up to 2' in a 2" x
	Minimum Charge	Job		242		242	6" stud wall 8' high.
3' Wide							
	Demolish	Ea.		17.30		17.30	Includes material and labor to install
	Install	Ea.	51	20		71	header, double studs each side of
	Demolish and Install	Ea.	51	37.30		88.30	opening, cripples, blocking, nails and
	Reinstall	Ea.		16.13		16.13	sub-sills for opening up to 3' in a 2" x
	Minimum Charge	Job		242		242	6" stud wall 8' high.
4' Wide							
	Demolish	Ea.		17.30		17.30	Includes material and labor to install
	Install	Ea.	55.50	20		75.50	header, double studs each side of
	Demolish and Install	Ea.	55.50	37.30		92.80	opening, cripples, blocking, nails and
	Reinstall	Ea.		16.13		16.13	sub-sills for opening up to 4' in a 2" x
	Minimum Charge	Job		242		242	6" stud wall 8' high.
5' Wide							
	Demolish	Ea.		17.30		17.30	Includes material and labor to install
	Install	Ea.	63.50	20		83.50	header, double studs each side of
	Demolish and Install	Ea.	63.50	37.30		100.80	opening, cripples, blocking, nails and
	Reinstall	Ea.		16.13		16.13	sub-sills for opening up to 5' in a 2" x
	Minimum Charge	Job		242		242	6" stud wall 8' high.
6' Wide							
	Demolish	Ea.		17.30		17.30	Includes material and labor to install
	Install	Ea.	70	20		90	header, double studs each side of
	Demolish and Install	Ea.	70	37.30		107.30	opening, cripples, blocking, nails and
	Reinstall	Ea.		16.13		16.13	sub-sills for opening up to 6' in a 2" x
	Minimum Charge	Job		242		242	6" stud wall 8' high.
7' Wide							
	Demolish	Ea.		17.30		17.30	Includes material and labor to install
	Install	Ea.	84.50	20		104.50	header, double studs each side of
	Demolish and Install	Ea.	84.50	37.30		121.80	opening, cripples, blocking, nails and
	Reinstall	Ea.		16.13		16.13	sub-sills for opening up to 7' in a 2" x
	Minimum Charge	Job		242		242	6" stud wall 8' high.

Rough Frame / Structure

Rough-in Opening

	Unit	Material	Labor	Equip.	Total	Specification
8' Wide						
Demolish	Ea.		18.90		18.90	Includes material and labor to install
Install	Ea.	92.50	22		114.50	header, double studs each side of
Demolish and Install	Ea.	92.50	40.90		133.40	opening, cripples, blocking, nails and
Reinstall	Ea.		17.60		17.60	sub-sills for opening up to 8' in a 2" x
Minimum Charge	Job		242		242	6" stud wall 8' high.
10' Wide						
Demolish	Ea.		18.90		18.90	Includes material and labor to install
Install	Ea.	122	22		144	header, double studs each side of
Demolish and Install	Ea.	122	40.90		162.90	opening, cripples, blocking, nails and
Reinstall	Ea.		17.60		17.60	sub-sills for opening up to 10' in a 2"
Minimum Charge	Job		242		242	x 6" stud wall 8' high.
12' Wide						
Demolish	Ea.		18.90		18.90	Includes material and labor to install
Install	Ea.	208	22		230	header, double studs each side of
Demolish and Install	Ea.	208	40.90		248.90	opening, cripples, blocking, nails and
Reinstall	Ea.		17.60		17.60	sub-sills for opening up to 12' in a 2"
Minimum Charge	Job		242		242	x 6" stud wall 8' high.

Glue-Laminated Beams

	Unit	Material	Labor	Equip.	Total	Specification
3-1/2" x 6"						
Demolish	L.F.		3.09		3.09	Includes material and labor to install
Install	L.F.	8.15	3.75	3.64	15.54	glue-laminated wood beam.
Demolish and Install	L.F.	8.15	6.84	3.64	18.63	
Reinstall	L.F.		3	2.91	5.91	
Clean	L.F.	.06	.47		.53	
Paint	L.F.	.57	1.19		1.76	
Minimum Charge	Job		242		242	
3-1/2" x 9"						
Demolish	L.F.		3.09		3.09	Includes material and labor to install
Install	L.F.	10.90	3.75	3.64	18.29	glue-laminated wood beam.
Demolish and Install	L.F.	10.90	6.84	3.64	21.38	
Reinstall	L.F.		3	2.91	5.91	
Clean	L.F.	.07	.48		.55	
Paint	L.F.	.74	1.54		2.28	
Minimum Charge	Job		242		242	
3-1/2" x 12"						
Demolish	L.F.		4.64		4.64	Includes material and labor to install
Install	L.F.	14.50	3.75	3.64	21.89	glue-laminated wood beam.
Demolish and Install	L.F.	14.50	8.39	3.64	26.53	
Reinstall	L.F.		3	2.91	5.91	
Clean	L.F.	.08	.50		.58	
Paint	L.F.	.91	1.92		2.83	
Minimum Charge	Job		242		242	
3-1/2" x 15"						
Demolish	L.F.		4.64		4.64	Includes material and labor to install
Install	L.F.	18.15	3.88	3.76	25.79	glue-laminated wood beam.
Demolish and Install	L.F.	18.15	8.52	3.76	30.43	
Reinstall	L.F.		3.10	3.01	6.11	
Clean	L.F.	.10	.52		.62	
Paint	L.F.	1.08	2.24		3.32	
Minimum Charge	Job		242		242	

For customer support on your Contractor's Pricing Guide: Residential Repair & Remodeling, call 800.448.8182.

Glue-Laminated Beams

Glue-Laminated Beams	Unit	Material	Labor	Equip.	Total	Specification
3-1/2″ x 18″						
Demolish	L.F.		5.80		5.80	Includes material and labor to install
Install	L.F.	24	3.88	3.76	31.64	glue-laminated wood beam.
Demolish and Install	L.F.	24	9.68	3.76	37.44	
Reinstall	L.F.		3.10	3.01	6.11	
Clean	L.F.	.11	.53		.64	
Paint	L.F.	1.27	2.60		3.87	
Minimum Charge	Job		242		242	
5-1/8″ x 6″						
Demolish	L.F.		4.64		4.64	Includes material and labor to install
Install	L.F.	11.90	3.75	3.64	19.29	glue-laminated wood beam.
Demolish and Install	L.F.	11.90	8.39	3.64	23.93	
Reinstall	L.F.		3	2.91	5.91	
Clean	L.F.	.06	.47		.53	
Paint	L.F.	.62	1.30		1.92	
Minimum Charge	Job		242		242	
5-1/8″ x 9″						
Demolish	L.F.		5.80		5.80	Includes material and labor to install
Install	L.F.	17.90	3.75	3.64	25.29	glue-laminated wood beam.
Demolish and Install	L.F.	17.90	9.55	3.64	31.09	
Reinstall	L.F.		3	2.91	5.91	
Clean	L.F.	.07	.48		.55	
Paint	L.F.	.79	1.67		2.46	
Minimum Charge	Job		242		242	
5-1/8″ x 12″						
Demolish	L.F.		6.90		6.90	Includes material and labor to install
Install	L.F.	23.50	3.75	3.64	30.89	glue-laminated wood beam.
Demolish and Install	L.F.	23.50	10.65	3.64	37.79	
Reinstall	L.F.		3	2.91	5.91	
Clean	L.F.	.09	.50		.59	
Paint	L.F.	.98	2.02		3	
Minimum Charge	Job		242		242	
5-1/8″ x 18″						
Demolish	L.F.		11.60		11.60	Includes material and labor to install
Install	L.F.	36	3.88	3.76	43.64	glue-laminated wood beam.
Demolish and Install	L.F.	36	15.48	3.76	55.24	
Reinstall	L.F.		3.10	3.01	6.11	
Clean	L.F.	.12	.53		.65	
Paint	L.F.	1.13	2.73		3.86	
Minimum Charge	Job		242		242	
6-3/4″ x 12″						
Demolish	L.F.		6.90		6.90	Includes material and labor to install
Install	L.F.	31.50	3.88	3.76	39.14	glue-laminated wood beam.
Demolish and Install	L.F.	31.50	10.78	3.76	46.04	
Reinstall	L.F.		3.10	3.01	6.11	
Clean	L.F.	.10	.52		.62	
Paint	L.F.	1.08	2.24		3.32	
Minimum Charge	Job		242		242	
6-3/4″ x 15″						
Demolish	L.F.		11.60		11.60	Includes material and labor to install
Install	L.F.	39	3.88	3.76	46.64	glue-laminated wood beam.
Demolish and Install	L.F.	39	15.48	3.76	58.24	
Reinstall	L.F.		3.10	3.01	6.11	
Clean	L.F.	.11	.53		.64	
Paint	L.F.	1.27	2.60		3.87	
Minimum Charge	Job		242		242	

For customer support on your Contractor's Pricing Guide: Residential Repair & Remodeling, call 800.448.8182.

Rough Frame / Structure

Glue-Laminated Beams

	Unit	Material	Labor	Equip.	Total	Specification
6-3/4" x 18"						
Demolish	L.F.		11.60		11.60	Includes material and labor to install
Install	L.F.	47.50	4.02	3.90	55.42	glue-laminated wood beam.
Demolish and Install	L.F.	47.50	15.62	3.90	67.02	
Reinstall	L.F.		3.21	3.12	6.33	
Clean	L.F.	.13	.55		.68	
Paint	L.F.	1.43	2.97		4.40	
Minimum Charge	Job		242		242	
Hardware						
5-1/4" Glue-Lam Seat						
Install	Ea.	230	2.69		232.69	Includes labor and material to install
Minimum Charge	Job		242		242	beam hangers.
6-3/4" Glue-Lam Seat						
Install	Ea.	230	2.69		232.69	Includes labor and material to install
Minimum Charge	Job		242		242	beam hangers.
Earthquake Strapping						
Install	Ea.	2.68	3.03		5.71	Includes labor and material to replace
Minimum Charge	Job		242		242	an earthquake strap.
Hurricane Clips						
Install	Ea.	1.42	3.34		4.76	Includes labor and material to install a
Minimum Charge	Job		242		242	hurricane clip.

Metal Stud Framing

	Unit	Material	Labor	Equip.	Total	Specification
16" O.C. System						
4", 16 Ga.						
Demolish	S.F.		1.52		1.52	Includes material and labor to install
Install	S.F.	1.35	1.49		2.84	load bearing cold rolled metal stud
Demolish and Install	S.F.	1.35	3.01		4.36	walls, to 10' high, including studs, top
Reinstall	S.F.		1.19		1.19	and bottom track and screws.
Minimum Charge	Job		242		242	
6", 16 Ga.						
Demolish	S.F.		1.52		1.52	Includes material and labor to install
Install	S.F.	1.72	1.51		3.23	load bearing cold rolled metal stud
Demolish and Install	S.F.	1.72	3.03		4.75	walls, to 10' high, including studs, top
Reinstall	S.F.		1.21		1.21	and bottom track and screws.
Minimum Charge	Job		242		242	
4", 25 Ga.						
Demolish	S.F.		1.05		1.05	Includes material and labor to install
Install	S.F.	.80	1.02		1.82	cold rolled metal stud walls, to 10'
Demolish and Install	S.F.	.80	2.07		2.87	high, including studs, top and bottom
Reinstall	S.F.		.82		.82	track and screws.
Minimum Charge	Job		242		242	
6", 25 Ga.						
Demolish	S.F.		1.05		1.05	Includes material and labor to install
Install	S.F.	.86	1.03		1.89	cold rolled metal stud walls, to 10'
Demolish and Install	S.F.	.86	2.08		2.94	high, including studs, top and bottom
Reinstall	S.F.		.82		.82	track and screws.
Minimum Charge	Job		242		242	
24" O.C. System						
4", 25 Ga.						
Demolish	S.F.		.89		.89	Includes material and labor to install
Install	S.F.	.58	.65		1.23	cold rolled metal stud walls, to 10'
Demolish and Install	S.F.	.58	1.54		2.12	high, including studs, top and bottom
Reinstall	S.F.		.52		.52	track and screws.
Minimum Charge	Job		242		242	

For customer support on your Contractor's Pricing Guide: Residential Repair & Remodeling, call 800.448.8182.

Rough Frame / Structure

Metal Stud Framing

	Unit	Material	Labor	Equip.	Total	Specification
6", 25 Ga.						
Demolish	S.F.		.89		.89	Includes material and labor to install
Install	S.F.	.63	.67		1.30	cold rolled metal stud walls, to 10'
Demolish and Install	S.F.	.63	1.56		2.19	high, including studs, top and bottom
Reinstall	S.F.		.53		.53	track and screws.
Minimum Charge	Job		242		242	
4", 16 Ga.						
Demolish	L.F.		10.20		10.20	Includes material and labor to install
Install	S.F.	.97	1.08		2.05	load bearing cold rolled metal stud
Demolish and Install	S.F.	.97	11.28		12.25	walls, to 10' high, including studs, top
Reinstall	S.F.		.86		.86	and bottom track and screws.
Minimum Charge	Job		242		242	
6", 16 Ga.						
Demolish	L.F.		10.20		10.20	Includes material and labor to install
Install	S.F.	1.25	1.10		2.35	load bearing cold rolled metal stud
Demolish and Install	S.F.	1.25	11.30		12.55	walls, to 10' high, including studs, top
Reinstall	S.F.		.88		.88	and bottom track and screws.
Minimum Charge	Job		242		242	

Metal Joist

	Unit	Material	Labor	Equip.	Total	Specification
Bar Joist						
18K9						
Demolish	L.F.		1.22	1.16	2.38	Includes material, labor and
Install	L.F.	10.40	2.50	1.53	14.43	equipment to install open web joist.
Demolish and Install	L.F.	10.40	3.72	2.69	16.81	
Reinstall	L.F.		2	1.22	3.22	
Minimum Charge	Job		2225	2175	4400	
16K6						
Demolish	L.F.		1.22	1.16	2.38	Includes material, labor and
Install	L.F.	5.35	1.67	.12	7.14	equipment to install open web joist.
Demolish and Install	L.F.	5.35	2.89	1.28	9.52	
Reinstall	L.F.		1.34	.10	1.44	
Minimum Charge	Job		2225	2175	4400	

Exterior Sheathing

	Unit	Material	Labor	Equip.	Total	Specification
CDX Plywood						
5/16"						
Demolish	S.F.		.71		.71	Cost includes material and labor to
Install	S.F.	1	.61		1.61	install 5/16" CDX plywood sheathing.
Demolish and Install	S.F.	1	1.32		2.32	
Minimum Charge	Job		242		242	
3/8"						
Demolish	S.F.		.72		.72	Cost includes material and labor to
Install	S.F.	.75	.81		1.56	install 3/8" CDX plywood sheathing.
Demolish and Install	S.F.	.75	1.53		2.28	
Minimum Charge	Job		242		242	
1/2"						
Demolish	S.F.		.74		.74	Cost includes material and labor to
Install	S.F.	.84	.86		1.70	install 1/2" CDX plywood sheathing.
Demolish and Install	S.F.	.84	1.60		2.44	
Minimum Charge	Job		242		242	

Rough Frame / Structure

Exterior Sheathing

Exterior Sheathing	Unit	Material	Labor	Equip.	Total	Specification
5/8″						
Demolish	S.F.		.76		.76	Cost includes material and labor to
Install	S.F.	.90	.92		1.82	install 5/8″ CDX plywood sheathing.
Demolish and Install	S.F.	.90	1.68		2.58	
Minimum Charge	Job		242		242	
3/4″						
Demolish	S.F.		.77		.77	Cost includes material and labor to
Install	S.F.	1.03	.99		2.02	install 3/4″ CDX plywood sheathing.
Demolish and Install	S.F.	1.03	1.76		2.79	
Minimum Charge	Job		242		242	
OSB						
1/2″						
Demolish	S.F.		.74		.74	Cost includes material and labor to
Install	S.F.	.52	.69		1.21	install 4′ x 8′ x 1/2″ OSB sheathing.
Demolish and Install	S.F.	.52	1.43		1.95	
Minimum Charge	Job		242		242	
5/8″						
Demolish	S.F.		.76		.76	Cost includes material and labor to
Install	S.F.	.52	.56		1.08	install 4′ x 8′ x 5/8″ OSB sheathing.
Demolish and Install	S.F.	.52	1.32		1.84	
Minimum Charge	Job		242		242	
Vapor Barrier						
Black Paper						
Install	S.F.	.04	.13		.17	Includes material and labor to install
Minimum Charge	Job		242		242	#15 felt building paper.

Plywood Sheathing

Plywood Sheathing	Unit	Material	Labor	Equip.	Total	Specification
Finish Plywood						
5/16″						
Demolish	S.F.		.71		.71	Includes material and labor to install
Install	S.F.	2.27	.30		2.57	exterior 5/16″ AC plywood on walls.
Demolish and Install	S.F.	2.27	1.01		3.28	
Minimum Charge	Job		242		242	
3/8″						
Demolish	S.F.		.72		.72	Includes material and labor to install
Install	S.F.	2.27	.40		2.67	exterior 3/8″ AC plywood on walls.
Demolish and Install	S.F.	2.27	1.12		3.39	
Minimum Charge	Job		242		242	
1/2″						
Demolish	S.F.		.74		.74	Includes material and labor to install
Install	S.F.	3.03	.43		3.46	exterior 1/2″ AC plywood on walls.
Demolish and Install	S.F.	3.03	1.17		4.20	
Minimum Charge	Job		242		242	
5/8″						
Demolish	S.F.		.76		.76	Includes material and labor to install
Install	S.F.	3.09	.43		3.52	exterior 5/8″ AC plywood on walls.
Demolish and Install	S.F.	3.09	1.19		4.28	
Reinstall	S.F.		.43		.43	
Minimum Charge	Job		242		242	
3/4″						
Demolish	S.F.		.77		.77	Includes material and labor to install
Install	S.F.	3.39	.43		3.82	exterior 3/4″ AC plywood on walls.
Demolish and Install	S.F.	3.39	1.20		4.59	
Reinstall	S.F.		.43		.43	
Minimum Charge	Job		242		242	

For customer support on your Contractor's Pricing Guide: Residential Repair & Remodeling, call 800.448.8182.

Rough Frame / Structure

Stairs

	Unit	Material	Labor	Equip.	Total	Specification
Job-built						
Treads and Risers						Includes material and labor to install
Demolish	Ea.		52		52	three 2" x 12" stringers, treads and
Install	Ea.	1125	485		1610	risers of 3/4" CDX plywood, installed
Demolish and Install	Ea.	1125	537		1662	in a straight or "L" shaped run
Clean	Flight	1.52	31.50		33.02	including carpet.
Paint	Ea.	.26	80.50		80.76	
Minimum Charge	Job		242		242	
Landing						
Demolish	S.F.		.42		.42	Includes material and labor to install
Install	S.F.	61	3.36		64.36	landing framing, 3/4" CDX plywood
Demolish and Install	S.F.	61	3.78		64.78	surface and carpet.
Clean	S.F.		.47		.47	
Paint	S.F.	.63	1		1.63	
Minimum Charge	Job		242		242	

Ceiling Framing

	Unit	Material	Labor	Equip.	Total	Specification
16" O.C. System						
2" x 6" Joists						Cost includes material and labor to
Demolish	S.F.		.64		.64	install 2" x 6" joists including end and
Install	S.F.	.81	.80		1.61	header joist installed 16" O.C. Does
Demolish and Install	S.F.	.81	1.44		2.25	not include beams, ledger strips,
Reinstall	S.F.		.64		.64	blocking or bridging.
Minimum Charge	Job		242		242	
2" x 8" Joists						Cost includes material and labor to
Demolish	S.F.		.66		.66	install 2" x 8" joists including end and
Install	S.F.	1.20	.98		2.18	header joist installed 16" O.C. Does
Demolish and Install	S.F.	1.20	1.64		2.84	not include beams, ledger strips,
Reinstall	S.F.		.78		.78	blocking or bridging.
Minimum Charge	Job		242		242	
2" x 10" Joists						Cost includes material and labor to
Demolish	S.F.		.68		.68	install 2" x 10" joists including end
Install	S.F.	1.66	1.16		2.82	and header joist installed 16" O.C.
Demolish and Install	S.F.	1.66	1.84		3.50	Does not include beams, ledger strips,
Reinstall	S.F.		.93		.93	blocking or bridging.
Minimum Charge	Job		242		242	
2" x 12" Joists						Cost includes material and labor to
Demolish	S.F.		.71		.71	install 2" x 12" joists including end
Install	S.F.	2.32	1.42		3.74	and header joist installed 16" O.C.
Demolish and Install	S.F.	2.32	2.13		4.45	Does not include beams, ledger strips,
Reinstall	S.F.		1.13		1.13	blocking or bridging.
Minimum Charge	Job		242		242	
Block / Bridge						
Demolish	Ea.		1.30		1.30	Includes material and labor to install
Install	Ea.	2.92	2.72		5.64	set of cross bridging or per block of
Demolish and Install	Ea.	2.92	4.02		6.94	solid bridging for 2" x 10" joists cut to
Minimum Charge	Job		242		242	size on site.

For customer support on your Contractor's Pricing Guide: Residential Repair & Remodeling, call 800.448.8182.

Ceiling Framing

Ceiling Framing	Unit	Material	Labor	Equip.	Total	Specification
24″ O.C. System						
2″ x 6″ Joists						
Demolish	S.F.		.43		.43	Cost includes material and labor to
Install	S.F.	.54	.53		1.07	install 2″ x 6″ joists including end and
Demolish and Install	S.F.	.54	.96		1.50	header joist installed 24″ O.C. Does
Reinstall	S.F.		.42		.42	not include beams, ledger strips,
Minimum Charge	Job		242		242	blocking or bridging.
2″ x 8″ Joists						
Demolish	S.F.		.44		.44	Cost includes material and labor to
Install	S.F.	.80	.65		1.45	install 2″ x 8″ joists including end and
Demolish and Install	S.F.	.80	1.09		1.89	header joist installed 24″ O.C. Does
Reinstall	S.F.		.52		.52	not include beams, ledger strips,
Minimum Charge	Job		242		242	blocking or bridging.
2″ x 10″ Joists						
Demolish	S.F.		.46		.46	Cost includes material and labor to
Install	S.F.	1.11	.77		1.88	install 2″ x 10″ joists including end
Demolish and Install	S.F.	1.11	1.23		2.34	and header joist installed 24″ O.C.
Reinstall	S.F.		.62		.62	Does not include beams, ledger strips,
Minimum Charge	Job		242		242	blocking or bridging.
2″ x 12″ Joists						
Demolish	S.F.		.47		.47	Cost includes material and labor to
Install	S.F.	1.56	.94		2.50	install 2″ x 12″ joists including end
Demolish and Install	S.F.	1.56	1.41		2.97	and header joist installed 24″ O.C.
Reinstall	S.F.		.75		.75	Does not include beams, ledger strips,
Minimum Charge	Job		242		242	blocking or bridging.
Block / Bridge						
Demolish	Ea.		1.30		1.30	Includes material and labor to install
Install	Ea.	2.92	2.72		5.64	set of cross bridging or per block of
Demolish and Install	Ea.	2.92	4.02		6.94	solid bridging for 2″ x 10″ joists cut to
Minimum Charge	Job		242		242	size on site.
Ledger Strips						
1″ x 2″						
Demolish	L.F.		.35		.35	Cost includes material and labor to
Install	L.F.	.33	1.40		1.73	install 1″ x 2″ ledger strip nailed to the
Demolish and Install	L.F.	.33	1.75		2.08	face of studs, beams or joist.
Minimum Charge	Job		242		242	
1″ x 3″						
Demolish	L.F.		.35		.35	Includes material and labor to install
Install	L.F.	.69	1.61		2.30	up to 1″ x 4″ ledger strip nailed to the
Demolish and Install	L.F.	.69	1.96		2.65	face of studs, beams or joist.
Minimum Charge	Job		242		242	
1″ x 4″						
Demolish	L.F.		.35		.35	Includes material and labor to install
Install	L.F.	.69	1.61		2.30	up to 1″ x 4″ ledger strip nailed to the
Demolish and Install	L.F.	.69	1.96		2.65	face of studs, beams or joist.
Minimum Charge	Job		242		242	
2″ x 2″						
Demolish	L.F.		.38		.38	Cost includes material and labor to
Install	L.F.	.51	1.47		1.98	install 2″ x 2″ ledger strip nailed to the
Demolish and Install	L.F.	.51	1.85		2.36	face of studs, beams or joist.
Minimum Charge	Job		242		242	
2″ x 4″						
Demolish	L.F.		.42		.42	Cost includes material and labor to
Install	L.F.	.73	1.94		2.67	install 2″ x 4″ ledger strip nailed to the
Demolish and Install	L.F.	.73	2.36		3.09	face of studs, beams or joist.
Reinstall	L.F.		1.55		1.55	
Minimum Charge	Job		242		242	

Rough Frame / Structure

Ceiling Framing

Ceiling Framing	Unit	Material	Labor	Equip.	Total	Specification
Ledger Boards						
2″ x 4″						
Demolish	L.F.		.42		.42	Cost includes material and labor to
Install	L.F.	3.97	2.69		6.66	install 2″ x 4″ ledger board fastened to
Demolish and Install	L.F.	3.97	3.11		7.08	a wall, joists or studs.
Reinstall	L.F.		2.15		2.15	
Minimum Charge	Job		242		242	
4″ x 6″						
Demolish	L.F.		.69		.69	Cost includes material and labor to
Install	L.F.	8.70	3.46		12.16	install 4″ x 6″ ledger board fastened to
Demolish and Install	L.F.	8.70	4.15		12.85	a wall, joists or studs.
Reinstall	L.F.		2.77		2.77	
Minimum Charge	Job		242		242	
4″ x 8″						
Demolish	L.F.		.92		.92	Cost includes material and labor to
Install	L.F.	10.30	4.03		14.33	install 4″ x 8″ ledger board fastened to
Demolish and Install	L.F.	10.30	4.95		15.25	a wall, joists or studs.
Reinstall	L.F.		3.23		3.23	
Minimum Charge	Job		242		242	
Earthquake Strapping						
Install	Ea.	2.68	3.03		5.71	Includes labor and material to replace
Minimum Charge	Job		242		242	an earthquake strap.
Hurricane Clips						
Install	Ea.	1.42	3.34		4.76	Includes labor and material to install a
Minimum Charge	Job		242		242	hurricane clip.

Roof Framing

Roof Framing	Unit	Material	Labor	Equip.	Total	Specification
16″ O.C. System						
2″ x 4″ Rafters						
Demolish	S.F.		.94		.94	Cost includes material and labor to
Install	S.F.	.54	.86		1.40	install 2″ x 4″, 16″ O.C. rafter framing
Demolish and Install	S.F.	.54	1.80		2.34	for flat, shed or gable roofs, 25′ span,
Reinstall	S.F.		.69		.69	up to 5/12 slope, per S.F. of roof.
Minimum Charge	Job		242		242	
2″ x 6″ Rafters						
Demolish	S.F.		.99		.99	Cost includes material and labor to
Install	S.F.	.80	.91		1.71	install 2″ x 6″, 16″ O.C. rafter framing
Demolish and Install	S.F.	.80	1.90		2.70	for flat, shed or gable roofs, 25′ span,
Reinstall	S.F.		.73		.73	up to 5/12 slope, per S.F. of roof.
Minimum Charge	Job		242		242	
2″ x 8″ Rafters						
Demolish	S.F.		1.01		1.01	Cost includes material and labor to
Install	S.F.	1.19	.97		2.16	install 2″ x 8″, 16″ O.C. rafter framing
Demolish and Install	S.F.	1.19	1.98		3.17	for flat, shed or gable roofs, 25′ span,
Reinstall	S.F.		.77		.77	up to 5/12 slope, per S.F. of roof.
Minimum Charge	Job		242		242	
2″ x 10″ Rafters						
Demolish	S.F.		1.01		1.01	Cost includes material and labor to
Install	S.F.	1.64	1.47		3.11	install 2″ x 10″, 16″ O.C. rafter
Demolish and Install	S.F.	1.64	2.48		4.12	framing for flat, shed or gable roofs,
Reinstall	S.F.		1.17		1.17	25′ span, up to 5/12 slope, per S.F.
Minimum Charge	Job		242		242	of roof.

For customer support on your Contractor's Pricing Guide: Residential Repair & Remodeling, call 800.448.8182.

Rough Frame / Structure

Roof Framing

	Unit	Material	Labor	Equip.	Total	Specification
2" x 12" Rafters						
Demolish	S.F.		1.03		1.03	Cost includes material and labor to
Install	S.F.	2.29	1.59		3.88	install 2" x 12", 16" O.C. rafter
Demolish and Install	S.F.	2.29	2.62		4.91	framing for flat, shed or gable roofs,
Reinstall	S.F.		1.28		1.28	25' span, up to 5/12 slope, per S.F.
Minimum Charge	Job		242		242	of roof.
24" O.C. System						
2" x 4" Rafters						
Demolish	S.F.		.71		.71	Cost includes material and labor to
Install	S.F.	.36	.65		1.01	install 2" x 4", 24" O.C. rafter framing
Demolish and Install	S.F.	.36	1.36		1.72	for flat, shed or gable roofs, 25' span,
Reinstall	S.F.		.52		.52	up to 5/12 slope, per S.F. of roof.
Minimum Charge	Job		242		242	
2" x 6" Rafters						
Demolish	S.F.		.74		.74	Cost includes material and labor to
Install	S.F.	.53	.68		1.21	install 2" x 6", 24" O.C. rafter framing
Demolish and Install	S.F.	.53	1.42		1.95	for flat, shed or gable roofs, 25' span,
Reinstall	S.F.		.55		.55	up to 5/12 slope, per S.F. of roof.
Minimum Charge	Job		242		242	
2" x 8" Rafters						
Demolish	S.F.		.76		.76	Cost includes material and labor to
Install	S.F.	.79	.73		1.52	install 2" x 8", 24" O.C. rafter framing
Demolish and Install	S.F.	.79	1.49		2.28	for flat, shed or gable roofs, 25' span,
Reinstall	S.F.		.58		.58	up to 5/12 slope, per S.F. of roof.
Minimum Charge	Job		242		242	
2" x 10" Rafters						
Demolish	S.F.		.76		.76	Cost includes material and labor to
Install	S.F.	1.09	1.10		2.19	install 2" x 10", 24" O.C. rafter
Demolish and Install	S.F.	1.09	1.86		2.95	framing for flat, shed or gable roofs,
Reinstall	S.F.		.88		.88	25' span, up to 5/12 slope, per S.F.
Minimum Charge	Job		242		242	of roof.
2" x 12" Rafters						
Demolish	S.F.		.77		.77	Cost includes material and labor to
Install	S.F.	1.53	1.20		2.73	install 2" x 12", 24" O.C. rafter
Demolish and Install	S.F.	1.53	1.97		3.50	framing for flat, shed or gable roofs,
Reinstall	S.F.		.96		.96	25' span, up to 5/12 slope, per S.F.
Minimum Charge	Job		242		242	of roof.
Rafter						
2" x 4"						
Demolish	L.F.		.96		.96	Cost includes material and labor to
Install	L.F.	.73	1.21		1.94	install 2" x 4" rafters for flat, shed or
Demolish and Install	L.F.	.73	2.17		2.90	gable roofs, up to 5/12 slope, 25'
Reinstall	L.F.		.97		.97	span, per L.F.
Minimum Charge	Job		242		242	
2" x 6"						
Demolish	L.F.		.98		.98	Cost includes material and labor to
Install	L.F.	1.07	1.21		2.28	install 2" x 6" rafters for flat, shed or
Demolish and Install	L.F.	1.07	2.19		3.26	gable roofs, up to 5/12 slope, 25'
Reinstall	L.F.		.97		.97	span, per L.F.
Minimum Charge	Job		242		242	
2" x 8"						
Demolish	L.F.		.99		.99	Cost includes material and labor to
Install	L.F.	1.57	1.29		2.86	install 2" x 8" rafters for flat, shed or
Demolish and Install	L.F.	1.57	2.28		3.85	gable roofs, up to 5/12 slope, 25'
Reinstall	L.F.		1.03		1.03	span, per L.F.
Minimum Charge	Job		242		242	

For customer support on your Contractor's Pricing Guide: Residential Repair & Remodeling, call 800.448.8182.

Roof Framing		Unit	Material	Labor	Equip.	Total	Specification
2″ x 10″							
	Demolish	L.F.		1.01		1.01	Cost includes material and labor to
	Install	L.F.	2.19	1.96		4.15	install 2″ x 10″ rafters for flat, shed or
	Demolish and Install	L.F.	2.19	2.97		5.16	gable roofs, up to 5/12 slope, 25′
	Reinstall	L.F.		1.56		1.56	span, per L.F.
	Minimum Charge	Job		242		242	
2″ x 12″							
	Demolish	L.F.		1.02		1.02	Cost includes material and labor to
	Install	L.F.	3.05	2.13		5.18	install 2″ x 12″ rafters for flat, shed or
	Demolish and Install	L.F.	3.05	3.15		6.20	gable roofs, up to 5/12 slope, 25′
	Reinstall	L.F.		1.70		1.70	span, per L.F.
	Minimum Charge	Job		242		242	
2″x 4″ Valley / Jack							
	Demolish	L.F.		.96		.96	Includes material and labor to install
	Install	L.F.	1.07	2.04		3.11	up to 2″ x 6″ valley/jack rafters for
	Demolish and Install	L.F.	1.07	3		4.07	flat, shed or gable roofs, up to 5/12
	Reinstall	L.F.		1.63		1.63	slope, 25′ span, per L.F.
	Minimum Charge	Job		242		242	
2″x 6″ Valley / Jack							
	Demolish	L.F.		.98		.98	Includes material and labor to install
	Install	L.F.	1.07	2.04		3.11	up to 2″ x 6″ valley/jack rafters for
	Demolish and Install	L.F.	1.07	3.02		4.09	flat, shed or gable roofs, up to 5/12
	Reinstall	L.F.		1.63		1.63	slope, 25′ span, per L.F.
	Minimum Charge	Job		242		242	
Ridgeboard							
2″ x 4″							
	Demolish	L.F.		.92		.92	Cost includes material and labor to
	Install	L.F.	.73	1.76		2.49	install 2″ x 4″ ridgeboard for flat, shed
	Demolish and Install	L.F.	.73	2.68		3.41	or gable roofs, up to 5/12 slope, 25′
	Reinstall	L.F.		1.41		1.41	span, per L.F.
	Minimum Charge	Job		242		242	
2″ x 6″							
	Demolish	L.F.		.95		.95	Cost includes material and labor to
	Install	L.F.	1.07	1.94		3.01	install 2″ x 6″ ridgeboard for flat, shed
	Demolish and Install	L.F.	1.07	2.89		3.96	or gable roofs, up to 5/12 slope, 25′
	Reinstall	L.F.		1.55		1.55	span, per L.F.
	Minimum Charge	Job		242		242	
2″ x 8″							
	Demolish	L.F.		.98		.98	Cost includes material and labor to
	Install	L.F.	1.57	2.15		3.72	install 2″ x 8″ ridgeboard for flat, shed
	Demolish and Install	L.F.	1.57	3.13		4.70	or gable roofs, up to 5/12 slope, 25′
	Reinstall	L.F.		1.72		1.72	span, per L.F.
	Minimum Charge	Job		242		242	
2″ x 10″							
	Demolish	L.F.		1.01		1.01	Cost includes material and labor to
	Install	L.F.	2.19	2.42		4.61	install 2″ x 10″ ridgeboard for flat,
	Demolish and Install	L.F.	2.19	3.43		5.62	shed or gable roofs, up to 5/12
	Reinstall	L.F.		1.94		1.94	slope, 25′ span, per L.F.
	Minimum Charge	Job		242		242	
2″ x 12″							
	Demolish	L.F.		1.04		1.04	Cost includes material and labor to
	Install	L.F.	3.05	1.38		4.43	install 2″ x 12″ ridgeboard for flat,
	Demolish and Install	L.F.	3.05	2.42		5.47	shed or gable roofs, up to 5/12
	Reinstall	L.F.		1.11		1.11	slope, 25′ span, per L.F.
	Minimum Charge	Job		242		242	

Roof Framing		Unit	Material	Labor	Equip.	Total	Specification
Collar Beam							
1" x 6"							
	Demolish	L.F.		.73		.73	Cost includes material and labor to
	Install	L.F.	1.08	.61		1.69	install 1" x 6" collar beam or tie for
	Demolish and Install	L.F.	1.08	1.34		2.42	roof framing.
	Reinstall	L.F.		.48		.48	
	Minimum Charge	Job		242		242	
2" x 6"							
	Demolish	L.F.		.86		.86	Cost includes material and labor to
	Install	L.F.	1.07	.61		1.68	install 2" x 6" collar beam or tie for
	Demolish and Install	L.F.	1.07	1.47		2.54	roof framing.
	Reinstall	L.F.		.48		.48	
	Minimum Charge	Job		242		242	
Purlins							
2" x 6"							
	Demolish	L.F.		.83		.83	Cost includes material and labor to
	Install	L.F.	1.07	.54		1.61	install 2" x 6" purlins below roof
	Demolish and Install	L.F.	1.07	1.37		2.44	rafters.
	Reinstall	L.F.		.43		.43	
	Minimum Charge	Job		242		242	
2" x 8"							
	Demolish	L.F.		.84		.84	Cost includes material and labor to
	Install	L.F.	1.57	.54		2.11	install 2" x 8" purlins below roof
	Demolish and Install	L.F.	1.57	1.38		2.95	rafters.
	Reinstall	L.F.		.44		.44	
	Minimum Charge	Job		242		242	
2" x 10"							
	Demolish	L.F.		.85		.85	Cost includes material and labor to
	Install	L.F.	2.19	.55		2.74	install 2" x 10" purlins below roof
	Demolish and Install	L.F.	2.19	1.40		3.59	rafters.
	Reinstall	L.F.		.44		.44	
	Minimum Charge	Job		242		242	
2" x 12"							
	Demolish	L.F.		.86		.86	Cost includes material and labor to
	Install	L.F.	3.05	.56		3.61	install 2" x 12" purlins below roof
	Demolish and Install	L.F.	3.05	1.42		4.47	rafters.
	Reinstall	L.F.		.45		.45	
	Minimum Charge	Job		242		242	
4" x 6"							
	Demolish	L.F.		.86		.86	Cost includes material and labor to
	Install	L.F.	4.57	.56		5.13	install 4" x 6" purlins below roof
	Demolish and Install	L.F.	4.57	1.42		5.99	rafters.
	Reinstall	L.F.		.45		.45	
	Minimum Charge	Job		242		242	
4" x 8"							
	Demolish	L.F.		.87		.87	Cost includes material and labor to
	Install	L.F.	6.15	.57		6.72	install 4" x 8" purlins below roof
	Demolish and Install	L.F.	6.15	1.44		7.59	rafters.
	Reinstall	L.F.		.46		.46	
	Minimum Charge	Job		242		242	
Ledger Board							
2" x 6"							
	Demolish	L.F.		.42		.42	Cost includes material and labor to
	Install	L.F.	1.07	2.18		3.25	install 2" x 6" ledger board nailed to
	Demolish and Install	L.F.	1.07	2.60		3.67	the face of a rafter.
	Reinstall	L.F.		1.74		1.74	
	Minimum Charge	Job		242		242	

Rough Frame / Structure

Roof Framing		Unit	Material	Labor	Equip.	Total	Specification
4" x 6"							
	Demolish	L.F.		.69		.69	Cost includes material and labor to
	Install	L.F.	4.57	3.21		7.78	install 4" x 6" ledger board nailed to
	Demolish and Install	L.F.	4.57	3.90		8.47	the face of a rafter.
	Reinstall	L.F.		2.56		2.56	
	Minimum Charge	Job		242		242	
4" x 8"							
	Demolish	L.F.		.92		.92	Cost includes material and labor to
	Install	L.F.	6.15	3.69		9.84	install 4" x 8" ledger board nailed to
	Demolish and Install	L.F.	6.15	4.61		10.76	the face of a rafter.
	Reinstall	L.F.		2.96		2.96	
	Minimum Charge	Job		242		242	
Outriggers							
2" x 4"							
	Demolish	L.F.		.29		.29	Cost includes material and labor to
	Install	L.F.	.73	1.94		2.67	install 2" x 4" outrigger rafters for flat,
	Demolish and Install	L.F.	.73	2.23		2.96	shed or gabled roofs, up to 5/12
	Reinstall	L.F.		1.55		1.55	slope, per L.F.
	Minimum Charge	Job		242		242	
2" x 6"							
	Demolish	L.F.		.43		.43	Cost includes material and labor to
	Install	L.F.	1.07	1.94		3.01	install 2" x 6" outrigger rafters for flat,
	Demolish and Install	L.F.	1.07	2.37		3.44	shed or gabled roofs, up to 5/12
	Reinstall	L.F.		1.55		1.55	slope, per L.F.
	Minimum Charge	Job		242		242	
Lookout Rafter							
2" x 4"							
	Demolish	L.F.		.96		.96	Cost includes material and labor to
	Install	L.F.	.73	.81		1.54	install 2" x 4" lookout rafters for flat,
	Demolish and Install	L.F.	.73	1.77		2.50	shed or gabled roofs, up to 5/12
	Reinstall	L.F.		.65		.65	slope, per L.F.
	Minimum Charge	Job		242		242	
2" x 6"							
	Demolish	L.F.		.98		.98	Cost includes material and labor to
	Install	L.F.	1.07	.82		1.89	install 2" x 6" lookout rafters for flat,
	Demolish and Install	L.F.	1.07	1.80		2.87	shed or gabled roofs, up to 5/12
	Reinstall	L.F.		.66		.66	slope, per L.F.
	Minimum Charge	Job		242		242	
Fly Rafter							
2" x 4"							
	Demolish	L.F.		.96		.96	Cost includes material and labor to
	Install	L.F.	.73	.81		1.54	install 2" x 4" fly rafters for flat, shed
	Demolish and Install	L.F.	.73	1.77		2.50	or gabled roofs, up to 5/12 slope,
	Reinstall	L.F.		.65		.65	per L.F.
	Minimum Charge	Job		242		242	
2" x 6"							
	Demolish	L.F.		.98		.98	Cost includes material and labor to
	Install	L.F.	1.06	.82		1.88	install 2" x 6" fly rafters for flat, shed
	Demolish and Install	L.F.	1.06	1.80		2.86	or gabled roofs, up to 5/12 slope,
	Reinstall	L.F.		.66		.66	per L.F.
	Minimum Charge	Job		242		242	

For customer support on your Contractor's Pricing Guide: Residential Repair & Remodeling, call 800.448.8182.

Rough Frame / Structure

Residential Trusses

	Unit	Material	Labor	Equip.	Total	Specification
W or Fink Truss						
24' Span						
Demolish	Ea.		38	36.50	74.50	Includes material, labor and
Install	Ea.	133	37.50	36.50	207	equipment to install wood gang-nailed
Demolish and Install	Ea.	133	75.50	73	281.50	residential truss, up to 24' span.
Reinstall	Ea.		37.27	36.35	73.62	
Minimum Charge	Job		242		242	
28' Span						
Demolish	Ea.		40.50	39	79.50	Includes material, labor and
Install	Ea.	183	42	41	266	equipment to install wood gang-nailed
Demolish and Install	Ea.	183	82.50	80	345.50	residential truss, up to 28' span.
Reinstall	Ea.		42.20	41.15	83.35	
Minimum Charge	Job		242		242	
32' Span						
Demolish	Ea.		44.50	43.50	88	Includes material, labor and
Install	Ea.	231	44.50	43.50	319	equipment to install wood gang-nailed
Demolish and Install	Ea.	231	89	87	407	residential truss, up to 32' span.
Reinstall	Ea.		44.73	43.62	88.35	
Minimum Charge	Job		242		242	
36' Span						
Demolish	Ea.		48	46.50	94.50	Includes material, labor and
Install	Ea.	246	48.50	47.50	342	equipment to install wood gang-nailed
Demolish and Install	Ea.	246	96.50	94	436.50	residential truss, up to 36' span.
Reinstall	Ea.		48.62	47.42	96.04	
Minimum Charge	Job		242		242	
Gable End						
24' Span						
Demolish	Ea.		38	36.50	74.50	Includes material, labor and
Install	Ea.	188	40	39	267	equipment to install wood gang-nailed
Demolish and Install	Ea.	188	78	75.50	341.50	residential gable end truss, up to 24'
Reinstall	Ea.		39.94	38.95	78.89	span.
Clean	Ea.	.64	50		50.64	
Paint	Ea.	7.15	54		61.15	
Minimum Charge	Job		242		242	
28' Span						
Demolish	Ea.		40.50	39	79.50	Includes material, labor and
Install	Ea.	240	47.50	46.50	334	equipment to install wood gang-nailed
Demolish and Install	Ea.	240	88	85.50	413.50	residential gable end truss, up to 30'
Reinstall	Ea.		47.58	46.41	93.99	span.
Clean	Ea.	.66	68		68.66	
Paint	Ea.	7.50	73.50		81	
Minimum Charge	Job		242		242	
32' Span						
Demolish	Ea.		44.50	43.50	88	Includes material, labor and
Install	Ea.	261	48.50	47.50	357	equipment to install wood gang-nailed
Demolish and Install	Ea.	261	93	91	445	residential gable end truss, up to 32'
Reinstall	Ea.		48.62	47.42	96.04	span.
Clean	Ea.	.69	89		89.69	
Paint	Ea.	7.80	96		103.80	
Minimum Charge	Job		242		242	
36' Span						
Demolish	Ea.		48	46.50	94.50	Includes material, labor and
Install	Ea.	285	56	54.50	395.50	equipment to install wood gang-nailed
Demolish and Install	Ea.	285	104	101	490	residential gable end truss, up to 36'
Reinstall	Ea.		55.91	54.53	110.44	span.
Clean	Ea.	.73	112		112.73	
Paint	Ea.	8.20	121		129.20	
Minimum Charge	Job		242		242	

For customer support on your Contractor's Pricing Guide: Residential Repair & Remodeling, call 800.448.8182.

Rough Frame / Structure

Residential Trusses

Residential Trusses		Unit	Material	Labor	Equip.	Total	Specification
Truss Hardware							
5-1/4" Glue-Lam Seat							
	Install	Ea.	230	2.69		232.69	Includes labor and material to install
	Minimum Charge	Job		242		242	beam hangers.
6-3/4" Glue-Lam Seat							
	Install	Ea.	230	2.69		232.69	Includes labor and material to install
	Minimum Charge	Job		242		242	beam hangers.
2" x 4" Joist Hanger							
	Install	Ea.	1.20	2.77		3.97	Includes labor and material to install
	Minimum Charge	Job		242		242	joist and beam hangers.
2" x 12" Joist Hanger							
	Install	Ea.	2.15	2.93		5.08	Cost includes labor and material to
	Minimum Charge	Job		242		242	install 2" x 12" joist hanger.

Commercial Trusses

Commercial Trusses		Unit	Material	Labor	Equip.	Total	Specification
Engineered Lumber, Truss / Joist							
9-1/2"							
	Demolish	S.F.		.68		.68	Includes material and labor to install
	Install	S.F.	4.20	.48		4.68	engineered lumber truss / joists per
	Demolish and Install	S.F.	4.20	1.16		5.36	S.F. of floor area based on joists 16"
	Reinstall	S.F.		.39		.39	O.C. Beams, supports and bridging
	Minimum Charge	Job		242		242	are not included.
11-7/8"							
	Demolish	S.F.		.71		.71	Includes material and labor to install
	Install	S.F.	4.42	.49		4.91	engineered lumber truss / joists per
	Demolish and Install	S.F.	4.42	1.20		5.62	S.F. of floor area based on joists 16"
	Reinstall	S.F.		.40		.40	O.C. Beams, supports and bridging
	Minimum Charge	Job		242		242	are not included.
14"							
	Demolish	S.F.		.73		.73	Includes material and labor to install
	Install	S.F.	5.10	.53		5.63	engineered lumber truss / joists per
	Demolish and Install	S.F.	5.10	1.26		6.36	S.F. of floor area based on joists 16"
	Reinstall	S.F.		.43		.43	O.C. Beams, supports and bridging
	Minimum Charge	Job		242		242	are not included.
16"							
	Demolish	S.F.		.76		.76	Includes material and labor to install
	Install	S.F.	7.45	.56		8.01	engineered lumber truss / joists per
	Demolish and Install	S.F.	7.45	1.32		8.77	S.F. of floor area based on joists 16"
	Reinstall	S.F.		.45		.45	O.C. Beams, supports and bridging
	Minimum Charge	Job		242		242	are not included.
Bowstring Truss							
100' Clear Span							
	Demolish	SF Flr.		.69	.74	1.43	Includes material, labor and
	Install	SF Flr.	12.55	.56	.62	13.73	equipment to install bow string truss
	Demolish and Install	SF Flr.	12.55	1.25	1.36	15.16	system up to 100' clear span per S.F.
	Minimum Charge	Job		242		242	of floor area.
120' Clear Span							
	Demolish	SF Flr.		.77	.83	1.60	Includes material, labor and
	Install	SF Flr.	13.30	.62	.69	14.61	equipment to install bow string truss
	Demolish and Install	SF Flr.	13.30	1.39	1.52	16.21	system up to 120' clear span per S.F.
	Minimum Charge	Job		242		242	of floor area.

For customer support on your Contractor's Pricing Guide: Residential Repair & Remodeling, call 800.448.8182.

Rough Frame / Structure

Roof Sheathing

Roof Sheathing	Unit	Material	Labor	Equip.	Total	Specification
Plywood						
3/8"						
Demolish	S.F.		.54		.54	Cost includes material and labor to
Install	S.F.	.75	.63		1.38	install 3/8" CDX plywood roof
Demolish and Install	S.F.	.75	1.17		1.92	sheathing.
Reinstall	S.F.		.63		.63	
Minimum Charge	Job		242		242	
1/2"						
Demolish	S.F.		.54		.54	Cost includes material and labor to
Install	S.F.	.84	.69		1.53	install 1/2" CDX plywood roof
Demolish and Install	S.F.	.84	1.23		2.07	sheathing.
Minimum Charge	Job		242		242	
5/8"						
Demolish	S.F.		.54		.54	Cost includes material and labor to
Install	S.F.	.90	.74		1.64	install 5/8" CDX plywood roof
Demolish and Install	S.F.	.90	1.28		2.18	sheathing.
Minimum Charge	Job		242		242	
3/4"						
Demolish	S.F.		.54		.54	Cost includes material and labor to
Install	S.F.	1.03	.81		1.84	install 3/4" CDX plywood roof
Demolish and Install	S.F.	1.03	1.35		2.38	sheathing.
Minimum Charge	Job		242		242	
OSB						
1/2"						
Demolish	S.F.		.54		.54	Cost includes material and labor to
Install	S.F.	.52	.69		1.21	install 4' x 8' x 1/2" OSB sheathing.
Demolish and Install	S.F.	.52	1.23		1.75	
Minimum Charge	Job		242		242	
5/8"						
Demolish	S.F.		.54		.54	Cost includes material and labor to
Install	S.F.	.52	.56		1.08	install 4' x 8' x 5/8" OSB sheathing.
Demolish and Install	S.F.	.52	1.10		1.62	
Minimum Charge	Job		242		242	
Plank						
Demolish	S.F.		.54		.54	Cost includes material and labor to
Install	S.F.	2.38	1.34		3.72	install 1" x 6" or 1" x 8" utility T&G
Demolish and Install	S.F.	2.38	1.88		4.26	board sheathing laid diagonal.
Reinstall	S.F.		1.34		1.34	
Minimum Charge	Job		242		242	
Skip-type						
1" x 4"						
Demolish	S.F.		.83		.83	Includes material and labor to install
Install	S.F.	.86	.40		1.26	sheathing board material 1" x 4", 7"
Demolish and Install	S.F.	.86	1.23		2.09	O.C.
Minimum Charge	Job		242		242	
1" x 6"						
Demolish	S.F.		.83		.83	Includes material and labor to install
Install	S.F.	1.07	.33		1.40	sheathing board material 1" x 6", 7"
Demolish and Install	S.F.	1.07	1.16		2.23	or 9" O.C.
Minimum Charge	Job		242		242	

Wood Deck		Unit	Material	Labor	Equip.	Total	Specification
Stairs							
	Demolish	Ea.		480		480	Includes material and labor to install
	Install	Ea.	232	860		1092	wood steps. 5 - 7 risers included.
	Demolish and Install	Ea.	232	1340		1572	
	Reinstall	Ea.		689.02		689.02	
	Clean	Flight	1.52	31.50		33.02	
	Paint	Flight	34.50	54		88.50	
	Minimum Charge	Job		242		242	
Bench Seating							
	Demolish	L.F.		6.95		6.95	Includes material and labor to install
	Install	L.F.	3.58	24		27.58	pressure treated bench seating.
	Demolish and Install	L.F.	3.58	30.95		34.53	
	Reinstall	L.F.		19.36		19.36	
	Clean	L.F.	.07	1.87		1.94	
	Paint	L.F.	1.02	3.23		4.25	
	Minimum Charge	Job		242		242	
Privacy Wall							
	Demolish	SF Flr.		1.66		1.66	Includes material, labor and
	Install	SF Flr.	1.49	.90	.32	2.71	equipment to install privacy wall.
	Demolish and Install	SF Flr.	1.49	2.56	.32	4.37	
	Reinstall	SF Flr.		.72	.26	.98	
	Clean	SF Flr.	.03	.38		.41	
	Paint	S.F.	.17	.31		.48	
	Minimum Charge	Job		242		242	
Treated Column or Post							
4″ x 4″							
	Demolish	L.F.		1.04		1.04	Includes material and labor to install
	Install	L.F.	1.79	2.48		4.27	4″ x 4″ treated column or post.
	Demolish and Install	L.F.	1.79	3.52		5.31	
	Reinstall	L.F.		1.99		1.99	
	Clean	L.F.	.06	.33		.39	
	Paint	L.F.	.35	1.01		1.36	
	Minimum Charge	Job		242		242	
4″ x 6″							
	Demolish	L.F.		1.51		1.51	Includes material and labor to install
	Install	L.F.	3.11	3.52		6.63	4″ x 6″ treated column or post.
	Demolish and Install	L.F.	3.11	5.03		8.14	
	Reinstall	L.F.		2.82		2.82	
	Clean	L.F.	.07	.47		.54	
	Paint	L.F.	.35	1.01		1.36	
	Minimum Charge	Job		242		242	
Cedar Columns or Posts							
4″ x 4″							
	Demolish	L.F.		1.04		1.04	Includes material and labor to install
	Install	L.F.	5.55	2.48		8.03	4″ x 4″ cedar columns or posts.
	Demolish and Install	L.F.	5.55	3.52		9.07	
	Reinstall	L.F.		1.99		1.99	
	Clean	L.F.	.06	.33		.39	
	Paint	L.F.	.35	1.01		1.36	
	Minimum Charge	Job		242		242	
4″ x 6″							
	Demolish	L.F.		1.51		1.51	Includes material and labor to install
	Install	L.F.	9	3.52		12.52	4″ x 6″ cedar columns or posts.
	Demolish and Install	L.F.	9	5.03		14.03	
	Reinstall	L.F.		2.82		2.82	
	Clean	L.F.	.07	.47		.54	
	Paint	L.F.	.35	1.01		1.36	
	Minimum Charge	Job		242		242	

Wood Deck	Unit	Material	Labor	Equip.	Total	Specification
Redwood Columns or Posts						
4″ x 4″						
Demolish	L.F.		1.04		1.04	Includes material and labor to install
Install	L.F.	13.60	2.48		16.08	4″ x 4″ redwood columns or posts.
Demolish and Install	L.F.	13.60	3.52		17.12	
Reinstall	L.F.		1.99		1.99	
Clean	L.F.	.06	.33		.39	
Paint	L.F.	.35	1.01		1.36	
Minimum Charge	Job		242		242	
4″ x 6″						
Demolish	L.F.		1.51		1.51	Includes material and labor to install
Install	L.F.	21.50	3.52		25.02	4″ x 6″ redwood columns or posts.
Demolish and Install	L.F.	21.50	5.03		26.53	
Reinstall	L.F.		2.82		2.82	
Clean	L.F.	.07	.47		.54	
Paint	L.F.	.35	1.01		1.36	
Minimum Charge	Job		242		242	
Treated Girder - Single						
4″ x 8″						
Demolish	L.F.		3.09		3.09	Includes material and labor to install a
Install	L.F.	7.45	1.84		9.29	treated lumber girder consisting of one
Demolish and Install	L.F.	7.45	4.93		12.38	4″ x 8″ for porch or deck construction.
Reinstall	L.F.		1.48		1.48	
Clean	L.F.	.07	.40		.47	
Minimum Charge	Job		242		242	
Treated Girder - Double						
2″ x 8″						
Demolish	L.F.		3.09		3.09	Includes material and labor to install a
Install	L.F.	3.58	1.68		5.26	treated lumber girder consisting of two
Demolish and Install	L.F.	3.58	4.77		8.35	2″ x 8″s for porch or deck
Reinstall	L.F.		1.35		1.35	construction.
Clean	L.F.	.07	.40		.47	
Minimum Charge	Job		242		242	
2″ x 10″						
Demolish	L.F.		3.87		3.87	Includes material and labor to install a
Install	L.F.	4.15	1.76		5.91	treated lumber girder consisting of two
Demolish and Install	L.F.	4.15	5.63		9.78	2″ x 10″s for porch or deck
Reinstall	L.F.		1.41		1.41	construction.
Clean	L.F.	.07	.47		.54	
Minimum Charge	Job		242		242	
2″ x 12″						
Demolish	L.F.		4.64		4.64	Includes material and labor to install a
Install	L.F.	6.55	1.84		8.39	treated lumber girder consisting of two
Demolish and Install	L.F.	6.55	6.48		13.03	2″ x 12″s for porch or deck
Reinstall	L.F.		1.48		1.48	construction.
Clean	L.F.	.08	.53		.61	
Minimum Charge	Job		242		242	
Treated Girder - Triple						
2″ x 8″						
Demolish	L.F.		4.64		4.64	Includes material and labor to install a
Install	L.F.	5.35	1.84		7.19	treated lumber girder consisting of
Demolish and Install	L.F.	5.35	6.48		11.83	three 2″ x 8″s for porch or deck
Reinstall	L.F.		1.48		1.48	construction.
Clean	L.F.	.07	.47		.54	
Minimum Charge	Job		242		242	

Rough Frame / Structure

Wood Deck		Unit	Material	Labor	Equip.	Total	Specification
2″ x 10″							
	Demolish	L.F.		5.80		5.80	Includes material and labor to install a
	Install	L.F.	6.20	1.94		8.14	treated lumber girder consisting of
	Demolish and Install	L.F.	6.20	7.74		13.94	three 2″ x 10″s for porch or deck
	Reinstall	L.F.		1.55		1.55	construction.
	Clean	L.F.	.08	.53		.61	
	Minimum Charge	Job		242		242	
2″ x 12″							
	Demolish	L.F.		6.90		6.90	Includes material and labor to install a
	Install	L.F.	9.85	2.04		11.89	treated lumber girder consisting of
	Demolish and Install	L.F.	9.85	8.94		18.79	three 2″ x 12″s for porch or deck
	Reinstall	L.F.		1.63		1.63	construction.
	Clean	L.F.	.09	.60		.69	
	Minimum Charge	Job		242		242	
Treated Ledger - Bolted							
2″ x 8″							
	Demolish	L.F.		.52		.52	Includes material and labor to install a
	Install	L.F.	2.08	2.48		4.56	treated lumber 2″ x 8″ ledger with
	Demolish and Install	L.F.	2.08	3		5.08	carriage bolts 4′ on center for porch
	Reinstall	L.F.		1.99		1.99	or deck construction.
	Clean	L.F.	.01	.27		.28	
	Minimum Charge	Job		242		242	
2″ x 10″							
	Demolish	L.F.		.52		.52	Includes material and labor to install a
	Install	L.F.	2.37	2.51		4.88	treated lumber 2″ x 10″ ledger with
	Demolish and Install	L.F.	2.37	3.03		5.40	carriage bolts 4′ on center for porch
	Reinstall	L.F.		2.01		2.01	or deck construction.
	Clean	L.F.	.02	.31		.33	
	Minimum Charge	Job		242		242	
2″ x 12″							
	Demolish	L.F.		.52		.52	Includes material and labor to install a
	Install	L.F.	3.55	2.55		6.10	treated lumber 2″ x 12″ ledger with
	Demolish and Install	L.F.	3.55	3.07		6.62	carriage bolts 4′ on center for porch
	Reinstall	L.F.		2.04		2.04	or deck construction.
	Clean	L.F.	.02	.31		.33	
	Minimum Charge	Job		242		242	
Treated Joists							
2″ x 8″							
	Demolish	L.F.		.88		.88	Includes material and labor to install
	Install	L.F.	1.84	.88		2.72	treated lumber 2″ x 8″ deck or floor
	Demolish and Install	L.F.	1.84	1.76		3.60	joists for porch or deck construction.
	Reinstall	L.F.		.70		.70	
	Clean	L.F.	.06	.33		.39	
	Minimum Charge	Job		242		242	
2″ x 10″							
	Demolish	L.F.		.91		.91	Includes material and labor to install
	Install	L.F.	2.13	1.08		3.21	treated lumber 2″ x 10″ deck or floor
	Demolish and Install	L.F.	2.13	1.99		4.12	joists for porch or deck construction.
	Reinstall	L.F.		.86		.86	
	Clean	L.F.	.07	.40		.47	
	Minimum Charge	Job		242		242	
2″ x 12″							
	Demolish	L.F.		.94		.94	Includes material and labor to install
	Install	L.F.	2.56	1.11		3.67	treated lumber 2″ x 12″ deck or floor
	Demolish and Install	L.F.	2.56	2.05		4.61	joists for porch or deck construction.
	Reinstall	L.F.		.89		.89	
	Clean	L.F.	.07	.47		.54	
	Minimum Charge	Job		242		242	

For customer support on your Contractor's Pricing Guide: Residential Repair & Remodeling, call 800.448.8182.

Wood Deck	Unit	Material	Labor	Equip.	Total	Specification
Treated Railings and Trim						
2″ x 2″						
Demolish	L.F.		.35		.35	Includes material and labor to install
Install	L.F.	.88	1.61		2.49	treated lumber 2″ x 2″ railings and
Demolish and Install	L.F.	.88	1.96		2.84	trim for porch or deck construction.
Reinstall	L.F.		1.29		1.29	
Clean	L.F.		.25		.25	
Paint	L.F.	.35	1.01		1.36	
Minimum Charge	Job		242		242	
2″ x 4″						
Demolish	L.F.		.35		.35	Includes material and labor to install
Install	L.F.	.95	1.61		2.56	treated lumber 2″ x 4″ railings and
Demolish and Install	L.F.	.95	1.96		2.91	trim for porch or deck construction.
Reinstall	L.F.		1.29		1.29	
Clean	L.F.	.01	.27		.28	
Paint	L.F.	.35	1.01		1.36	
Minimum Charge	Job		242		242	
2″ x 6″						
Demolish	L.F.		.35		.35	Includes material and labor to install
Install	L.F.	1.33	1.61		2.94	treated lumber 2″ x 6″ railings and
Demolish and Install	L.F.	1.33	1.96		3.29	trim for porch or deck construction.
Reinstall	L.F.		1.29		1.29	
Clean	L.F.	.01	.31		.32	
Paint	L.F.	.35	1.01		1.36	
Minimum Charge	Job		242		242	
Cedar Railings and Trim						
1″ x 4″						
Demolish	L.F.		.35		.35	Includes material and labor to install
Install	L.F.	5.05	1.61		6.66	cedar 1″ x 4″ railings and trim for
Demolish and Install	L.F.	5.05	1.96		7.01	porch or deck construction.
Reinstall	L.F.		1.29		1.29	
Clean	L.F.		.25		.25	
Paint	L.F.	.35	1.01		1.36	
Minimum Charge	Job		242		242	
2″ x 4″						
Demolish	L.F.		.35		.35	Includes material and labor to install
Install	L.F.	10.45	1.61		12.06	cedar 2″ x 4″ railings and trim for
Demolish and Install	L.F.	10.45	1.96		12.41	porch or deck construction.
Reinstall	L.F.		1.29		1.29	
Clean	L.F.	.01	.27		.28	
Paint	L.F.	.35	1.01		1.36	
Minimum Charge	Job		242		242	
2″ x 6″						
Demolish	L.F.		.35		.35	Includes material and labor to install
Install	L.F.	15.75	1.61		17.36	cedar 2″ x 6″ railings and trim for
Demolish and Install	L.F.	15.75	1.96		17.71	porch or deck construction.
Reinstall	L.F.		1.29		1.29	
Clean	L.F.	.01	.31		.32	
Paint	L.F.	.35	1.01		1.36	
Minimum Charge	Job		242		242	
Redwood Railings and Trim						
1″ x 4″						
Demolish	L.F.		.35		.35	Includes material and labor to install
Install	L.F.	3.62	1.61		5.23	redwood 1″ x 4″ railings and trim for
Demolish and Install	L.F.	3.62	1.96		5.58	porch or deck construction.
Reinstall	L.F.		1.29		1.29	
Clean	L.F.		.25		.25	
Paint	L.F.	.35	1.01		1.36	
Minimum Charge	Job		242		242	

For customer support on your Contractor's Pricing Guide: Residential Repair & Remodeling, call 800.448.8182.

Wood Deck		Unit	Material	Labor	Equip.	Total	Specification
2" x 6"							
	Demolish	L.F.		.35		.35	Includes material and labor to install
	Install	L.F.	22.50	1.61		24.11	redwood 2" x 6" railings and trim for
	Demolish and Install	L.F.	22.50	1.96		24.46	porch or deck construction.
	Reinstall	L.F.		1.29		1.29	
	Clean	L.F.	.01	.31		.32	
	Paint	L.F.	.35	1.01		1.36	
	Minimum Charge	Job		242		242	
Treated Decking							
1" x 4"							
	Demolish	L.F.		.50		.50	Includes material and labor to install
	Install	S.F.	2.93	1.76		4.69	treated lumber 1" x 4" decking for
	Demolish and Install	S.F.	2.93	2.26		5.19	porch or deck construction.
	Reinstall	S.F.		1.41		1.41	
	Clean	S.F.	.03	.38		.41	
	Paint	S.F.	.34	.81		1.15	
	Minimum Charge	Job		242		242	
2" x 4"							
	Demolish	L.F.		.50		.50	Includes material and labor to install
	Install	S.F.	3.20	1.61		4.81	treated lumber 2" x 4" decking for
	Demolish and Install	S.F.	3.20	2.11		5.31	porch or deck construction.
	Reinstall	S.F.		1.29		1.29	
	Clean	S.F.	.03	.38		.41	
	Paint	S.F.	.34	.81		1.15	
	Minimum Charge	Job		242		242	
2" x 6"							
	Demolish	L.F.		.50		.50	Includes material and labor to install
	Install	S.F.	2.87	1.51		4.38	treated lumber 2" x 6" decking for
	Demolish and Install	S.F.	2.87	2.01		4.88	porch or deck construction.
	Reinstall	S.F.		1.21		1.21	
	Clean	S.F.	.03	.38		.41	
	Paint	S.F.	.34	.81		1.15	
	Minimum Charge	Job		242		242	
5/4" x 6"							
	Demolish	L.F.		.50		.50	Includes material and labor to install
	Install	S.F.	3.15	1.51		4.66	treated lumber 5/4" x 6" decking for
	Demolish and Install	S.F.	3.15	2.01		5.16	porch or deck construction.
	Reinstall	S.F.		1.21		1.21	
	Clean	S.F.	.03	.38		.41	
	Paint	S.F.	.34	.81		1.15	
	Minimum Charge	Job		242		242	
Wood / Plastic Composite Decking							
5/4" x 6"							
	Demolish	L.F.		.50		.50	Includes material and labor to install
	Install	L.F.	2.54	1.51		4.05	wood/plastic composite 5/4" x 6"
	Demolish and Install	L.F.	2.54	2.01		4.55	decking for porch or deck
	Reinstall	L.F.		1.21		1.21	construction.
	Clean	S.F.	.03	.38		.41	
	Paint	S.F.	.34	.81		1.15	
	Minimum Charge	Job		242		242	
Square Edge Fir Decking							
1" x 4"							
	Demolish	L.F.		.50		.50	Includes material and labor to install
	Install	L.F.	2.95	1.76		4.71	square edge fir 1" x 4" decking for
	Demolish and Install	L.F.	2.95	2.26		5.21	porch or deck construction.
	Reinstall	L.F.		1.41		1.41	
	Clean	S.F.	.03	.38		.41	
	Paint	S.F.	.34	.81		1.15	
	Minimum Charge	Job		242		242	

Wood Deck	Unit	Material	Labor	Equip.	Total	Specification
Tongue and Groove Fir Decking						
1″ x 4″						
Demolish	L.F.		.50		.50	Includes material and labor to install
Install	L.F.	2.18	2.15		4.33	treated lumber 1″ x 4″ decking for
Demolish and Install	L.F.	2.18	2.65		4.83	porch or deck construction.
Reinstall	L.F.		1.72		1.72	
Clean	S.F.	.03	.38		.41	
Paint	S.F.	.34	.81		1.15	
Minimum Charge	Job		242		242	
Mahogany Decking						
1″ x 4″						
Demolish	L.F.		.50		.50	Includes material and labor to install
Install	L.F.	2.33	1.76		4.09	treated lumber 1″ x 4″ decking for
Demolish and Install	L.F.	2.33	2.26		4.59	porch or deck construction.
Reinstall	L.F.		1.41		1.41	
Clean	S.F.	.03	.38		.41	
Paint	S.F.	.34	.81		1.15	
Minimum Charge	Job		242		242	
PVC Decking						
5/4″ x 6″						
Demolish	L.F.		.50		.50	Includes material and labor to install
Install	L.F.	3.63	1.76		5.39	PVC 5/4″ x 6″ decking for porch or
Demolish and Install	L.F.	3.63	2.26		5.89	deck construction.
Reinstall	L.F.		1.41		1.41	
Clean	S.F.	.03	.38		.41	
Minimum Charge	Job		242		242	
PVC Handrails and Balusters						
Straight						
Demolish	L.F.		3.46		3.46	Includes material and labor to install
Install	L.F.	28	5.05		33.05	straight PVC handrails and balusters
Demolish and Install	L.F.	28	8.51		36.51	for porch or deck construction.
Reinstall	L.F.		4.03		4.03	
Clean	L.F.	.31	2.92		3.23	
Minimum Charge	Job		242		242	
Angled						
Demolish	L.F.		3.46		3.46	Includes material and labor to install
Install	L.F.	28	6.70		34.70	angled PVC handrails and balusters
Demolish and Install	L.F.	28	10.16		38.16	for porch or deck construction.
Reinstall	L.F.		5.38		5.38	
Clean	L.F.	.31	2.92		3.23	
Minimum Charge	Job		242		242	
PVC Post Sleeve						
for 4″ x 4″ post						
Demolish	L.F.		1.04		1.04	Includes material and labor to install
Install	L.F.	30.50	5.05		35.55	PVC post sleeves over 4″ x 4″ porch
Demolish and Install	L.F.	30.50	6.09		36.59	or deck posts.
Reinstall	L.F.		4.03		4.03	
Clean	L.F.	.06	.33		.39	
Minimum Charge	Job		242		242	
Post Cap						
Demolish	Ea.		5.20		5.20	Includes material and labor to install
Install	Ea.	40	10.10		50.10	PVC post caps for porch or deck
Demolish and Install	Ea.	40	15.30		55.30	construction.
Reinstall	Ea.		8.07		8.07	
Clean	L.F.	.06	.33		.39	
Minimum Charge	Job		242		242	

For customer support on your Contractor's Pricing Guide: Residential Repair & Remodeling, call 800.448.8182.

Rough Frame / Structure

Porch		Unit	Material	Labor	Equip.	Total	Specification
Stairs							
	Demolish	Ea.		480		480	Includes material and labor to install
	Install	Ea.	232	860		1092	wood steps. 5 - 7 risers included.
	Demolish and Install	Ea.	232	1340		1572	
	Reinstall	Ea.		689.02		689.02	
	Clean	Flight	1.52	31.50		33.02	
	Paint	Flight	34.50	54		88.50	
	Minimum Charge	Job		242		242	
Bench Seating							
	Demolish	L.F.		6.95		6.95	Includes material and labor to install
	Install	L.F.	3.58	24		27.58	pressure treated bench seating.
	Demolish and Install	L.F.	3.58	30.95		34.53	
	Reinstall	L.F.		19.36		19.36	
	Clean	L.F.	.07	1.87		1.94	
	Paint	L.F.	1.02	3.23		4.25	
	Minimum Charge	Job		242		242	
Privacy Wall							
	Demolish	SF Flr.		1.66		1.66	Includes material, labor and
	Install	SF Flr.	1.49	.90	.32	2.71	equipment to install privacy wall.
	Demolish and Install	SF Flr.	1.49	2.56	.32	4.37	
	Reinstall	SF Flr.		.72	.26	.98	
	Clean	SF Flr.	.03	.38		.41	
	Paint	S.F.	.17	.31		.48	
	Minimum Charge	Job		242		242	
Wood Board Ceiling							
	Demolish	S.F.		.69		.69	Includes material and labor to install
	Install	S.F.	5.65	2.15		7.80	wood board on ceiling.
	Demolish and Install	S.F.	5.65	2.84		8.49	
	Clean	S.F.	.01	.22		.23	
	Paint	S.F.	.33	.50		.83	
	Minimum Charge	Job		242		242	
Re-screen							
	Install	S.F.	2.45	1.94		4.39	Includes labor and material to
	Minimum Charge	Job		242		242	re-screen wood frame.

Exterior Trim

Exterior Trim	Unit	Material	Labor	Equip.	Total	Specification
Stock Lumber						
1″ x 2″						
Demolish	L.F.		.35		.35	Cost includes material and labor to
Install	L.F.	.36	1.47		1.83	install 1″ x 2″ pine trim.
Demolish and Install	L.F.	.36	1.82		2.18	
Clean	L.F.		.25		.25	
Paint	L.F.	.04	.63		.67	
Minimum Charge	Job		242		242	
1″ x 3″						
Demolish	L.F.		.35		.35	Cost includes material and labor to
Install	L.F.	.37	1.67		2.04	install 1″ x 3″ pine trim.
Demolish and Install	L.F.	.37	2.02		2.39	
Clean	L.F.		.25		.25	
Paint	L.F.	.04	.63		.67	
Minimum Charge	Job		242		242	
1″ x 4″						
Demolish	L.F.		.35		.35	Cost includes material and labor to
Install	L.F.	.76	1.94		2.70	install 1″ x 4″ pine trim.
Demolish and Install	L.F.	.76	2.29		3.05	
Clean	L.F.		.25		.25	
Paint	L.F.	.04	.63		.67	
Minimum Charge	Job		242		242	
1″ x 6″						
Demolish	L.F.		.35		.35	Cost includes material and labor to
Install	L.F.	1.13	1.94		3.07	install 1″ x 6″ pine trim.
Demolish and Install	L.F.	1.13	2.29		3.42	
Clean	L.F.	.01	.25		.26	
Paint	L.F.	.14	.63		.77	
Minimum Charge	Job		242		242	
1″ x 8″						
Demolish	L.F.		.38		.38	Cost includes material and labor to
Install	L.F.	1.55	2.42		3.97	install 1″ x 8″ pine trim.
Demolish and Install	L.F.	1.55	2.80		4.35	
Reinstall	L.F.		1.94		1.94	
Clean	L.F.	.01	.27		.28	
Paint	L.F.	.14	.63		.77	
Minimum Charge	Job		242		242	

Exterior Trim		Unit	Material	Labor	Equip.	Total	Specification
Stock Lumber							
1" x 10"							
	Demolish	L.F.		.58		.58	Cost includes material and labor to
	Install	L.F.	2.13	2.69		4.82	install 1" x 10" pine trim.
	Demolish and Install	L.F.	2.13	3.27		5.40	
	Minimum Charge	Job		242		242	
1" x 12"							
	Demolish	L.F.		.58		.58	Cost includes material and labor to
	Install	L.F.	2.95	2.69		5.64	install 1" x 12" pine trim.
	Demolish and Install	L.F.	2.95	3.27		6.22	
	Minimum Charge	Job		242		242	
PVC							
1" x 4"							
	Demolish	L.F.		.35		.35	Includes material and labor to install
	Install	L.F.	2.72	1.94		4.66	1" x 4" PVC exterior trim.
	Demolish and Install	L.F.	2.72	2.29		5.01	
	Reinstall	L.F.		1.55		1.55	
	Clean	L.F.		.25		.25	
	Paint	L.F.	.04	.63		.67	
	Minimum Charge	Job		242		242	
1" x 6"							
	Demolish	L.F.		.35		.35	Includes material and labor to install
	Install	L.F.	3.85	1.94		5.79	1" x 6" PVC exterior trim.
	Demolish and Install	L.F.	3.85	2.29		6.14	
	Reinstall	L.F.		1.55		1.55	
	Clean	L.F.	.01	.25		.26	
	Paint	L.F.	.14	.63		.77	
	Minimum Charge	Job		242		242	
1" x 8"							
	Demolish	L.F.		.38		.38	Includes material and labor to install
	Install	L.F.	4.29	2.15		6.44	1" x 8" PVC exterior trim.
	Demolish and Install	L.F.	4.29	2.53		6.82	
	Reinstall	L.F.		1.72		1.72	
	Clean	L.F.	.01	.27		.28	
	Paint	L.F.	.14	.63		.77	
	Minimum Charge	Job		242		242	
1" x 10"							
	Demolish	L.F.		.40		.40	Includes material and labor to install
	Install	L.F.	5.50	2.15		7.65	1" x 10" PVC exterior trim.
	Demolish and Install	L.F.	5.50	2.55		8.05	
	Reinstall	L.F.		1.72		1.72	
	Clean	L.F.	.02	.31		.33	
	Paint	L.F.	.14	.63		.77	
	Minimum Charge	Job		242		242	
1" x 12"							
	Demolish	L.F.		.42		.42	Includes material and labor to install
	Install	L.F.	7.05	2.42		9.47	1" x 12" PVC exterior trim.
	Demolish and Install	L.F.	7.05	2.84		9.89	
	Reinstall	L.F.		1.94		1.94	
	Clean	L.F.	.02	.31		.33	
	Paint	L.F.	.14	.63		.77	
	Minimum Charge	Job		242		242	

Exterior Trim

Fascia		Unit	Material	Labor	Equip.	Total	Specification
Aluminum							
	Demolish	L.F.		.72		.72	Includes material and labor to install
	Install	S.F.	1.86	2.30		4.16	aluminum fascia.
	Demolish and Install	S.F.	1.86	3.02		4.88	
	Reinstall	S.F.		1.84		1.84	
	Clean	L.F.		.47		.47	
	Minimum Charge	Job		242		242	
Vinyl							
	Demolish	L.F.		.72		.72	Cost includes material and labor to
	Install	L.F.	4.17	2.77		6.94	install 6" vinyl fascia with 12" vinyl
	Demolish and Install	L.F.	4.17	3.49		7.66	soffit and J-channel.
	Reinstall	L.F.		2.21		2.21	
	Clean	L.F.		.47		.47	
	Minimum Charge	Job		242		242	
Plywood							
	Demolish	L.F.		.72		.72	Cost includes material and labor to
	Install	L.F.	3.49	1.08		4.57	install 12" wide plywood fascia.
	Demolish and Install	L.F.	3.49	1.80		5.29	
	Reinstall	L.F.		.86		.86	
	Clean	L.F.	.02	.31		.33	
	Paint	L.F.	.35	1.01		1.36	
	Minimum Charge	Job		242		242	
Cedar							
	Demolish	L.F.		.72		.72	Cost includes material and labor to
	Install	L.F.	5.40	1.08		6.48	install 12" wide cedar fascia.
	Demolish and Install	L.F.	5.40	1.80		7.20	
	Reinstall	L.F.		.86		.86	
	Clean	L.F.	.02	.31		.33	
	Paint	L.F.	.35	1.01		1.36	
	Minimum Charge	Job		242		242	
Redwood							
1" x 6"							
	Demolish	L.F.		.56		.56	Cost includes material and labor to
	Install	L.F.	5.35	1.94		7.29	install 1" x 6" redwood fascia board.
	Demolish and Install	L.F.	5.35	2.50		7.85	
	Reinstall	L.F.		1.55		1.55	
	Minimum Charge	Job		242		242	
1" x 8"							
	Demolish	L.F.		.63		.63	Cost includes material and labor to
	Install	L.F.	5.90	2.10		8	install 1" x 8" redwood fascia board.
	Demolish and Install	L.F.	5.90	2.73		8.63	
	Reinstall	L.F.		1.68		1.68	
	Minimum Charge	Job		242		242	
2" x 6"							
	Demolish	L.F.		.56		.56	Cost includes material and labor to
	Install	L.F.	1.07	3.87		4.94	install 2" x 6" exterior trim board.
	Demolish and Install	L.F.	1.07	4.43		5.50	
	Reinstall	L.F.		3.10		3.10	
	Minimum Charge	Job		242		242	
2" x 8"							
	Demolish	L.F.		.63		.63	Cost includes material and labor to
	Install	L.F.	1.57	4.30		5.87	install 2" x 8" exterior trim board.
	Demolish and Install	L.F.	1.57	4.93		6.50	
	Reinstall	L.F.		3.44		3.44	
	Minimum Charge	Job		242		242	

Exterior Trim

Fascia	Unit	Material	Labor	Equip.	Total	Specification
Hem-fir Std & Better						
2″ x 6″						
Demolish	L.F.		.56		.56	Cost includes material and labor to
Install	L.F.	1.07	3.87		4.94	install 2″ x 6″ exterior trim board.
Demolish and Install	L.F.	1.07	4.43		5.50	
Reinstall	L.F.		3.10		3.10	
Clean	L.F.	.01	.31		.32	
Paint	L.F.	.35	1.01		1.36	
Minimum Charge	Job		242		242	
2″ x 8″						
Demolish	L.F.		.63		.63	Cost includes material and labor to
Install	L.F.	1.57	4.30		5.87	install 2″ x 8″ exterior trim board.
Demolish and Install	L.F.	1.57	4.93		6.50	
Reinstall	L.F.		3.44		3.44	
Clean	L.F.	.01	.31		.32	
Paint	L.F.	.35	1.01		1.36	
Minimum Charge	Job		242		242	
2″ x 10″						
Demolish	L.F.		.72		.72	Cost includes material and labor to
Install	L.F.	2.19	5.40		7.59	install 2″ x 10″ exterior trim board.
Demolish and Install	L.F.	2.19	6.12		8.31	
Reinstall	L.F.		4.30		4.30	
Clean	L.F.	.02	.34		.36	
Paint	L.F.	.35	1.01		1.36	
Minimum Charge	Job		242		242	
Barge Rafter						
2″ x 6″						
Demolish	L.F.		1.11		1.11	Cost includes material and labor to
Install	L.F.	1.07	.54		1.61	install 2″ x 6″ barge rafters.
Demolish and Install	L.F.	1.07	1.65		2.72	
Reinstall	L.F.		.43		.43	
Clean	L.F.	.01	.31		.32	
Paint	L.F.	.35	1.01		1.36	
Minimum Charge	Job		242		242	
2″ x 8″						
Demolish	L.F.		1.13		1.13	Cost includes material and labor to
Install	L.F.	1.57	.57		2.14	install 2″ x 8″ barge rafters.
Demolish and Install	L.F.	1.57	1.70		3.27	
Reinstall	L.F.		.46		.46	
Clean	L.F.	.01	.31		.32	
Paint	L.F.	.35	1.01		1.36	
Minimum Charge	Job		242		242	
2″ x 10″						
Demolish	L.F.		1.15		1.15	Cost includes material and labor to
Install	L.F.	2.19	.59		2.78	install 2″ x 10″ barge rafters.
Demolish and Install	L.F.	2.19	1.74		3.93	
Reinstall	L.F.		.47		.47	
Clean	L.F.	.02	.34		.36	
Paint	L.F.	.35	1.01		1.36	
Minimum Charge	Job		242		242	

Soffit	Unit	Material	Labor	Equip.	Total	Specification
Aluminum						
12" w / Fascia						
Demolish	L.F.		.72		.72	Cost includes material and labor to
Install	L.F.	3.93	2.77		6.70	install 12" aluminum soffit with 6"
Demolish and Install	L.F.	3.93	3.49		7.42	fascia and J-channel.
Reinstall	L.F.		2.21		2.21	
Clean	L.F.		.47		.47	
Paint	L.F.	.39	1.01		1.40	
Minimum Charge	Job		242		242	
18" w / Fascia						
Demolish	L.F.		.72		.72	Cost includes material and labor to
Install	L.F.	4.73	3.23		7.96	install 18" aluminum soffit with 6"
Demolish and Install	L.F.	4.73	3.95		8.68	fascia and J-channel.
Reinstall	L.F.		2.58		2.58	
Clean	L.F.		.47		.47	
Paint	L.F.	.39	1.01		1.40	
Minimum Charge	Job		242		242	
24" w / Fascia						
Demolish	L.F.		.72		.72	Cost includes material and labor to
Install	L.F.	5.65	3.59		9.24	install 24" aluminum soffit with 6"
Demolish and Install	L.F.	5.65	4.31		9.96	fascia and J-channel.
Reinstall	L.F.		2.87		2.87	
Clean	L.F.		.68		.68	
Paint	L.F.	.39	1.01		1.40	
Minimum Charge	Job		242		242	
Vinyl						
Demolish	L.F.		.72		.72	Cost includes material and labor to
Install	L.F.	4.17	2.77		6.94	install 6" vinyl fascia with 12" vinyl
Demolish and Install	L.F.	4.17	3.49		7.66	soffit and J-channel.
Reinstall	L.F.		2.21		2.21	
Clean	L.F.		.47		.47	
Paint	L.F.	.39	1.01		1.40	
Minimum Charge	Job		242		242	
Plywood						
Demolish	L.F.		.80		.80	Cost includes material and labor to
Install	L.F.	3.39	1.38		4.77	install 12" wide plywood soffit.
Demolish and Install	L.F.	3.39	2.18		5.57	
Reinstall	L.F.		1.11		1.11	
Clean	L.F.	.02	.31		.33	
Paint	L.F.	.35	1.01		1.36	
Minimum Charge	Job		242		242	
Redwood						
Demolish	L.F.		.80		.80	Cost includes material and labor to
Install	L.F.	6.90	1.38		8.28	install 12" wide redwood soffit.
Demolish and Install	L.F.	6.90	2.18		9.08	
Reinstall	L.F.		1.11		1.11	
Clean	L.F.	.02	.31		.33	
Paint	L.F.	.35	1.01		1.36	
Minimum Charge	Job		242		242	
Cedar						
Demolish	L.F.		.80		.80	Cost includes material and labor to
Install	L.F.	5.30	1.38		6.68	install 12" wide cedar soffit.
Demolish and Install	L.F.	5.30	2.18		7.48	
Reinstall	L.F.		1.11		1.11	
Clean	L.F.	.02	.31		.33	
Paint	L.F.	.35	1.01		1.36	
Minimum Charge	Job		242		242	

Exterior Trim

Soffit	Unit	Material	Labor	Equip.	Total	Specification
Douglas Fir						
Demolish	L.F.		.80		.80	Cost includes material and labor to
Install	L.F.	2.84	1.38		4.22	install 12″ wide Douglas fir soffit.
Demolish and Install	L.F.	2.84	2.18		5.02	
Reinstall	L.F.		1.11		1.11	
Clean	L.F.	.02	.31		.33	
Paint	L.F.	.35	1.01		1.36	
Minimum Charge	Job		242		242	

For customer support on your Contractor's Pricing Guide: Residential Repair & Remodeling, call 800.448.8182.

Roofing and Flashing

Tile Roofing

Metal Flashing		Unit	Material	Labor	Equip.	Total	Specification
Galvanized							
6″							
	Demolish	L.F.		.28		.28	Includes material and labor to install
	Install	L.F.	.70	1.14		1.84	flat galvanized steel flashing, 6″ wide.
	Demolish and Install	L.F.	.70	1.42		2.12	
	Paint	L.F.	.21	.50		.71	
	Minimum Charge	Job		227		227	
14″							
	Demolish	L.F.		.28		.28	Includes material and labor to install
	Install	L.F.	1.63	1.14		2.77	flat galvanized steel flashing, 14″
	Demolish and Install	L.F.	1.63	1.42		3.05	wide.
	Paint	L.F.	.21	.50		.71	
	Minimum Charge	Job		227		227	
20″							
	Demolish	L.F.		.28		.28	Includes material and labor to install
	Install	L.F.	2.42	1.14		3.56	flat galvanized steel flashing, 20″
	Demolish and Install	L.F.	2.42	1.42		3.84	wide.
	Paint	L.F.	.21	.50		.71	
	Minimum Charge	Job		227		227	
Aluminum							
6″							
	Demolish	L.F.		.28		.28	Includes material and labor to install
	Install	L.F.	.47	1.14		1.61	flat aluminum flashing, 6″ wide.
	Demolish and Install	L.F.	.47	1.42		1.89	
	Paint	L.F.	.21	.50		.71	
	Minimum Charge	Job		227		227	
14″							
	Demolish	L.F.		.28		.28	Includes material and labor to install
	Install	L.F.	.83	1.14		1.97	flat aluminum flashing, 14″ wide.
	Demolish and Install	L.F.	.83	1.42		2.25	
	Paint	L.F.	.21	.50		.71	
	Minimum Charge	Job		227		227	
20″							
	Demolish	L.F.		.28		.28	Includes material and labor to install
	Install	L.F.	1.12	1.14		2.26	flat aluminum flashing, 20″ wide.
	Demolish and Install	L.F.	1.12	1.42		2.54	
	Paint	L.F.	.21	.50		.71	
	Minimum Charge	Job		227		227	

Roofing

Metal Flashing	Unit	Material	Labor	Equip.	Total	Specification
Copper						
6"						
Demolish	L.F.		.28		.28	Includes material and labor to install
Install	L.F.	7.80	1.51		9.31	flat copper flashing, 6" wide.
Demolish and Install	L.F.	7.80	1.79		9.59	
Minimum Charge	Job		267		267	
14"						
Demolish	L.F.		.28		.28	Includes material and labor to install
Install	L.F.	18.20	1.51		19.71	flat copper flashing, 14" wide.
Demolish and Install	L.F.	18.20	1.79		19.99	
Minimum Charge	Job		267		267	
18"						
Demolish	L.F.		.28		.28	Includes material and labor to install
Install	L.F.	26	1.51		27.51	flat copper flashing, 18" wide.
Demolish and Install	L.F.	26	1.79		27.79	
Minimum Charge	Job		267		267	
Aluminum Valley						
Demolish	L.F.		.28		.28	Includes material and labor to install
Install	L.F.	4.62	2.85		7.47	aluminum valley flashing, .024" thick.
Demolish and Install	L.F.	4.62	3.13		7.75	
Paint	L.F.	.21	.50		.71	
Minimum Charge	Job		242		242	
Gravel Stop						
Demolish	L.F.		.44		.44	Includes material and labor to install
Install	L.F.	4.69	3.68		8.37	aluminum gravel stop, .050" thick, 4"
Demolish and Install	L.F.	4.69	4.12		8.81	high, mill finish.
Reinstall	L.F.		2.94		2.94	
Paint	L.F.	.33	.50		.83	
Minimum Charge	Job		242		242	
Drip Edge						
Demolish	L.F.		.42		.42	Includes material and labor to install
Install	L.F.	.76	1.21		1.97	aluminum drip edge, .016" thick, 5"
Demolish and Install	L.F.	.76	1.63		2.39	wide, mill finish.
Reinstall	L.F.		.97		.97	
Paint	I.F.	.33	.50		.83	
Minimum Charge	Job		242		242	
End Wall						
Demolish	L.F.		1.28		1.28	Includes material and labor to install
Install	L.F.	3.52	2.85		6.37	aluminum end wall flashing, .024"
Demolish and Install	L.F.	3.52	4.13		7.65	thick.
Minimum Charge	Job		242		242	
Side Wall						
Demolish	L.F.		.25		.25	Includes material and labor to install
Install	L.F.	4.21	2.85		7.06	aluminum side wall flashing, .024"
Demolish and Install	L.F.	4.21	3.10		7.31	thick.
Minimum Charge	Job		242		242	
Coping and Wall Cap						
Demolish	L.F.		1.73		1.73	Includes material and labor to install
Install	L.F.	4.92	3.13		8.05	aluminum flashing, mill finish, 0.40
Demolish and Install	L.F.	4.92	4.86		9.78	thick, 12" wide.
Clean	L.F.	.03	1.17		1.20	
Minimum Charge	Job		227		227	

For customer support on your Contractor's Pricing Guide: Residential Repair & Remodeling, call 800.448.8182.

Roofing

Roof Shingles

Roof Shingles	Unit	Material	Labor	Equip.	Total	Specification
Underlayment						
15#						
Demolish	Sq.		13.85		13.85	Includes material and labor to install
Install	Sq.	4.87	7.10		11.97	#15 felt underlayment.
Demolish and Install	Sq.	4.87	20.95		25.82	
Minimum Charge	Job		227		227	
30#						
Demolish	Sq.		13.85		13.85	Includes material and labor to install
Install	Sq.	9.55	7.85		17.40	#30 felt underlayment.
Demolish and Install	Sq.	9.55	21.70		31.25	
Minimum Charge	Job		227		227	
Self Adhering						
Demolish	Sq.		34.50		34.50	Includes material and labor to install
Install	Sq.	72.50	20.50		93	self adhering ice barrier roofing
Demolish and Install	Sq.	72.50	55		127.50	underlayment.
Minimum Charge	Job		227		227	

Rolled Roofing

Rolled Roofing	Unit	Material	Labor	Equip.	Total	Specification
90 lb						
Demolish	Sq.		34.50		34.50	Cost includes material and labor to
Install	Sq.	47.50	38		85.50	install 90 lb mineral surface rolled
Demolish and Install	Sq.	47.50	72.50		120	roofing.
Minimum Charge	Job		227		227	

Composition Shingles

Composition Shingles	Unit	Material	Labor	Equip.	Total	Specification
Fiberglass (3-tab)						
25 Year						
Demolish	Sq.		37.50		37.50	Cost includes material and labor to
Install	Sq.	98	50.50		148.50	install 25 year fiberglass shingles.
Demolish and Install	Sq.	98	88		186	
Minimum Charge	Job		227		227	
30 Year						
Demolish	Sq.		37.50		37.50	Cost includes material and labor to
Install	Sq.	129	56.50		185.50	install 30 year fiberglass shingles.
Demolish and Install	Sq.	129	94		223	
Minimum Charge	Job		227		227	
Emergency Repairs						
Install	Ea.	216	227		443	Includes labor and material for
Minimum Charge	Job		227		227	emergency repairs.
Architectural						
Laminated, 25 Year						
Demolish	Sq.		37.50		37.50	Cost includes material and labor to
Install	Sq.	126	63		189	install 25 year fiberglass roof shingles.
Demolish and Install	Sq.	126	100.50		226.50	
Minimum Charge	Job		227		227	
Laminated, 30 Year						
Demolish	Sq.		37.50		37.50	Cost includes material and labor to
Install	Sq.	164	81		245	install 30 year laminated roof
Demolish and Install	Sq.	164	118.50		282.50	shingles.
Minimum Charge	Job		227		227	

Roofing

Composition Shingles		Unit	Material	Labor	Equip.	Total	Specification
Laminated, 40 Year							
	Demolish	Sq.		37.50		37.50	Cost includes material and labor to
	Install	Sq.	425	94.50		519.50	install 40 year laminated roof
	Demolish and Install	Sq.	425	132		557	shingles.
	Minimum Charge	Job		227		227	
Laminated Shake, 40 Year							
	Demolish	Sq.		37.50		37.50	Cost includes material and labor to
	Install	Sq.	425	94.50		519.50	install 40 year laminated roof
	Demolish and Install	Sq.	425	132		557	shingles.
	Minimum Charge	Job		227		227	
Emergency Repairs							
	Install	Ea.	216	227		443	Includes labor and material for
	Minimum Charge	Job		227		227	emergency repairs.
Roof Jack							
	Demolish	Ea.		7.55		7.55	Includes material and labor to install
	Install	Ea.	27	48.50		75.50	residential roof jack, w/bird screen,
	Demolish and Install	Ea.	27	56.05		83.05	backdraft damper, 3" & 4" dia. round
	Paint	Ea.	.11	5.05		5.16	duct.
	Minimum Charge	Job		284		284	
Ridge Vent							
	Demolish	L.F.		1.34		1.34	Includes material and labor to install a
	Install	L.F.	3.03	3.44		6.47	mill finish aluminum ridge vent strip.
	Demolish and Install	L.F.	3.03	4.78		7.81	
	Minimum Charge	Job		242		242	
Shingle Molding							
1" x 2"							
	Demolish	L.F.		.26		.26	Cost includes material and labor to
	Install	L.F.	.36	1.47		1.83	install 1" x 2" pine trim.
	Demolish and Install	L.F.	.36	1.73		2.09	
	Paint	L.F.	.04	.63		.67	
	Minimum Charge	Job		242		242	
1" x 3"							
	Demolish	L.F.		.26		.26	Cost includes material and labor to
	Install	L.F.	.37	1.67		2.04	install 1" x 3" pine trim.
	Demolish and Install	L.F.	.37	1.93		2.30	
	Paint	L.F.	.04	.63		.67	
	Minimum Charge	Job		242		242	
1" x 4"							
	Demolish	L.F.		.26		.26	Cost includes material and labor to
	Install	L.F.	.76	1.94		2.70	install 1" x 4" pine trim.
	Demolish and Install	L.F.	.76	2.20		2.96	
	Paint	L.F.	.04	.63		.67	
	Minimum Charge	Job		242		242	
Add for Steep Pitch							
	Install	Sq.		28.50		28.50	Includes labor and material added for
Add for Additional Story							2nd story or steep roofs.
	Install	Sq.		28.50		28.50	Includes labor and material added for 2nd story or steep roofs.

For customer support on your Contractor's Pricing Guide: Residential Repair & Remodeling, call 800.448.8182.

Roofing

Wood Shingles	Unit	Material	Labor	Equip.	Total	Specification
Cedar Wood Shingles						
16" Red Label (#2)						
Demolish	Sq.		92.50		92.50	Includes material and labor to install
Install	Sq.	219	194		413	wood shingles/shakes.
Demolish and Install	Sq.	219	286.50		505.50	
Minimum Charge	Job		242		242	
16" Blue Label (#1)						
Demolish	Sq.		92.50		92.50	Includes material and labor to install
Install	Sq.	510	149		659	wood shingles/shakes.
Demolish and Install	Sq.	510	241.50		751.50	
Minimum Charge	Job		242		242	
18" Red Label (#2)						
Demolish	Sq.		92.50		92.50	Includes material and labor to install
Install	Sq.	395	124		519	wood shingles/shakes.
Demolish and Install	Sq.	395	216.50		611.50	
Minimum Charge	Job		242		242	
18" Blue Label (#1)						
Demolish	Sq.		92.50		92.50	Includes material and labor to install
Install	Sq.	510	136		646	wood shingles/shakes.
Demolish and Install	Sq.	510	228.50		738.50	
Minimum Charge	Job		242		242	
Replace Shingle						
Install	Ea.	9.45	12.10		21.55	Includes labor and material to replace
Minimum Charge	Job		242		242	individual wood shingle/shake. Cost is per shingle.
Cedar Shake						
18" Medium Handsplit						
Demolish	Sq.		92.50		92.50	Includes material and labor to install
Install	Sq.	455	186		641	wood shingles/shakes.
Demolish and Install	Sq.	455	278.50		733.50	
Minimum Charge	Job		242		242	
18" Heavy Handsplit						
Demolish	Sq.		92.50		92.50	Includes material and labor to install
Install	Sq.	455	207		662	wood shingles/shakes.
Demolish and Install	Sq.	455	299.50		754.50	
Minimum Charge	Job		242		242	
24" Heavy Handsplit						
Demolish	Sq.		92.50		92.50	Includes material and labor to install
Install	Sq.	550	166		716	wood shingles/shakes.
Demolish and Install	Sq.	550	258.50		808.50	
Minimum Charge	Job		242		242	
24" Medium Handsplit						
Demolish	Sq.		92.50		92.50	Includes material and labor to install
Install	Sq.	550	149		699	wood shingles/shakes.
Demolish and Install	Sq.	550	241.50		791.50	
Minimum Charge	Job		242		242	
24" Straight Split						
Demolish	Sq.		92.50		92.50	Includes material and labor to install
Install	Sq.	430	220		650	wood shingles/shakes.
Demolish and Install	Sq.	430	312.50		742.50	
Minimum Charge	Job		242		242	
24" Tapersplit						
Demolish	Sq.		92.50		92.50	Includes material and labor to install
Install	Sq.	325	220		545	wood shingles/shakes.
Demolish and Install	Sq.	325	312.50		637.50	
Minimum Charge	Job		242		242	

For customer support on your Contractor's Pricing Guide: Residential Repair & Remodeling, call 800.448.8182.

Roofing

Wood Shingles

Wood Shingles	Unit	Material	Labor	Equip.	Total	Specification
18″ Straight Split						
Demolish	Sq.		92.50		92.50	Includes material and labor to install
Install	Sq.	550	220		770	wood shingles/shakes.
Demolish and Install	Sq.	550	312.50		862.50	
Minimum Charge	Job		242		242	
Replace Shingle						
Install	Ea.	9.45	12.10		21.55	Includes labor and material to replace
Minimum Charge	Job		242		242	individual wood shingle/shake. Cost is per shingle.

Slate Tile Roofing

Slate Tile Roofing	Unit	Material	Labor	Equip.	Total	Specification
Unfading Green						
Demolish	Sq.		83		83	Includes material and labor to install
Install	Sq.	695	260		955	clear Vermont slate tile.
Demolish and Install	Sq.	695	343		1038	
Reinstall	Sq.		207.73		207.73	
Minimum Charge	Job		227		227	
Unfading Purple						
Demolish	Sq.		83		83	Includes material and labor to install
Install	Sq.	680	260		940	clear Vermont slate tile.
Demolish and Install	Sq.	680	343		1023	
Reinstall	Sq.		207.73		207.73	
Minimum Charge	Job		227		227	
Replace Slate Tiles						
Install	Ea.	9.80	24		33.80	Includes labor and material to replace
Minimum Charge	Job		227		227	individual slate tiles. Cost is per tile.
Variegated Purple						
Demolish	Sq.		83		83	Includes material and labor to install
Install	Sq.	635	260		895	clear Vermont slate tile.
Demolish and Install	Sq.	635	343		978	
Reinstall	Sq.		207.73		207.73	
Minimum Charge	Job		227		227	
Unfading Grey / Black						
Demolish	Sq.		83		83	Includes material and labor to install
Install	Sq.	595	260		855	clear Vermont slate tile.
Demolish and Install	Sq.	595	343		938	
Reinstall	Sq.		207.73		207.73	
Minimum Charge	Job		227		227	
Unfading Red						
Demolish	Sq.		83		83	Includes material and labor to install
Install	Sq.	1625	260		1885	clear Vermont slate tile.
Demolish and Install	Sq.	1625	343		1968	
Reinstall	Sq.		207.73		207.73	
Minimum Charge	Job		227		227	
Weathering Black						
Demolish	Sq.		83		83	Includes material and labor to install
Install	Sq.	680	260		940	clear Pennsylvania slate tile.
Demolish and Install	Sq.	680	343		1023	
Reinstall	Sq.		207.73		207.73	
Minimum Charge	Job		227		227	
Weathering Green						
Demolish	Sq.		83		83	Includes material and labor to install
Install	Sq.	555	260		815	clear Vermont slate tile.
Demolish and Install	Sq.	555	343		898	
Reinstall	Sq.		207.73		207.73	
Minimum Charge	Job		227		227	

Roofing

Flat Clay Tile Roofing

Flat Clay Tile Roofing	Unit	Material	Labor	Equip.	Total	Specification
Glazed						
Demolish	Sq.		92.50		92.50	Includes material and labor to install
Install	Sq.	725	227		952	tile roofing. Cost based on terra cotta
Demolish and Install	Sq.	725	319.50		1044.50	red flat clay tile and includes ridge
Reinstall	Sq.		181.76		181.76	and rake tiles.
Minimum Charge	Job		227		227	
Terra Cotta Red						
Demolish	Sq.		92.50		92.50	Includes material and labor to install
Install	Sq.	725	227		952	tile roofing. Cost based on terra cotta
Demolish and Install	Sq.	725	319.50		1044.50	red flat clay tile and includes ridge
Reinstall	Sq.		181.76		181.76	and rake tiles.
Minimum Charge	Job		227		227	

Mission Tile Roofing

Mission Tile Roofing	Unit	Material	Labor	Equip.	Total	Specification
Glazed Red						
Demolish	Sq.		92.50		92.50	Includes material and labor to install
Install	Sq.	625	248		873	tile roofing. Cost based on glazed red
Demolish and Install	Sq.	625	340.50		965.50	tile and includes birdstop, ridge and
Reinstall	Sq.		198.28		198.28	rake tiles.
Minimum Charge	Job		227		227	
Unglazed Red						
Demolish	Sq.		92.50		92.50	Includes material and labor to install
Install	Sq.	485	248		733	tile roofing. Cost based on unglazed
Demolish and Install	Sq.	485	340.50		825.50	red tile and includes birdstop, ridge
Reinstall	Sq.		198.28		198.28	and rake tiles.
Minimum Charge	Job		227		227	
Glazed Blue						
Demolish	Sq.		92.50		92.50	Includes material and labor to install
Install	Sq.	415	248		663	tile roofing. Cost based on glazed
Demolish and Install	Sq.	415	340.50		755.50	blue tile and includes birdstop, ridge
Reinstall	Sq.		198.28		198.28	and rake tiles.
Minimum Charge	Job		227		227	
Glazed White						
Demolish	Sq.		92.50		92.50	Includes material and labor to install
Install	Sq.	420	248		668	tile roofing. Cost based on glazed
Demolish and Install	Sq.	420	340.50		760.50	white tile and includes birdstop, ridge
Reinstall	Sq.		198.28		198.28	and rake tiles.
Minimum Charge	Job		227		227	
Unglazed White						
Demolish	Sq.		92.50		92.50	Includes material and labor to install
Install	Sq.	296	248		544	tile roofing. Cost based on unglazed
Demolish and Install	Sq.	296	340.50		636.50	white tile and includes birdstop, ridge
Reinstall	Sq.		198.28		198.28	and rake tiles.
Minimum Charge	Job		227		227	
Color Blend						
Demolish	Sq.		92.50		92.50	Includes material and labor to install
Install	Sq.	299	248		547	tile roofing. Cost based on color blend
Demolish and Install	Sq.	299	340.50		639.50	tile and includes birdstop, ridge and
Reinstall	Sq.		198.28		198.28	rake tiles.
Minimum Charge	Job		227		227	

For customer support on your Contractor's Pricing Guide: Residential Repair & Remodeling, call 800.448.8182.

Roofing

Mission Tile Roofing

Mission Tile Roofing		Unit	Material	Labor	Equip.	Total	Specification
Accessories / Extras							
Add for Cap Furring Strips	Install	Sq.	101	284		385	Includes labor and material to install vertically placed wood furring strips under cap tiles.
Add for Tile Adhesive	Install	Sq.	21.50	16.60		38.10	Includes material and labor for adhesive material placed between tiles.
Add for Wire Attachment	Install	Sq.	15.25	198		213.25	Includes labor and material to install single wire anchors for attaching tile.
Add for Braided Runners	Install	Sq.	37	189		226	Includes labor and material to install braided wire runners for secure attachment in high wind areas.
Add for Stainless Steel Nails	Install	Sq.	23.50			23.50	Includes material only for stainless steel nails.
Add for Brass Nails	Install	Sq.	42.50			42.50	Includes material only for brass nails.
Add for Hurricane Clips	Install	Sq.	24	267		291	Includes material and labor for hurricane or wind clips for high wind areas.
Add for Copper Nails	Install	Sq.	42.50			42.50	Includes material only for copper nails.
Add for Vertical Furring Strips	Install	Sq.	20	53		73	Includes labor and material to install vertically placed wood furring strips under tiles.
Add for Horizontal Furring Strips	Install	Sq.	13.80	23.50		37.30	Includes labor and material to install horizontally placed wood furring strips under tiles.

Spanish Tile Roofing

Spanish Tile Roofing		Unit	Material	Labor	Equip.	Total	Specification
Glazed Red	Demolish	Sq.		92.50		92.50	Includes material and labor to install tile roofing. Cost based on glazed red tile and includes birdstop, ridge and rake tiles.
	Install	Sq.	615	189		804	
	Demolish and Install	Sq.	615	281.50		896.50	
	Reinstall	Sq.		151.47		151.47	
	Minimum Charge	Job		227		227	
Glazed White	Demolish	Sq.		92.50		92.50	Includes material and labor to install tile roofing. Cost based on glazed white tile and includes birdstop, ridge and rake tiles.
	Install	Sq.	350	252		602	
	Demolish and Install	Sq.	350	344.50		694.50	
	Reinstall	Sq.		201.96		201.96	
	Minimum Charge	Job		227		227	

For customer support on your Contractor's Pricing Guide: Residential Repair & Remodeling, call 800.448.8182.

Spanish Tile Roofing	Unit	Material	Labor	Equip.	Total	Specification
Glazed Blue						
Demolish	Sq.		92.50		92.50	Includes material and labor to install
Install	Sq.	555	189		744	tile roofing. Cost based on glazed
Demolish and Install	Sq.	555	281.50		836.50	blue tile and includes birdstop, ridge
Reinstall	Sq.		151.47		151.47	and rake tiles.
Minimum Charge	Job		227		227	
Unglazed Red						
Demolish	Sq.		92.50		92.50	Includes material and labor to install
Install	Sq.	665	189		854	tile roofing. Cost based on unglazed
Demolish and Install	Sq.	665	281.50		946.50	red tile and includes birdstop, ridge
Reinstall	Sq.		151.47		151.47	and rake tiles.
Minimum Charge	Job		227		227	
Unglazed White						
Demolish	Sq.		92.50		92.50	Includes material and labor to install
Install	Sq.	580	189		769	tile roofing. Cost based on unglazed
Demolish and Install	Sq.	580	281.50		861.50	white tile and includes birdstop, ridge
Reinstall	Sq.		151.47		151.47	and rake tiles.
Minimum Charge	Job		227		227	
Color Blend						
Demolish	Sq.		92.50		92.50	Includes material and labor to install
Install	Sq.	705	189		894	tile roofing. Cost based on color blend
Demolish and Install	Sq.	705	281.50		986.50	tile and includes birdstop, ridge and
Reinstall	Sq.		151.47		151.47	rake tiles.
Minimum Charge	Job		227		227	
Accessories / Extras						
Add for Vertical Furring Strips						
Install	Sq.	20	53		73	Includes labor and material to install vertically placed wood furring strips under tiles.
Add for Wire Attachment						
Install	Sq.	7.55	151		158.55	Includes labor and material to install single wire ties.
Add for Braided Runners						
Install	Sq.	28.50	130		158.50	Includes labor and material to install braided wire runners for secure attachment in high wind areas.
Add for Stainless Steel Nails						
Install	Sq.	23.50			23.50	Includes material only for stainless steel nails.
Add for Copper Nails						
Install	Sq.	36.50			36.50	Includes material only for copper nails.
Add for Brass Nails						
Install	Sq.	32.50			32.50	Includes material only for brass nails.
Add for Hurricane Clips						
Install	Sq.	11.85	157		168.85	Includes material and labor to install hurricane or wind clips for high wind areas.
Add for Mortar Set Tiles						
Install	Sq.	53	227		280	Includes material and labor to install setting tiles by using mortar material placed between tiles.

Roofing

Spanish Tile Roofing

Spanish Tile Roofing	Unit	Material	Labor	Equip.	Total	Specification
Add for Tile Adhesive						
Install	Sq.	14.15	14.75		28.90	Includes material and labor for adhesive placed between tiles.
Add for Cap Furring Strips						
Install	Sq.	101	284		385	Includes labor and material to install vertically placed wood furring strips under cap tiles.
Add for Horizontal Furring Strips						
Install	Sq.	13.80	23.50		37.30	Includes labor and material to install horizontally placed wood furring strips under tiles.

Concrete Tile Roofing

Concrete Tile Roofing	Unit	Material	Labor	Equip.	Total	Specification
Corrugated Red / Brown						
Demolish	Sq.		92.50		92.50	Includes material and labor to install tile roofing. Cost includes birdstop, booster, ridge and rake tiles.
Install	Sq.	131	335		466	
Demolish and Install	Sq.	131	427.50		558.50	
Reinstall	Sq.		269.27		269.27	
Minimum Charge	Job		227		227	
Corrugated Gray						
Demolish	Sq.		92.50		92.50	Includes material and labor to install tile roofing. Cost includes birdstop, booster, ridge and rake tiles.
Install	Sq.	118	335		453	
Demolish and Install	Sq.	118	427.50		545.50	
Reinstall	Sq.		269.27		269.27	
Minimum Charge	Job		227		227	
Corrugated Bright Red						
Demolish	Sq.		92.50		92.50	Includes material and labor to install tile roofing. Cost includes birdstop, booster, ridge and rake tiles.
Install	Sq.	130	335		465	
Demolish and Install	Sq.	130	427.50		557.50	
Reinstall	Sq.		269.27		269.27	
Minimum Charge	Job		227		227	
Corrugated Black						
Demolish	Sq.		92.50		92.50	Includes material and labor to install tile roofing. Cost includes birdstop, booster, ridge and rake tiles.
Install	Sq.	130	335		465	
Demolish and Install	Sq.	130	427.50		557.50	
Reinstall	Sq.		269.27		269.27	
Minimum Charge	Job		227		227	
Corrugated Green						
Demolish	Sq.		92.50		92.50	Includes material and labor to install tile roofing. Cost includes birdstop, booster, ridge and rake tiles.
Install	Sq.	130	335		465	
Demolish and Install	Sq.	130	427.50		557.50	
Reinstall	Sq.		269.27		269.27	
Minimum Charge	Job		227		227	
Corrugated Blue						
Demolish	Sq.		92.50		92.50	Includes material and labor to install tile roofing. Cost includes birdstop, booster, ridge and rake tiles.
Install	Sq.	130	335		465	
Demolish and Install	Sq.	130	427.50		557.50	
Reinstall	Sq.		269.27		269.27	
Minimum Charge	Job		227		227	
Flat Natural Gray						
Demolish	Sq.		92.50		92.50	Includes material and labor to install tile roofing. Cost includes birdstop, booster, ridge and rake tiles.
Install	Sq.	179	335		514	
Demolish and Install	Sq.	179	427.50		606.50	
Reinstall	Sq.		269.27		269.27	
Minimum Charge	Job		227		227	

For customer support on your Contractor's Pricing Guide: Residential Repair & Remodeling, call 800.448.8182.

Concrete Tile Roofing		Unit	Material	Labor	Equip.	Total	Specification
Flat Red / Brown							
	Demolish	Sq.		92.50		92.50	Includes material and labor to install
	Install	Sq.	179	335		514	tile roofing. Cost includes birdstop,
	Demolish and Install	Sq.	179	427.50		606.50	booster, ridge and rake tiles.
	Reinstall	Sq.		269.27		269.27	
	Minimum Charge	Job		227		227	
Flat Bright Red							
	Demolish	Sq.		92.50		92.50	Includes material and labor to install
	Install	Sq.	179	335		514	tile roofing. Cost includes birdstop,
	Demolish and Install	Sq.	179	427.50		606.50	booster, ridge and rake tiles.
	Reinstall	Sq.		269.27		269.27	
	Minimum Charge	Job		227		227	
Flat Green							
	Demolish	Sq.		92.50		92.50	Includes material and labor to install
	Install	Sq.	179	335		514	tile roofing. Cost includes birdstop,
	Demolish and Install	Sq.	179	427.50		606.50	booster, ridge and rake tiles.
	Minimum Charge	Job		227		227	
Flat Blue							
	Demolish	Sq.		92.50		92.50	Includes material and labor to install
	Install	Sq.	179	335		514	tile roofing. Cost includes birdstop,
	Demolish and Install	Sq.	179	427.50		606.50	booster, ridge and rake tiles.
	Reinstall	Sq.		269.27		269.27	
	Minimum Charge	Job		227		227	
Flat Black							
	Demolish	Sq.		92.50		92.50	Includes material and labor to install
	Install	Sq.	179	335		514	tile roofing. Cost includes birdstop,
	Demolish and Install	Sq.	179	427.50		606.50	booster, ridge and rake tiles.
	Reinstall	Sq.		269.27		269.27	
	Minimum Charge	Job		227		227	

Aluminum Sheet		Unit	Material	Labor	Equip.	Total	Specification
Corrugated							
.019" Thick Natural							
	Demolish	S.F.		.98		.98	Includes material and labor to install
	Install	S.F.	1.51	2.22		3.73	aluminum sheet metal roofing.
	Demolish and Install	S.F.	1.51	3.20		4.71	
	Reinstall	S.F.		1.78		1.78	
	Minimum Charge	Job		535		535	
.016" Thick Natural							
	Demolish	S.F.		.98		.98	Includes material and labor to install
	Install	S.F.	.87	2.22		3.09	aluminum sheet metal roofing.
	Demolish and Install	S.F.	.87	3.20		4.07	
	Reinstall	S.F.		1.78		1.78	
	Minimum Charge	Job		535		535	
.016" Colored Finish							
	Demolish	S.F.		.98		.98	Includes material and labor to install
	Install	S.F.	1.28	2.22		3.50	aluminum sheet metal roofing.
	Demolish and Install	S.F.	1.28	3.20		4.48	
	Reinstall	S.F.		1.78		1.78	
	Minimum Charge	Job		535		535	
.019" Colored Finish							
	Demolish	S.F.		.98		.98	Includes material and labor to install
	Install	S.F.	1.87	2.22		4.09	aluminum sheet metal roofing.
	Demolish and Install	S.F.	1.87	3.20		5.07	
	Reinstall	S.F.		1.78		1.78	
	Minimum Charge	Job		535		535	

Aluminum Sheet	Unit	Material	Labor	Equip.	Total	Specification
Ribbed						
.019″ Thick Natural						
Demolish	S.F.		.98		.98	Includes material and labor to install
Install	S.F.	1.51	2.22		3.73	aluminum sheet metal roofing.
Demolish and Install	S.F.	1.51	3.20		4.71	
Reinstall	S.F.		1.78		1.78	
Minimum Charge	Job		535		535	
.016″ Thick Natural						
Demolish	S.F.		.98		.98	Includes material and labor to install
Install	S.F.	.87	2.22		3.09	aluminum sheet metal roofing.
Demolish and Install	S.F.	.87	3.20		4.07	
Reinstall	S.F.		1.78		1.78	
Minimum Charge	Job		535		535	
.050″ Colored Finish						
Demolish	S.F.		.98		.98	Includes material and labor to install
Install	S.F.	5.55	2.22		7.77	aluminum sheet metal roofing.
Demolish and Install	S.F.	5.55	3.20		8.75	
Reinstall	S.F.		1.78		1.78	
Minimum Charge	Job		535		535	
.032″ Thick Natural						
Demolish	S.F.		.98		.98	Includes material and labor to install
Install	S.F.	3.10	2.22		5.32	aluminum sheet metal roofing.
Demolish and Install	S.F.	3.10	3.20		6.30	
Reinstall	S.F.		1.78		1.78	
Minimum Charge	Job		535		535	
.040″ Thick Natural						
Demolish	S.F.		.98		.98	Includes material and labor to install
Install	S.F.	4.52	2.22		6.74	aluminum sheet metal roofing.
Demolish and Install	S.F.	4.52	3.20		7.72	
Reinstall	S.F.		1.78		1.78	
Minimum Charge	Job		535		535	
.050″ Thick Natural						
Demolish	S.F.		.98		.98	Includes material and labor to install
Install	S.F.	4.64	2.22		6.86	aluminum sheet metal roofing.
Demolish and Install	S.F.	4.64	3.20		7.84	
Reinstall	S.F.		1.78		1.78	
Minimum Charge	Job		535		535	
.016″ Colored Finish						
Demolish	S.F.		.98		.98	Includes material and labor to install
Install	S.F.	1.28	2.22		3.50	aluminum sheet metal roofing.
Demolish and Install	S.F.	1.28	3.20		4.48	
Reinstall	S.F.		1.78		1.78	
Minimum Charge	Job		535		535	
.019″ Colored Finish						
Demolish	S.F.		.98		.98	Includes material and labor to install
Install	S.F.	1.87	2.22		4.09	aluminum sheet metal roofing.
Demolish and Install	S.F.	1.87	3.20		5.07	
Reinstall	S.F.		1.78		1.78	
Minimum Charge	Job		535		535	
.032″ Colored Finish						
Demolish	S.F.		.98		.98	Includes material and labor to install
Install	S.F.	5.05	2.22		7.27	aluminum sheet metal roofing.
Demolish and Install	S.F.	5.05	3.20		8.25	
Reinstall	S.F.		1.78		1.78	
Minimum Charge	Job		535		535	

For customer support on your Contractor's Pricing Guide: Residential Repair & Remodeling, call 800.448.8182.

Roofing

Aluminum Sheet

	Unit	Material	Labor	Equip.	Total	Specification
.040" Colored Finish						
Demolish	S.F.		.98		.98	Includes material and labor to install
Install	S.F.	6.05	2.22		8.27	aluminum sheet metal roofing.
Demolish and Install	S.F.	6.05	3.20		9.25	
Reinstall	S.F.		1.78		1.78	
Minimum Charge	Job		535		535	

Fiberglass Sheet

	Unit	Material	Labor	Equip.	Total	Specification
12 Ounce Corrugated						
Demolish	S.F.		1.76		1.76	Cost includes material and labor to
Install	S.F.	5.25	1.82		7.07	install 12 ounce corrugated fiberglass
Demolish and Install	S.F.	5.25	3.58		8.83	sheet roofing.
Reinstall	S.F.		1.45		1.45	
Minimum Charge	Job		227		227	
8 Ounce Corrugated						
Demolish	S.F.		1.76		1.76	Cost includes material and labor to
Install	S.F.	3.78	1.82		5.60	install 8 ounce corrugated fiberglass
Demolish and Install	S.F.	3.78	3.58		7.36	roofing.
Reinstall	S.F.		1.45		1.45	
Minimum Charge	Job		227		227	

Galvanized Steel

	Unit	Material	Labor	Equip.	Total	Specification
Corrugated						
24 Gauge						
Demolish	S.F.		.98		.98	Includes material and labor to install
Install	S.F.	3.47	1.91		5.38	galvanized steel sheet metal roofing.
Demolish and Install	S.F.	3.47	2.89		6.36	Cost based on 24 gauge corrugated
Reinstall	S.F.		1.53		1.53	or ribbed steel roofing.
Minimum Charge	Job		227		227	
26 Gauge						
Demolish	S.F.		.98		.98	Includes material and labor to install
Install	S.F.	2.20	1.82		4.02	galvanized steel sheet metal roofing.
Demolish and Install	S.F.	2.20	2.80		5	Cost based on 26 gauge corrugated
Reinstall	S.F.		1.45		1.45	or ribbed steel roofing.
Minimum Charge	Job		227		227	
28 Gauge						
Demolish	S.F.		.98		.98	Includes material and labor to install
Install	S.F.	1.78	1.73		3.51	galvanized steel sheet metal roofing.
Demolish and Install	S.F.	1.78	2.71		4.49	Cost based on 28 gauge corrugated
Reinstall	S.F.		1.38		1.38	or ribbed steel roofing.
Minimum Charge	Job		227		227	
30 Gauge						
Demolish	S.F.		.98		.98	Includes material and labor to install
Install	S.F.	1.31	1.65		2.96	galvanized steel sheet metal roofing.
Demolish and Install	S.F.	1.31	2.63		3.94	Cost based on 30 gauge corrugated
Reinstall	S.F.		1.32		1.32	or ribbed steel roofing.
Minimum Charge	Job		227		227	

For customer support on your Contractor's Pricing Guide: Residential Repair & Remodeling, call 800.448.8182.

Roofing

Galvanized Steel		Unit	Material	Labor	Equip.	Total	Specification
Ribbed							
24 Gauge							
	Demolish	S.F.		.98		.98	Includes material and labor to install
	Install	S.F.	3.47	1.91		5.38	galvanized steel sheet metal roofing.
	Demolish and Install	S.F.	3.47	2.89		6.36	Cost based on 24 gauge corrugated
	Reinstall	S.F.		1.53		1.53	or ribbed steel roofing.
	Minimum Charge	Job		227		227	
26 Gauge							
	Demolish	S.F.		.98		.98	Includes material and labor to install
	Install	S.F.	2.20	1.82		4.02	galvanized steel sheet metal roofing.
	Demolish and Install	S.F.	2.20	2.80		5	Cost based on 26 gauge corrugated
	Reinstall	S.F.		1.45		1.45	or ribbed steel roofing.
	Minimum Charge	Job		227		227	
28 Gauge							
	Demolish	S.F.		.98		.98	Includes material and labor to install
	Install	S.F.	1.78	1.73		3.51	galvanized steel sheet metal roofing.
	Demolish and Install	S.F.	1.78	2.71		4.49	Cost based on 28 gauge corrugated
	Reinstall	S.F.		1.38		1.38	or ribbed steel roofing.
	Minimum Charge	Job		227		227	
30 Gauge							
	Demolish	S.F.		.98		.98	Includes material and labor to install
	Install	S.F.	1.31	1.65		2.96	galvanized steel sheet metal roofing.
	Demolish and Install	S.F.	1.31	2.63		3.94	Cost based on 30 gauge corrugated
	Reinstall	S.F.		1.32		1.32	or ribbed steel roofing.
	Minimum Charge	Job		227		227	

Standing Seam		Unit	Material	Labor	Equip.	Total	Specification
Copper							
16 Ounce							
	Demolish	Sq.		98		98	Includes material and labor to install
	Install	Sq.	1950	410		2360	copper standing seam metal roofing.
	Demolish and Install	Sq.	1950	508		2458	Cost based on 16 ounce copper
	Reinstall	Sq.		328.37		328.37	roofing.
	Minimum Charge	Job		267		267	
18 Ounce							
	Demolish	Sq.		98		98	Includes material and labor to install
	Install	Sq.	2175	445		2620	copper standing seam metal roofing.
	Demolish and Install	Sq.	2175	543		2718	Cost based on 18 ounce copper
	Reinstall	Sq.		355.73		355.73	roofing.
	Minimum Charge	Job		267		267	
20 Ounce							
	Demolish	Sq.		98		98	Includes material and labor to install
	Install	Sq.	2375	485		2860	copper standing seam metal roofing.
	Demolish and Install	Sq.	2375	583		2958	Cost based on 20 ounce copper
	Reinstall	Sq.		388.07		388.07	roofing.
Stainless Steel							
26 Gauge							
	Demolish	Sq.		98		98	Cost includes material and labor to
	Install	Sq.	675	465		1140	install 26 gauge stainless steel
	Demolish and Install	Sq.	675	563		1238	standing seam metal roofing.
	Reinstall	Sq.		371.20		371.20	
	Minimum Charge	Job		267		267	

For customer support on your Contractor's Pricing Guide: Residential Repair & Remodeling, call 800.448.8182.

Roofing

Standing Seam

	Unit	Material	Labor	Equip.	Total	Specification
28 Gauge						
Demolish	Sq.		98		98	Includes material and labor to install
Install	Sq.	545	445		990	standing seam metal roofing. Cost
Demolish and Install	Sq.	545	543		1088	based on 28 gauge stainless steel
Reinstall	Sq.		355.73		355.73	roofing.
Minimum Charge	Job		267		267	

Elastomeric Roofing

	Unit	Material	Labor	Equip.	Total	Specification
Neoprene						
1/16"						
Demolish	Sq.		34		34	Includes material and labor to install
Install	Sq.	223	217	44	484	neoprene membrane, fully adhered,
Demolish and Install	Sq.	223	251	44	518	1/16" thick.
Minimum Charge	Job		1025	91.50	1116.50	
GRM Membrane						
1/16"						
Demolish	Sq.		34		34	Includes material and labor to install
Install	Sq.	77	345	30.50	452.50	coal tar based elastomeric roofing
Demolish and Install	Sq.	77	379	30.50	486.50	membrane, polyester fiber reinforced.
Minimum Charge	Job		1025	91.50	1116.50	
Cant Strips						
3"						
Demolish	L.F.		1.05		1.05	Includes material and labor to install
Install	L.F.	1.07	1.40		2.47	cants 3" x 3", treated timber, cut
Demolish and Install	L.F.	1.07	2.45		3.52	diagonally.
Minimum Charge	Job		114		114	
4"						
Demolish	L.F.		1.05		1.05	Includes material and labor to install
Install	L.F.	2.16	1.40		3.56	cants 4" x 4", treated timber, cut
Demolish and Install	L.F.	2.16	2.45		4.61	diagonally.
Minimum Charge	Job		114		114	
Polyvinyl Chloride (PVC)						
45 mil (loose laid)						
Demolish	Sq.		34		34	Cost includes material and labor to
Install	Sq.	104	40.50	3.59	148.09	install 45 mil PVC elastomeric roofing,
Demolish and Install	Sq.	104	74.50	3.59	182.09	loose-laid with stone.
Minimum Charge	Job		1025	91.50	1116.50	
48 mil (loose laid)						
Demolish	Sq.		34		34	Cost includes material and labor to
Install	Sq.	136	40.50	3.59	180.09	install 48 mil PVC elastomeric roofing,
Demolish and Install	Sq.	136	74.50	3.59	214.09	loose-laid with stone.
Minimum Charge	Job		1025	91.50	1116.50	
60 mil (loose laid)						
Demolish	Sq.		34		34	Cost includes material and labor to
Install	Sq.	149	40.50	3.59	193.09	install 60 mil PVC elastomeric roofing,
Demolish and Install	Sq.	149	74.50	3.59	227.09	loose-laid with stone.
Minimum Charge	Job		1025	91.50	1116.50	
45 mil (attached)						
Demolish	Sq.		54		54	Cost includes material and labor to
Install	Sq.	90	79.50	7.05	176.55	install 45 mil PVC elastomeric roofing
Demolish and Install	Sq.	90	133.50	7.05	230.55	fully attached to roof deck.
Minimum Charge	Job		1025	91.50	1116.50	

For customer support on your Contractor's Pricing Guide: Residential Repair & Remodeling, call 800.448.8182.

Elastomeric Roofing	Unit	Material	Labor	Equip.	Total	Specification
48 mil (attached)						Cost includes material and labor to install 48 mil PVC elastomeric roofing fully attached to roof deck.
Demolish	Sq.		54		54	
Install	Sq.	121	79.50	7.05	207.55	
Demolish and Install	Sq.	121	133.50	7.05	261.55	
Minimum Charge	Job		1025	91.50	1116.50	
60 mil (attached)						Cost includes material and labor to install 60 mil PVC elastomeric roofing fully attached to roof deck.
Demolish	Sq.		54		54	
Install	Sq.	132	79.50	7.05	218.55	
Demolish and Install	Sq.	132	133.50	7.05	272.55	
Minimum Charge	Job		1025	91.50	1116.50	

CSPE Type

	Unit	Material	Labor	Equip.	Total	Specification
45 mil (loose laid w / stone)						Cost includes material and labor to install 45 mil CSPE (chlorosulfonated polyethylene-hypalon) roofing material loose laid on deck and ballasted with stone.
Demolish	Sq.		34		34	
Install	Sq.	305	79.50	7.05	391.55	
Demolish and Install	Sq.	305	113.50	7.05	425.55	
Minimum Charge	Job		1025	91.50	1116.50	
45 mil (attached at seams)						Cost includes material and labor to install 45 mil CSPE (chlorosulfonated polyethylene-hypalon) roofing material attached to deck at seams with batten strips.
Demolish	Sq.		40.50		40.50	
Install	Sq.	400	59	5.25	464.25	
Demolish and Install	Sq.	400	99.50	5.25	504.75	
Minimum Charge	Job		1025	91.50	1116.50	
45 mil (fully attached)						Cost includes material and labor to install 45 mil CSPE (chlorosulfonated polyethylene-hypalon) roofing material fully attached to deck.
Demolish	Sq.		54		54	
Install	Sq.	293	59	5.25	357.25	
Demolish and Install	Sq.	293	113	5.25	411.25	
Minimum Charge	Job		1025	91.50	1116.50	

EPDM Type

	Unit	Material	Labor	Equip.	Total	Specification
45 mil (attached at seams)						Cost includes material and labor to install 45 mil EPDM (ethylene propylene diene monomer) roofing material attached to deck at seams.
Demolish	Sq.		40.50		40.50	
Install	Sq.	105	59	5.25	169.25	
Demolish and Install	Sq.	105	99.50	5.25	209.75	
Minimum Charge	Job		1025	91.50	1116.50	
45 mil (fully attached)						Cost includes material and labor to install 45 mil EPDM (ethylene propylene diene monomer) roofing material fully attached to deck.
Demolish	Sq.		54		54	
Install	Sq.	99	79.50	7.05	185.55	
Demolish and Install	Sq.	99	133.50	7.05	239.55	
Minimum Charge	Job		1025	91.50	1116.50	
45 mil (loose laid w / stone)						Cost includes material and labor to install 45 mil EPDM (ethylene propylene diene monomer) roofing material loose laid on deck and ballasted with stones (approx. 1/2 ton of stone per square).
Demolish	Sq.		34		34	
Install	Sq.	111	40.50	3.59	155.09	
Demolish and Install	Sq.	111	74.50	3.59	189.09	
Minimum Charge	Job		1025	91.50	1116.50	
60 mil (attached at seams)						Cost includes material and labor to install 60 mil EPDM (ethylene propylene diene monomer) roofing material attached to deck at seams.
Demolish	Sq.		40.50		40.50	
Install	Sq.	121	59	5.25	185.25	
Demolish and Install	Sq.	121	99.50	5.25	225.75	
Minimum Charge	Job		1025	91.50	1116.50	

Roofing

Elastomeric Roofing

	Unit	Material	Labor	Equip.	Total	Specification
60 mil (fully attached)						Cost includes material and labor to
Demolish	Sq.		54		54	install 60 mil EPDM (ethylene
Install	Sq.	116	79.50	7.05	202.55	propylene diene monomer) roofing
Demolish and Install	Sq.	116	133.50	7.05	256.55	material fully attached to deck.
Minimum Charge	Job		1025	91.50	1116.50	
60 mil (loose laid w / stone)						Cost includes material and labor to
Demolish	Sq.		34		34	install 60 mil EPDM (ethylene
Install	Sq.	130	40.50	3.59	174.09	propylene diene monomer) roofing
Demolish and Install	Sq.	130	74.50	3.59	208.09	material loose laid on deck and
Minimum Charge	Job		1025	91.50	1116.50	ballasted with stones (approx. 1/2 ton of stone per square).

Modified Bitumen Roofing

	Unit	Material	Labor	Equip.	Total	Specification
Fully Attached						
120 mil						Cost includes material and labor to
Demolish	Sq.		81		81	install 120 mil modified bitumen
Install	Sq.	66	10.10	2.05	78.15	roofing, fully attached to deck by
Demolish and Install	Sq.	66	91.10	2.05	159.15	torch.
Minimum Charge	Job		114		114	
150 mil						Cost includes material and labor to
Demolish	Sq.		81		81	install 150 mil modified bitumen
Install	Sq.	109	25	5.05	139.05	roofing, fully attached to deck by
Demolish and Install	Sq.	109	106	5.05	220.05	torch.
Minimum Charge	Job		114		114	
160 mil						Cost includes material and labor to
Demolish	Sq.		81		81	install 160 mil modified bitumen
Install	Sq.	97.50	26	5.25	128.75	roofing, fully attached to deck by
Demolish and Install	Sq.	97.50	107	5.25	209.75	torch.
Minimum Charge	Job		114		114	

Built-up Roofing

	Unit	Material	Labor	Equip.	Total	Specification
Emergency Repairs						
Install	Ea.	286	2125	435	2846	Minimum charge for work requiring
Minimum Charge	Job		1500	305	1805	hot asphalt.
Asphalt						
3-ply Asphalt						
Demolish	Sq.		84.50		84.50	Cost includes material and labor to
Install	Sq.	75.50	124	25.50	225	install 3-ply asphalt built-up roof
Demolish and Install	Sq.	75.50	208.50	25.50	309.50	including one 30 lb and two 15 lb
Minimum Charge	Job		1500	305	1805	layers of felt.
4-ply Asphalt						
Demolish	Sq.		100		100	Cost includes material and labor to
Install	Sq.	120	136	27.50	283.50	install 4-ply asphalt built-up roof
Demolish and Install	Sq.	120	236	27.50	383.50	including one 30 lb and three 15 lb
Minimum Charge	Job		1500	305	1805	layers of felt.

87

Roofing

Built-up Roofing		Unit	Material	Labor	Equip.	Total	Specification
5-ply Asphalt							Cost includes material and labor to
	Demolish	Sq.		132		132	install 5-ply asphalt built-up roof
	Install	Sq.	147	149	30.50	326.50	including one 30 lb and four 15 lb
	Demolish and Install	Sq.	147	281	30.50	458.50	layers of felt.
	Minimum Charge	Job		1500	305	1805	
Components / Accessories							
Flood Coat							
	Install	Sq.	30.50	70		100.50	Includes labor and material to install
R / R Gravel and Flood Coat							flood coat.
	Install	Sq.	37	227		264	Includes labor and material to install
Aluminum UV Coat							flood coat and gravel.
	Install	Sq.	37	259	23	319	Includes labor and material to install
Gravel Coat							aluminum-based UV coating.
	Demolish	Sq.		42		42	Includes labor and material to install
	Install	Sq.	6.80	38		44.80	gravel coat.
	Demolish and Install	Sq.	6.80	80		86.80	
Mineral Surface Cap Sheet							
	Install	Sq.	81	7.85		88.85	Cost includes labor and material to
Fibrated Aluminum UV Coating							install mineral surfaced cap sheet.
	Install	Sq.	31.50	70		101.50	Includes labor and material to install
Impregnated Walk							fibered aluminum-based UV coating.
1″							
	Install	S.F.	3.49	2.42		5.91	Cost includes labor and material to
1/2″							install 1″ layer of asphalt coating.
	Install	S.F.	2.60	1.91		4.51	Cost includes labor and material to
3/4″							install 1/2″ layer of asphalt coating.
	Install	S.F.	3.15	2.02		5.17	Cost includes labor and material to
							install 3/4″ layer of asphalt coating.

For customer support on your Contractor's Pricing Guide: Residential Repair & Remodeling, call 800.448.8182.

Doors	Roof Window	Double Hung Window

Exterior Type Door	Unit	Material	Labor	Equip.	Total	Specification
Entry						
Flush (1-3/4″)						
Demolish	Ea.		45		45	Cost includes material and labor to
Install	Ea.	252	30.50		282.50	install 3′ x 6′ 8″ pre-hung 1-3/4″
Demolish and Install	Ea.	252	75.50		327.50	door, including casing and stop,
Reinstall	Ea.		30.25		30.25	hinges, jamb, aluminum sill,
Clean	Ea.	1.52	15.60		17.12	weatherstripped.
Paint	Ea.	7.70	57.50		65.20	
Minimum Charge	Job		242		242	
Raised Panel						
Demolish	Ea.		45		45	Cost includes material and labor to
Install	Ea.	580	27		607	install 3′ x 6′ 8″ pre-hung 1-3/4″
Demolish and Install	Ea.	580	72		652	door, jamb, including casing and
Reinstall	Ea.		26.89		26.89	stop, hinges, aluminum sill,
Clean	Ea.	1.52	15.60		17.12	weatherstripping.
Paint	Ea.	7.70	57.50		65.20	
Minimum Charge	Job		242		242	
Entry w / Lights						
Good Quality						
Demolish	Ea.		45		45	Cost includes material and labor to
Install	Ea.	500	60.50		560.50	install 3′ x 6′ 8″ pre-hung 1-3/4″
Demolish and Install	Ea.	500	105.50		605.50	door, solid wood lower panel, with up
Reinstall	Ea.		60.50		60.50	to 9-light glass panels, casing, sill,
Clean	Ea.	1.52	15.60		17.12	weatherstripped.
Paint	Ea.	12.25	27		39.25	
Minimum Charge	Job		242		242	
Better Quality						
Demolish	Ea.		45		45	Cost includes material and labor to
Install	Ea.	515	60.50		575.50	install 3′ x 6′ 8″ pre-hung 1-3/4″ Ash
Demolish and Install	Ea.	515	105.50		620.50	door, solid wood, raised panel door
Reinstall	Ea.		60.50		60.50	without glass panels.
Clean	Ea.	1.52	15.60		17.12	
Paint	Ea.	7.70	57.50		65.20	
Minimum Charge	Job		242		242	
Premium Quality						
Demolish	Ea.		45		45	Cost includes material and labor to
Install	Ea.	1050	28.50		1078.50	install 3′ x 6′ 8″ pre-hung 1-3/4″ Oak
Demolish and Install	Ea.	1050	73.50		1123.50	door, solid wood, raised panel door
Reinstall	Ea.		28.47		28.47	without glass panels.
Clean	Ea.	1.52	15.60		17.12	
Paint	Ea.	7.70	57.50		65.20	
Minimum Charge	Job		242		242	

Exterior Type Door		Unit	Material	Labor	Equip.	Total	Specification
Custom Quality							
	Demolish	Ea.		45		45	Cost includes material and labor to
	Install	Ea.	585	64.50		649.50	install 3' x 6' 8" pre-hung 1-3/4"
	Demolish and Install	Ea.	585	109.50		694.50	hand-made, solid wood, raised panel
	Reinstall	Ea.		64.53		64.53	door with custom stained glass.
	Clean	Ea.	.11	15.60		15.71	
	Paint	Ea.	5.85	135		140.85	
	Minimum Charge	Job		242		242	
Miami Style							
	Demolish	Ea.		45		45	Cost includes material and labor to
	Install	Ea.	239	28.50		267.50	install 3' x 6' 8" pre-hung 1-3/4"
	Demolish and Install	Ea.	239	73.50		312.50	lauan door of solid wood.
	Reinstall	Ea.		28.47		28.47	
	Clean	Ea.	.11	15.60		15.71	
	Paint	Ea.	7.70	57.50		65.20	
	Minimum Charge	Job		242		242	
French Style							
1-Light							
	Demolish	Ea.		45		45	Cost includes material and labor to
	Install	Ea.	1375	80.50		1455.50	install 3' x 6' 8" pre-hung 1-3/4"
	Demolish and Install	Ea.	1375	125.50		1500.50	French style door with 1 light.
	Reinstall	Ea.		80.67		80.67	
	Clean	Ea.	.96	23.50		24.46	
	Paint	Ea.	22	101		123	
	Minimum Charge	Job		242		242	
9-Light							
	Demolish	Ea.		45		45	Cost includes material and labor to
	Install	Ea.	895	34.50		929.50	install 3' x 6' 8" pre-hung 1-3/4"
	Demolish and Install	Ea.	895	79.50		974.50	French style door with 9 lights.
	Reinstall	Ea.		34.57		34.57	
	Clean	Ea.	.21	31		31.21	
	Paint	Ea.	22	101		123	
	Minimum Charge	Job		242		242	
15-Light							
	Demolish	Ea.		52		52	Cost includes material and labor to
	Install	Ea.	1025	69		1094	install 3' x 6' 8" pre-hung 1-3/4"
	Demolish and Install	Ea.	1025	121		1146	French style door with 15 lights.
	Reinstall	Ea.		69.14		69.14	
	Clean	Ea.	.21	31		31.21	
	Paint	Ea.	22	101		123	
	Minimum Charge	Job		242		242	
Jamb							
Wood							
	Demolish	Ea.		33.50		33.50	Includes material and labor to install
	Install	Ea.	142	13.10		155.10	flat pine jamb with square cut heads
	Demolish and Install	Ea.	142	46.60		188.60	and rabbeted sides for 6' 8" high and
	Clean	Ea.	.04	9.35		9.39	3-9/16" door including trim sets for
	Paint	Ea.	12.25	27		39.25	both sides.
	Minimum Charge	Job		242		242	
Metal 6' 8" High							
	Demolish	Ea.		27		27	Cost includes material and labor to
	Install	Ea.	175	30.50		205.50	install 18 gauge hollow metal door
	Demolish and Install	Ea.	175	57.50		232.50	frame to fit a 4-1/2" jamb 6' 8" high
	Reinstall	Ea.		30.25		30.25	and up to 3' 6" opening.
	Clean	Ea.	.32	7.80		8.12	
	Paint	Ea.	2.52	8.05		10.57	
	Minimum Charge	Job		242		242	

For customer support on your Contractor's Pricing Guide: Residential Repair & Remodeling, call 800.448.8182.

Exterior Doors and Windows

Exterior Type Door		Unit	Material	Labor	Equip.	Total	Specification
Metal 7' High							
	Demolish	Ea.		27		27	Cost includes material and labor to
	Install	Ea.	188	32.50		220.50	install 18 gauge hollow metal door
	Demolish and Install	Ea.	188	59.50		247.50	frame to fit a 4-1/2" jamb 7' high and
	Reinstall	Ea.		32.27		32.27	up to 3' 6" opening.
	Clean	Ea.	.32	7.80		8.12	
	Paint	Ea.	2.52	8.05		10.57	
	Minimum Charge	Job		242		242	
Metal 8' High							
	Demolish	Ea.		27		27	Cost includes material and labor to
	Install	Ea.	211	40.50		251.50	install 18 gauge hollow metal door
	Demolish and Install	Ea.	211	67.50		278.50	frame to fit a 4-1/2" jamb 8' high and
	Reinstall	Ea.		40.33		40.33	up to 3' 6" opening.
	Clean	Ea.	.45	9.35		9.80	
	Paint	Ea.	2.82	8.95		11.77	
	Minimum Charge	Job		242		242	
Metal 9' High							
	Demolish	Ea.		27		27	Cost includes material and labor to
	Install	Ea.	264	48.50		312.50	install 18 gauge hollow metal door
	Demolish and Install	Ea.	264	75.50		339.50	frame to fit a 4-1/2" jamb 9' high and
	Reinstall	Ea.		48.40		48.40	up to 3' 6" opening.
	Clean	Ea.	.45	9.35		9.80	
	Paint	Ea.	2.82	8.95		11.77	
	Minimum Charge	Job		242		242	
Casing Trim							
Single Width							
	Demolish	Ea.		8.55		8.55	Cost includes material and labor to
	Install	Opng.	27	40.50		67.50	install 11/16" x 2-1/2" pine ranch
	Demolish and Install	Opng.	27	49.05		76.05	style casing for one side of a standard
	Reinstall	Opng.		32.27		32.27	door opening.
	Clean	Opng.	.22	3.74		3.96	
	Paint	Ea.	7.45	8.05		15.50	
	Minimum Charge	Job		242		242	
Double Width							
	Demolish	Ea.		10.15		10.15	Cost includes material and labor to
	Install	Opng.	31.50	48.50		80	install 11/16" x 2-1/2" pine ranch
	Demolish and Install	Opng.	31.50	58.65		90.15	style door casing for one side of a
	Reinstall	Opng.		38.72		38.72	double door opening.
	Clean	Opng.	.25	7.50		7.75	
	Paint	Ea.	7.45	9.15		16.60	
	Minimum Charge	Job		242		242	

Residential Metal Door		Unit	Material	Labor	Equip.	Total	Specification
Pre-hung, Steel-clad							
2' 8" x 6' 8"							
	Demolish	Ea.		45		45	Cost includes material and labor to
	Install	Ea.	530	57		587	install 2' 6" x 6' 8" pre-hung 1-3/4"
	Demolish and Install	Ea.	530	102		632	24 gauge steel clad door with 4-1/2"
	Reinstall	Ea.		56.94		56.94	wood frame, trims, hinges, aluminum
	Clean	Ea.	1.52	15.60		17.12	sill, weatherstripped.
	Paint	Ea.	22	57.50		79.50	
	Minimum Charge	Job		242		242	

For customer support on your Contractor's Pricing Guide: Residential Repair & Remodeling, call 800.448.8182.

Exterior Doors and Windows

Residential Metal Door

	Unit	Material	Labor	Equip.	Total	Specification
3' x 6' 8"						
Demolish	Ea.		45		45	Cost includes material and labor to
Install	Ea.	530	64.50		594.50	install 3' x 6' 8" pre-hung 1-3/4" 24
Demolish and Install	Ea.	530	109.50		639.50	gauge steel clad door with 4-1/2"
Reinstall	Ea.		64.53		64.53	wood frame, trims, hinges, aluminum
Clean	Ea.	1.52	15.60		17.12	sill, weatherstripped.
Paint	Ea.	22	57.50		79.50	
Minimum Charge	Job		242		242	

Metal Door

	Unit	Material	Labor	Equip.	Total	Specification
Steel, Door Only						
2' 8" x 6' 8", 18 Gauge						
Demolish	Ea.		9.60		9.60	Cost includes material and labor to
Install	Ea.	720	30.50		750.50	install 18 gauge 2' 8" x 6' 8" steel
Demolish and Install	Ea.	720	40.10		760.10	door only, 1-3/4" thick, hinges.
Reinstall	Ea.		30.25		30.25	
Clean	Ea.	.64	11.70		12.34	
Paint	Ea.	9.15	33.50		42.65	
Minimum Charge	Job		242		242	
3' x 6' 8", 18 Gauge						
Demolish	Ea.		9.60		9.60	Cost includes material and labor to
Install	Ea.	735	32.50		767.50	install 18 gauge 3' x 6' 8" steel door
Demolish and Install	Ea.	735	42.10		777.10	only, 1-3/4" thick, hinges.
Reinstall	Ea.		32.27		32.27	
Clean	Ea.	.64	11.70		12.34	
Paint	Ea.	9.15	33.50		42.65	
Minimum Charge	Job		242		242	
2' 8" x 7', 18 Gauge						
Demolish	Ea.		9.60		9.60	Cost includes material and labor to
Install	Ea.	735	32.50		767.50	install 18 gauge 2' 8" x 7' steel door
Demolish and Install	Ea.	735	42.10		777.10	only, 1-3/4" thick, hinges.
Reinstall	Ea.		32.27		32.27	
Clean	Ea.	.64	11.70		12.34	
Paint	Ea.	9.15	33.50		42.65	
Minimum Charge	Job		242		242	
3' x 7', 18 Gauge						
Demolish	Ea.		9.60		9.60	Cost includes material and labor to
Install	Ea.	735	34.50		769.50	install 18 gauge 3' x 7' steel door
Demolish and Install	Ea.	735	44.10		779.10	only, 1-3/4" thick, hinges.
Reinstall	Ea.		34.57		34.57	
Clean	Ea.	.64	11.70		12.34	
Paint	Ea.	9.15	33.50		42.65	
Minimum Charge	Job		242		242	
3' 6" x 7', 18 Gauge						
Demolish	Ea.		9.60		9.60	Cost includes material and labor to
Install	Ea.	915	37		952	install 18 gauge 3' 6" x 7' steel door
Demolish and Install	Ea.	915	46.60		961.60	only, 1-3/4" thick, hinges.
Reinstall	Ea.		37.23		37.23	
Clean	Ea.	.64	11.70		12.34	
Paint	Ea.	9.15	33.50		42.65	
Minimum Charge	Job		242		242	

Exterior Doors and Windows

Metal Door	Unit	Material	Labor	Equip.	Total	Specification
Casing Trim						
Single Width						Cost includes material and labor to
Demolish	Ea.		8.55		8.55	install 11/16" x 2-1/2" pine ranch
Install	Opng.	27	40.50		67.50	style casing for one side of a standard
Demolish and Install	Opng.	27	49.05		76.05	door opening.
Reinstall	Opng.		32.27		32.27	
Clean	Opng.	.22	3.74		3.96	
Paint	Ea.	7.45	8.05		15.50	
Minimum Charge	Job		242		242	
Double Width						Cost includes material and labor to
Demolish	Ea.		10.15		10.15	install 11/16" x 2-1/2" pine ranch
Install	Opng.	31.50	48.50		80	style door casing for one side of a
Demolish and Install	Opng.	31.50	58.65		90.15	double door opening.
Reinstall	Opng.		38.72		38.72	
Clean	Opng.	.25	7.50		7.75	
Paint	Ea.	7.45	9.15		16.60	
Minimum Charge	Job		242		242	

Solid Core Door	Unit	Material	Labor	Equip.	Total	Specification
Exterior Type						
Flush (1-3/4")						Cost includes material and labor to
Demolish	Ea.		45		45	install 3' x 6' 8" pre-hung 1-3/4"
Install	Ea.	490	60.50		550.50	door, including casing and stop,
Demolish and Install	Ea.	490	105.50		595.50	hinges, jamb, aluminum sill,
Reinstall	Ea.		60.50		60.50	weatherstripped.
Clean	Ea.	1.52	15.60		17.12	
Paint	Ea.	22	57.50		79.50	
Minimum Charge	Job		242		242	
Raised Panel						Cost includes material and labor to
Demolish	Ea.		45		45	install 3' x 6' 8" pre-hung 1-3/4"
Install	Ea.	580	27		607	door, jamb, including casing and
Demolish and Install	Ea.	580	72		652	stop, hinges, aluminum sill,
Reinstall	Ea.		26.89		26.89	weatherstripping.
Clean	Ea.	1.52	15.60		17.12	
Paint	Ea.	22	135		157	
Minimum Charge	Job		242		242	
Entry w / Lights						Cost includes material and labor to
Demolish	Ea.		45		45	install 3' x 6' 8" pre-hung 1-3/4"
Install	Ea.	500	60.50		560.50	door, solid wood lower panel, with up
Demolish and Install	Ea.	500	105.50		605.50	to 9-light glass panels, casing, sill,
Reinstall	Ea.		60.50		60.50	weatherstripped.
Clean	Ea.	1.52	15.60		17.12	
Paint	Ea.	5.85	135		140.85	
Minimum Charge	Job		242		242	
Custom Entry						Cost includes material and labor to
Demolish	Ea.		45		45	install prehung, fir, single glazed with
Install	Ea.	1675	97		1772	lead caming, 1-3/4" x 6' 8" x 3'
Demolish and Install	Ea.	1675	142		1817	
Reinstall	Ea.		96.80		96.80	
Clean	Ea.	1.52	15.60		17.12	
Paint	Ea.	9.15	33.50		42.65	
Minimum Charge	Job		242		242	

Solid Core Door		Unit	Material	Labor	Equip.	Total	Specification
Dutch Style							
	Demolish	Ea.		45		45	Includes material and labor to install
	Install	Ea.	1300	80.50		1380.50	dutch door, pine, 1-3/4" x 6' 8" x 2'
	Demolish and Install	Ea.	1300	125.50		1425.50	8" wide.
	Reinstall	Ea.		80.67		80.67	
	Clean	Ea.	1.52	15.60		17.12	
	Paint	Ea.	5.85	135		140.85	
	Minimum Charge	Job		242		242	
Miami Style							
	Demolish	Ea.		45		45	Cost includes material and labor to
	Install	Ea.	239	28.50		267.50	install 3' x 6' 8" pre-hung 1-3/4"
	Demolish and Install	Ea.	239	73.50		312.50	lauan door of solid wood.
	Reinstall	Ea.		28.47		28.47	
	Clean	Ea.	1.52	15.60		17.12	
	Paint	Ea.	7.70	57.50		65.20	
	Minimum Charge	Job		242		242	
French Style							
	Demolish	Ea.		52		52	Cost includes material and labor to
	Install	Ea.	1025	69		1094	install 3' x 6' 8" pre-hung 1-3/4"
	Demolish and Install	Ea.	1025	121		1146	French style door with 15 lights.
	Reinstall	Ea.		69.14		69.14	
	Clean	Ea.	.11	15.60		15.71	
	Paint	Ea.	5.20	202		207.20	
	Minimum Charge	Job		242		242	
Double							
	Demolish	Ea.		34.50		34.50	Includes material and labor to install a
	Install	Ea.	1425	121		1546	pre-hung double entry door including
	Demolish and Install	Ea.	1425	155.50		1580.50	frame and exterior trim.
	Clean	Ea.	.06	59.50		59.56	
	Paint	Ea.	44	202		246	
	Minimum Charge	Job		121		121	
Commercial Grade							
3' x 6' 8"							
	Demolish	Ea.		45		45	Includes material and labor to install
	Install	Ea.	735	28.50		763.50	flush solid core 1-3/4" hardwood
	Demolish and Install	Ea.	735	73.50		808.50	veneer 3' x 6' 8" door only.
	Reinstall	Ea.		28.47		28.47	
	Clean	Ea.	1.52	15.60		17.12	
	Paint	Ea.	15.25	33.50		48.75	
	Minimum Charge	Job		242		242	
3' 6" x 6' 8"							
	Demolish	Ea.		45		45	Includes material and labor to install
	Install	Ea.	835	28.50		863.50	flush solid core 1-3/4" hardwood
	Demolish and Install	Ea.	835	73.50		908.50	veneer 3' 6" x 6' 8" door only.
	Reinstall	Ea.		28.47		28.47	
	Clean	Ea.	1.52	15.60		17.12	
	Paint	Ea.	15.25	33.50		48.75	
	Minimum Charge	Job		242		242	
3' x 7'							
	Demolish	Ea.		45		45	Includes material and labor to install
	Install	Ea.	470	69		539	flush solid core 1-3/4" hardwood
	Demolish and Install	Ea.	470	114		584	veneer 3' x 7' door only.
	Reinstall	Ea.		69.14		69.14	
	Clean	Ea.	1.52	15.60		17.12	
	Paint	Ea.	15.25	33.50		48.75	
	Minimum Charge	Job		242		242	

Solid Core Door		Unit	Material	Labor	Equip.	Total	Specification
3' 6" x 7'							
	Demolish	Ea.		45		45	Includes material and labor to install
	Install	Ea.	1125	34.50		1159.50	flush solid core 1-3/4" hardwood
	Demolish and Install	Ea.	1125	79.50		1204.50	veneer 3' 6" x 7' door only.
	Reinstall	Ea.		34.57		34.57	
	Clean	Ea.	1.52	15.60		17.12	
	Paint	Ea.	15.25	33.50		48.75	
	Minimum Charge	Job		242		242	
3' x 8'							
	Demolish	Ea.		45		45	Includes material and labor to install
	Install	Ea.	825	60.50		885.50	flush solid core 1-3/4" hardwood
	Demolish and Install	Ea.	825	105.50		930.50	veneer 3' x 8' door only.
	Reinstall	Ea.		60.50		60.50	
	Clean	Ea.	1.83	23.50		25.33	
	Paint	Ea.	15.25	33.50		48.75	
	Minimum Charge	Job		242		242	
3' 6" x 8'							
	Demolish	Ea.		45		45	Includes material and labor to install
	Install	Ea.	930	60.50		990.50	flush solid core 1-3/4" hardwood
	Demolish and Install	Ea.	930	105.50		1035.50	veneer 3' 6" x 8' door only.
	Reinstall	Ea.		60.50		60.50	
	Clean	Ea.	1.83	23.50		25.33	
	Paint	Ea.	15.25	33.50		48.75	
	Minimum Charge	Job		242		242	
3' x 9'							
	Demolish	Ea.		45		45	Includes material and labor to install
	Install	Ea.	815	69		884	flush solid core 1-3/4" hardwood
	Demolish and Install	Ea.	815	114		929	veneer 3' x 9' door only.
	Reinstall	Ea.		69.14		69.14	
	Clean	Ea.	1.83	23.50		25.33	
	Paint	Ea.	15.25	33.50		48.75	
	Minimum Charge	Job		242		242	
Stain							
	Install	Ea.	13.80	45		58.80	Includes labor and material to stain
	Minimum Charge	Job		202		202	exterior type door and trim on both sides.
Door Only (SC)							
2' 4" x 6' 8"							
	Demolish	Ea.		9.60		9.60	Cost includes material and labor to
	Install	Ea.	271	27		298	install 2' 4" x 6' 8" lauan solid core
	Demolish and Install	Ea.	271	36.60		307.60	door slab.
	Reinstall	Ea.		26.89		26.89	
	Clean	Ea.	.64	11.70		12.34	
	Paint	Ea.	8.05	50.50		58.55	
	Minimum Charge	Job		242		242	
2' 8" x 6' 8"							
	Demolish	Ea.		9.60		9.60	Cost includes material and labor to
	Install	Ea.	285	27		312	install 2' 8" x 6' 8" lauan solid core
	Demolish and Install	Ea.	285	36.60		321.60	door slab.
	Reinstall	Ea.		26.89		26.89	
	Clean	Ea.	.64	11.70		12.34	
	Paint	Ea.	8.05	50.50		58.55	
	Minimum Charge	Job		242		242	

For customer support on your Contractor's Pricing Guide: Residential Repair & Remodeling, call 800.448.8182.

Solid Core Door		Unit	Material	Labor	Equip.	Total	Specification
3' x 6' 8"							
	Demolish	Ea.		9.60		9.60	Cost includes material and labor to
	Install	Ea.	285	28.50		313.50	install 3' x 6' 8" lauan solid core door
	Demolish and Install	Ea.	285	38.10		323.10	slab.
	Reinstall	Ea.		28.47		28.47	
	Clean	Ea.	.64	11.70		12.34	
	Paint	Ea.	8.05	50.50		58.55	
	Minimum Charge	Job		242		242	
3' 6" x 6' 8"							
	Demolish	Ea.		9.60		9.60	Cost includes material and labor to
	Install	Ea.	335	30.50		365.50	install 3' 6" x 6' 8" lauan solid core
	Demolish and Install	Ea.	335	40.10		375.10	door slab.
	Reinstall	Ea.		30.25		30.25	
	Clean	Ea.	.64	11.70		12.34	
	Paint	Ea.	8.05	50.50		58.55	
	Minimum Charge	Job		242		242	
Panel							
	Demolish	Ea.		9.60		9.60	Cost includes material and labor to
	Install	Ea.	405	57		462	install 3' x 6' 8" six (6) panel interior
	Demolish and Install	Ea.	405	66.60		471.60	door.
	Reinstall	Ea.		56.94		56.94	
	Clean	Ea.	.76	11.70		12.46	
	Paint	Ea.	8.05	50.50		58.55	
	Minimum Charge	Job		242		242	
Casing Trim							
Single Width							
	Demolish	Ea.		8.55		8.55	Cost includes material and labor to
	Install	Opng.	27	40.50		67.50	install 11/16" x 2-1/2" pine ranch
	Demolish and Install	Opng.	27	49.05		76.05	style casing for one side of a standard
	Reinstall	Opng.		32.27		32.27	door opening.
	Clean	Opng.	.22	3.74		3.96	
	Paint	Ea.	7.45	8.05		15.50	
	Minimum Charge	Job		242		242	
Double Width							
	Demolish	Ea.		10.15		10.15	Cost includes material and labor to
	Install	Opng.	31.50	48.50		80	install 11/16" x 2-1/2" pine ranch
	Demolish and Install	Opng.	31.50	58.65		90.15	style door casing for one side of a
	Reinstall	Opng.		38.72		38.72	double door opening.
	Clean	Opng.	.25	7.50		7.75	
	Paint	Ea.	7.45	9.15		16.60	
	Minimum Charge	Job		242		242	

Components		Unit	Material	Labor	Equip.	Total	Specification
Hardware							
Doorknob							
	Demolish	Ea.		19.20		19.20	Includes material and labor to install
	Install	Ea.	54	30.50		84.50	residential passage lockset, keyless
	Demolish and Install	Ea.	54	49.70		103.70	bored type.
	Reinstall	Ea.		24.20		24.20	
	Clean	Ea.	.15	11.70		11.85	
	Minimum Charge	Job		242		242	

For customer support on your Contractor's Pricing Guide: Residential Repair & Remodeling, call 800.448.8182.

Exterior Doors and Windows

Components		Unit	Material	Labor	Equip.	Total	Specification
Doorknob w / Lock							
	Demolish	Ea.		19.20		19.20	Includes material and labor to install
	Install	Ea.	55.50	30.50		86	privacy lockset.
	Demolish and Install	Ea.	55.50	49.70		105.20	
	Reinstall	Ea.		24.20		24.20	
	Clean	Ea.	.15	11.70		11.85	
	Minimum Charge	Job		242		242	
Lever Handle							
	Demolish	Ea.		19.20		19.20	Includes material and labor to install
	Install	Ea.	241	48.50		289.50	residential passage lockset, keyless
	Demolish and Install	Ea.	241	67.70		308.70	bored type with lever handle.
	Reinstall	Ea.		38.72		38.72	
	Clean	Ea.	.15	11.70		11.85	
	Minimum Charge	Job		242		242	
Deadbolt							
	Demolish	Ea.		16.80		16.80	Includes material and labor to install a
	Install	Ea.	90	34.50		124.50	deadbolt.
	Demolish and Install	Ea.	90	51.30		141.30	
	Reinstall	Ea.		27.66		27.66	
	Clean	Ea.	.15	11.70		11.85	
	Minimum Charge	Job		242		242	
Grip Handle Entry							
	Demolish	Ea.		15.35		15.35	Includes material and labor to install
	Install	Ea.	140	54		194	residential keyed entry lock set with
	Demolish and Install	Ea.	140	69.35		209.35	grip type handle.
	Reinstall	Ea.		43.02		43.02	
	Clean	Ea.	.15	11.70		11.85	
	Minimum Charge	Job		242		242	
Panic Device							
	Demolish	Ea.		41.50		41.50	Includes material and labor to install
	Install	Ea.	430	80.50		510.50	touch bar, low profile, exit only.
	Demolish and Install	Ea.	430	122		552	
	Reinstall	Ea.		64.53		64.53	
	Clean	Ea.	.08	21		21.08	
	Minimum Charge	Job		242		242	
Closer							
	Demolish	Ea.		10.10		10.10	Includes material and labor to install
	Install	Ea.	292	80.50		372.50	pneumatic heavy duty closer for
	Demolish and Install	Ea.	292	90.60		382.60	exterior type doors.
	Reinstall	Ea.		64.53		64.53	
	Clean	Ea.	.08	15.60		15.68	
	Minimum Charge	Job		242		242	
Peep-hole							
	Demolish	Ea.		9.25		9.25	Includes material and labor to install
	Install	Ea.	14.25	15.15		29.40	peep-hole for door.
	Demolish and Install	Ea.	14.25	24.40		38.65	
	Reinstall	Ea.		12.10		12.10	
	Clean	Ea.	.01	4.59		4.60	
	Minimum Charge	Job		242		242	
Exit Sign							
	Demolish	Ea.		28.50		28.50	Includes material and labor to install a
	Install	Ea.	116	71		187	wall mounted interior electric exit sign.
	Demolish and Install	Ea.	116	99.50		215.50	
	Reinstall	Ea.		56.84		56.84	
	Clean	Ea.	.01	9.35		9.36	
	Minimum Charge	Job		242		242	

For customer support on your Contractor's Pricing Guide: Residential Repair & Remodeling, call 800.448.8182.

Components		Unit	Material	Labor	Equip.	Total	Specification
Kickplate							
	Demolish	Ea.		14.95		14.95	Includes material and labor to install
	Install	Ea.	29	32.50		61.50	aluminum kickplate, 10" x 28".
	Demolish and Install	Ea.	29	47.45		76.45	
	Reinstall	Ea.		25.81		25.81	
	Clean	Ea.	.15	11.70		11.85	
	Minimum Charge	Job		242		242	
Threshold							
	Install	L.F.	3.03	4.84		7.87	Includes labor and material to install
	Minimum Charge	Job		121		121	oak threshold.
Weatherstripping							
	Install	L.F.	.77	1.21		1.98	Includes labor and material to install
	Minimum Charge	Job		121		121	rubber weatherstripping for door.
Metal Jamb							
6' 8" High							
	Demolish	Ea.		27		27	Cost includes material and labor to
	Install	Ea.	175	30.50		205.50	install 18 gauge hollow metal door
	Demolish and Install	Ea.	175	57.50		232.50	frame to fit a 4-1/2" jamb 6' 8" high
	Reinstall	Ea.		30.25		30.25	and up to 3' 6" opening.
	Clean	Ea.	.32	7.80		8.12	
	Paint	Ea.	2.52	8.05		10.57	
	Minimum Charge	Job		242		242	
7' High							
	Demolish	Ea.		27		27	Cost includes material and labor to
	Install	Ea.	188	32.50		220.50	install 18 gauge hollow metal door
	Demolish and Install	Ea.	188	59.50		247.50	frame to fit a 4-1/2" jamb 7' high and
	Reinstall	Ea.		32.27		32.27	up to 3' 6" opening.
	Clean	Ea.	.32	7.80		8.12	
	Paint	Ea.	2.52	8.05		10.57	
	Minimum Charge	Job		242		242	
8' High							
	Demolish	Ea.		27		27	Cost includes material and labor to
	Install	Ea.	211	40.50		251.50	install 18 gauge hollow metal door
	Demolish and Install	Ea.	211	67.50		278.50	frame to fit a 4-1/2" jamb 8' high and
	Reinstall	Ea.		40.33		40.33	up to 3' 6" opening.
	Clean	Ea.	.45	9.35		9.80	
	Paint	Ea.	2.82	8.95		11.77	
	Minimum Charge	Job		242		242	
9' High							
	Demolish	Ea.		27		27	Cost includes material and labor to
	Install	Ea.	264	48.50		312.50	install 18 gauge hollow metal door
	Demolish and Install	Ea.	264	75.50		339.50	frame to fit a 4-1/2" jamb 9' high and
	Reinstall	Ea.		48.40		48.40	up to 3' 6" opening.
	Clean	Ea.	.45	9.35		9.80	
	Paint	Ea.	2.82	8.95		11.77	
	Minimum Charge	Job		242		242	
Wood Jamb							
	Demolish	Ea.		33.50		33.50	Includes material and labor to install
	Install	Ea.	142	13.10		155.10	flat pine jamb with square cut heads
	Demolish and Install	Ea.	142	46.60		188.60	and rabbeted sides for 6' 8" high and
	Clean	Ea.	.04	9.35		9.39	3-9/16" door including trim sets for
	Paint	Ea.	12.25	27		39.25	both sides.
	Minimum Charge	Job		242		242	
Stain							
	Install	Ea.	9.50	33.50		43	Includes labor and material to stain
	Minimum Charge	Job		202		202	single door and trim on both sides.

Components		Unit	Material	Labor	Equip.	Total	Specification
Shave & Refit							
	Install	Ea.		40.50		40.50	Includes labor to shave and rework
	Minimum Charge	Job		121		121	door to fit opening at the job site.

Sliding Patio Door		Unit	Material	Labor	Equip.	Total	Specification
Aluminum							
6' Single Pane							
	Demolish	Ea.		67		67	Cost includes material and labor to
	Install	Ea.	510	242		752	install 6' x 6' 8" sliding anodized
	Demolish and Install	Ea.	510	309		819	aluminum patio door with single pane
	Reinstall	Ea.		193.60		193.60	tempered safety glass, anodized
	Clean	Ea.	.96	23.50		24.46	frame, screen, weatherstripped.
	Minimum Charge	Job		242		242	
8' Single Pane							
	Demolish	Ea.		67		67	Cost includes material and labor to
	Install	Ea.	645	325		970	install 8' x 6' 8" sliding aluminum
	Demolish and Install	Ea.	645	392		1037	patio door with single pane tempered
	Reinstall	Ea.		258.13		258.13	safety glass, anodized frame, screen,
	Clean	Ea.	.96	23.50		24.46	weatherstripped.
	Minimum Charge	Job		242		242	
10' Single Pane							
	Demolish	Ea.		67		67	Cost includes material and labor to
	Install	Ea.	760	485		1245	install 10' x 6' 8" sliding aluminum
	Demolish and Install	Ea.	760	552		1312	patio door with single pane tempered
	Reinstall	Ea.		387.20		387.20	safety glass, anodized frame, screen,
	Clean	Ea.	.96	23.50		24.46	weatherstripped.
	Minimum Charge	Job		242		242	
6' Insulated							
	Demolish	Ea.		67		67	Cost includes material and labor to
	Install	Ea.	1050	242		1292	install 6' x 6' 8" sliding aluminum
	Demolish and Install	Ea.	1050	309		1359	patio door with double insulated
	Reinstall	Ea.		193.60		193.60	tempered glass, baked-on enamel
	Clean	Ea.	.96	23.50		24.46	finish, screen, weatherstripped.
	Minimum Charge	Job		242		242	
8' Insulated							
	Demolish	Ea.		67		67	Cost includes material and labor to
	Install	Ea.	1250	325		1575	install 8' x 6' 8" sliding aluminum
	Demolish and Install	Ea.	1250	392		1642	patio door with insulated tempered
	Reinstall	Ea.		258.13		258.13	glass, baked-on enamel finish, screen,
	Clean	Ea.	.96	23.50		24.46	weatherstripped.
	Minimum Charge	Job		242		242	
10' Insulated							
	Demolish	Ea.		67		67	Cost includes material and labor to
	Install	Ea.	1425	485		1910	install 10' x 6' 8" sliding aluminum
	Demolish and Install	Ea.	1425	552		1977	patio door with double insulated
	Reinstall	Ea.		387.20		387.20	tempered glass, baked-on enamel
	Clean	Ea.	.96	23.50		24.46	finish, screen, weatherstripped.
	Minimum Charge	Job		242		242	

Sliding Patio Door		Unit	Material	Labor	Equip.	Total	Specification
Vinyl							
6' Single Pane							
	Demolish	Ea.		67		67	Cost includes material and labor to
	Install	Ea.	495	242		737	install 6' x 6' 8" sliding vinyl patio
	Demolish and Install	Ea.	495	309		804	door with single pane tempered safety
	Reinstall	Ea.		193.60		193.60	glass, anodized frame, screen,
	Clean	Ea.	.96	23.50		24.46	weatherstripped.
	Minimum Charge	Job		242		242	
8' Single Pane							
	Demolish	Ea.		67		67	Cost includes material and labor to
	Install	Ea.	660	325		985	install 8' x 6' 8" sliding vinyl patio
	Demolish and Install	Ea.	660	392		1052	door with single pane tempered safety
	Reinstall	Ea.		258.13		258.13	glass, anodized frame, screen,
	Clean	Ea.	.96	23.50		24.46	weatherstripped.
	Minimum Charge	Job		242		242	
10' Single Pane							
	Demolish	Ea.		67		67	Cost includes material and labor to
	Install	Ea.	710	485		1195	install 10' x 6' 8" sliding vinyl patio
	Demolish and Install	Ea.	710	552		1262	door with single pane tempered safety
	Reinstall	Ea.		387.20		387.20	glass, anodized frame, screen,
	Clean	Ea.	.96	23.50		24.46	weatherstripped.
	Minimum Charge	Job		242		242	
6' Insulated							
	Demolish	Ea.		67		67	Cost includes material and labor to
	Install	Ea.	1325	242		1567	install 6' x 6' 8" sliding vinyl patio
	Demolish and Install	Ea.	1325	309		1634	door with double insulated tempered
	Reinstall	Ea.		193.60		193.60	glass, anodized frame, screen,
	Clean	Ea.	.96	23.50		24.46	weatherstripped.
	Minimum Charge	Job		242		242	
8' Insulated							
	Demolish	Ea.		67		67	Cost includes material and labor to
	Install	Ea.	1125	325		1450	install 8' x 6' 8" sliding vinyl patio
	Demolish and Install	Ea.	1125	392		1517	door with double insulated tempered
	Reinstall	Ea.		258.13		258.13	glass, anodized frame, screen,
	Clean	Ea.	.96	23.50		24.46	weatherstripped.
	Minimum Charge	Job		242		242	
10' Insulated							
	Demolish	Ea.		67		67	Cost includes material and labor to
	Install	Ea.	1950	242		2192	install 10' x 6' 8" sliding vinyl patio
	Demolish and Install	Ea.	1950	309		2259	door with double insulated tempered
	Reinstall	Ea.		193.60		193.60	glass, anodized frame, screen,
	Clean	Ea.	.96	23.50		24.46	weatherstripped.
	Minimum Charge	Job		242		242	
Wood Framed							
6' Premium Insulated							
	Demolish	Ea.		67		67	Cost includes material and labor to
	Install	Ea.	1650	242		1892	install 6' x 6' 8" sliding wood patio
	Demolish and Install	Ea.	1650	309		1959	door with pine frame, double insulated
	Reinstall	Ea.		193.60		193.60	glass, screen, lock and dead bolt,
	Clean	Ea.	.96	23.50		24.46	weatherstripped. Add for casing and
	Minimum Charge	Job		242		242	finishing.

Exterior Doors and Windows

Sliding Patio Door		Unit	Material	Labor	Equip.	Total	Specification
8' Premium Insulated							
	Demolish	Ea.		67		67	Cost includes material and labor to
	Install	Ea.	2125	325		2450	install 8' x 6' 8" sliding wood patio
	Demolish and Install	Ea.	2125	392		2517	door with pine frame, double insulated
	Reinstall	Ea.		258.13		258.13	glass, screen, lock and dead bolt,
	Clean	Ea.	.96	23.50		24.46	weatherstripped. Add for casing and
	Minimum Charge	Job		242		242	finishing.
9' Premium Insulated							
	Demolish	Ea.		67		67	Cost includes material and labor to
	Install	Ea.	2125	325		2450	install 9' x 6' 8" sliding wood patio
	Demolish and Install	Ea.	2125	392		2517	door with pine frame, double insulated
	Reinstall	Ea.		258.13		258.13	glass, screen, lock and dead bolt,
	Clean	Ea.	.96	23.50		24.46	weatherstripped. Add for casing and
	Minimum Charge	Job		242		242	finishing.
Stain							
	Install	Ea.	13.80	45		58.80	Includes labor and material to stain
	Minimum Charge	Job		202		202	exterior type door and trim on both
							sides.
Casing Trim							
	Demolish	Ea.		10.15		10.15	Cost includes material and labor to
	Install	Opng.	31.50	48.50		80	install 11/16" x 2-1/2" pine ranch
	Demolish and Install	Opng.	31.50	58.65		90.15	style door casing for one side of a
	Reinstall	Opng.		38.72		38.72	double door opening.
	Clean	Opng.	.25	7.50		7.75	
	Paint	Ea.	7.45	9.15		16.60	
	Minimum Charge	Job		242		242	
Sliding Screen							
	Demolish	Ea.		9.60		9.60	Includes material and labor to install
	Install	Ea.	153	24		177	sliding screen door, 6' 8" x 3'.
	Demolish and Install	Ea.	153	33.60		186.60	
	Reinstall	Ea.		19.36		19.36	
	Clean	Ea.	.32	7.80		8.12	
	Minimum Charge	Job		242		242	

Screen Door		Unit	Material	Labor	Equip.	Total	Specification
Aluminum							
	Demolish	Ea.		45		45	Includes material and labor to install
	Install	Ea.	271	69		340	aluminum frame screen door with
	Demolish and Install	Ea.	271	114		385	fiberglass screen cloth, closer, hinges,
	Reinstall	Ea.		55.31		55.31	latch.
	Clean	Ea.	1.52	15.60		17.12	
	Paint	Ea.	7.15	16.15		23.30	
	Minimum Charge	Job		242		242	
Wood							
	Demolish	Ea.		45		45	Includes material and labor to install
	Install	Ea.	320	80.50		400.50	wood screen door with aluminum cloth
	Demolish and Install	Ea.	320	125.50		445.50	screen. Frame and hardware not
	Reinstall	Ea.		64.53		64.53	included.
	Clean	Ea.	1.52	15.60		17.12	
	Paint	Ea.	7.15	16.15		23.30	
	Minimum Charge	Job		242		242	
Shave & Refit							
	Install	Ea.		40.50		40.50	Includes labor to shave and rework
	Minimum Charge	Job		242		242	door to fit opening at the job site.

For customer support on your Contractor's Pricing Guide: Residential Repair & Remodeling, call 800.448.8182.

Storm Door

Storm Door	Unit	Material	Labor	Equip.	Total	Specification
Good Grade						
Demolish	Ea.		45		45	Includes material and labor to install
Install	Ea.	315	69		384	aluminum frame storm door with
Demolish and Install	Ea.	315	114		429	fiberglass screen cloth, upper glass
Reinstall	Ea.		55.31		55.31	panel, closer, hinges, latch.
Clean	Ea.	1.52	15.60		17.12	
Paint	Ea.	7.15	16.15		23.30	
Minimum Charge	Job		242		242	
Better Grade						
Demolish	Ea.		45		45	Includes material and labor to install
Install	Ea.	287	69		356	aluminum frame storm door with
Demolish and Install	Ea.	287	114		401	fiberglass screen cloth, upper and
Reinstall	Ea.		55.31		55.31	lower glass panel, closer, latch.
Clean	Ea.	1.52	15.60		17.12	
Paint	Ea.	7.15	16.15		23.30	
Minimum Charge	Job		242		242	
Premium Grade						
Demolish	Ea.		45		45	Includes material and labor to install
Install	Ea.	325	69		394	aluminum frame storm door with
Demolish and Install	Ea.	325	114		439	fiberglass screen cloth, upper and
Reinstall	Ea.		55.31		55.31	lower glass panel, closer, latch.
Clean	Ea.	1.52	15.60		17.12	
Paint	Ea.	7.15	16.15		23.30	
Minimum Charge	Job		242		242	
Insulated						
Demolish	Ea.		45		45	Includes material and labor to install
Install	Ea.	275	34.50		309.50	aluminum frame storm door with
Demolish and Install	Ea.	275	79.50		354.50	fiberglass screen cloth, upper and
Reinstall	Ea.		27.66		27.66	lower glass panel, closer, latch.
Clean	Ea.	1.52	15.60		17.12	
Paint	Ea.	7.15	16.15		23.30	
Minimum Charge	Job		242		242	

Garage Door

Garage Door	Unit	Material	Labor	Equip.	Total	Specification
9' One Piece Metal						
Demolish	Ea.		113		113	Includes material and labor to install
Install	Ea.	705	121		826	one piece steel garage door 9' x 7'
Demolish and Install	Ea.	705	234		939	primed including hardware.
Reinstall	Ea.		96.80		96.80	
Clean	Ea.	1.90	47		48.90	
Paint	Ea.	9.95	101		110.95	
Minimum Charge	Job		242		242	
16' One Piece Metal						
Demolish	Ea.		134		134	Includes material and labor to install
Install	Ea.	1350	161		1511	one piece steel garage door 16' x 7'
Demolish and Install	Ea.	1350	295		1645	primed including hardware.
Reinstall	Ea.		129.07		129.07	
Clean	Ea.	3.81	62.50		66.31	
Paint	Ea.	19.90	115		134.90	
Minimum Charge	Job		242		242	

Garage Door		Unit	Material	Labor	Equip.	Total	Specification
9' Metal Sectional							Includes material and labor to install
	Demolish	Ea.		113		113	sectional metal garage door 9' x 7'
	Install	Ea.	665	121		786	primed including hardware.
	Demolish and Install	Ea.	665	234		899	
	Reinstall	Ea.		96.80		96.80	
	Clean	Ea.	1.95	53.50		55.45	
	Paint	Ea.	19.90	115		134.90	
	Minimum Charge	Job		242		242	
16' Metal Sectional							Includes material and labor to install
	Demolish	Ea.		134		134	sectional metal garage door 16' x 7'
	Install	Ea.	1025	161		1186	primed including hardware.
	Demolish and Install	Ea.	1025	295		1320	
	Reinstall	Ea.		129.07		129.07	
	Clean	Ea.	3.81	62.50		66.31	
	Paint	Ea.	19.90	115		134.90	
	Minimum Charge	Job		242		242	
Metal / Steel Roll Up							
8' x 8' Steel							Cost includes material and labor to
	Demolish	Ea.		154		154	install 8' x 8' steel overhead roll-up,
	Install	Ea.	590	695		1285	chain hoist operated service door.
	Demolish and Install	Ea.	590	849		1439	
	Reinstall	Ea.		557.20		557.20	
	Clean	Ea.	1.95	53.50		55.45	
	Paint	Ea.	21	135		156	
	Minimum Charge	Job		1125		1125	
10' x 10' Steel							Cost includes material and labor to
	Demolish	Ea.		179		179	install 10' x 10' steel overhead roll-up,
	Install	Ea.	865	795		1660	chain hoist operated service door.
	Demolish and Install	Ea.	865	974		1839	
	Reinstall	Ea.		636.80		636.80	
	Clean	Ea.	3.05	75		78.05	
	Paint	Ea.	33	135		168	
	Minimum Charge	Job		1125		1125	
12' x 12' Steel							Cost includes material and labor to
	Demolish	Ea.		215		215	install 12' x 12' steel overhead roll-up,
	Install	Ea.	1150	930		2080	chain hoist operated service door.
	Demolish and Install	Ea.	1150	1145		2295	
	Reinstall	Ea.		742.93		742.93	
	Clean	Ea.	4.39	83		87.39	
	Paint	Ea.	48	161		209	
	Minimum Charge	Job		1125		1125	
14' x 14' Steel							Cost includes material and labor to
	Demolish	Ea.		269		269	install 14' x 14' steel overhead roll-up,
	Install	Ea.	2000	1400		3400	chain hoist operated service door.
	Demolish and Install	Ea.	2000	1669		3669	
	Reinstall	Ea.		1114.40		1114.40	
	Clean	Ea.	6	107		113	
	Paint	Ea.	55.50	202		257.50	
	Minimum Charge	Job		1125		1125	
18' x 18' Steel							Cost includes material and labor to
	Demolish	Ea.		299		299	install 18' x 18' steel overhead roll-up,
	Install	Opng.	3225	1850		5075	chain hoist operated service door.
	Demolish and Install	Opng.	3225	2149		5374	
	Reinstall	Opng.		1485.87		1485.87	
	Clean	Ea.	6	62.50		68.50	
	Paint	Ea.	107	405		512	
	Minimum Charge	Job		1125		1125	
Automatic Motor							Includes labor and material to install
	Install	Ea.	1075	48.50		1123.50	electric opener motor, 1/3 HP.
	Minimum Charge	Job		242		242	

103

Exterior Doors and Windows

Garage Door		Unit	Material	Labor	Equip.	Total	Specification
9' One Piece Wood							
	Demolish	Ea.		113		113	Includes material and labor to install
	Install	Ea.	865	121		986	one piece wood garage door 9' x 7'
	Demolish and Install	Ea.	865	234		1099	primed including hardware.
	Reinstall	Ea.		96.80		96.80	
	Clean	Ea.	1.90	47		48.90	
	Paint	Ea.	9.95	101		110.95	
	Minimum Charge	Job		242		242	
9' Wood Sectional							
	Demolish	Ea.		113		113	Includes material and labor to install
	Install	Ea.	2525	121		2646	sectional wood garage door 9' x 7'
	Demolish and Install	Ea.	2525	234		2759	unfinished including hardware.
	Reinstall	Ea.		96.80		96.80	
	Clean	Ea.	1.90	47		48.90	
	Paint	Ea.	9.95	101		110.95	
	Minimum Charge	Job		242		242	
16' Wood Sectional							
	Demolish	Ea.		134		134	Includes material and labor to install
	Install	Ea.	2400	161		2561	sectional fiberglass garage door 16' x
	Demolish and Install	Ea.	2400	295		2695	7' unfinished including hardware.
	Reinstall	Ea.		129.07		129.07	
	Clean	Ea.	3.81	62.50		66.31	
	Paint	Ea.	19.90	115		134.90	
	Minimum Charge	Job		242		242	
Electric Opener							
	Demolish	Ea.		83		83	Includes material and labor to install
	Install	Ea.	385	91		476	electric opener, 1/3 HP economy.
	Demolish and Install	Ea.	385	174		559	
	Reinstall	Ea.		72.65		72.65	
	Clean	Ea.	.08	21		21.08	
	Minimum Charge	Job		242		242	
Springs							
	Install	Pr.	197	121		318	Includes labor and material to replace
	Minimum Charge	Job		242		242	springs for garage door.

Commercial Metal Door		Unit	Material	Labor	Equip.	Total	Specification
Pre-hung Steel Door							
2' 8" x 6' 8"							
	Demolish	Ea.		67		67	Cost includes material and labor to
	Install	Ea.	860	30.50		890.50	install 2' 8" x 6' 8" pre-hung 1-3/4"
	Demolish and Install	Ea.	860	97.50		957.50	18 gauge steel door with steel frame,
	Reinstall	Ea.		30.25		30.25	hinges, aluminum sill,
	Clean	Ea.	1.52	15.60		17.12	weather-stripped.
	Paint	Ea.	9.15	33.50		42.65	
	Minimum Charge	Job		242		242	
3' x 6' 8"							
	Demolish	Ea.		67		67	Cost includes material and labor to
	Install	Ea.	1700	32.50		1732.50	install 3' x 6' 8" pre-hung 1-3/4" 18
	Demolish and Install	Ea.	1700	99.50		1799.50	gauge steel door with steel frame,
	Reinstall	Ea.		32.27		32.27	hinges, aluminum sill,
	Clean	Ea.	1.52	15.60		17.12	weather-stripped.
	Paint	Ea.	9.15	33.50		42.65	
	Minimum Charge	Job		242		242	

For customer support on your Contractor's Pricing Guide: Residential Repair & Remodeling, call 800.448.8182.

Commercial Metal Door		Unit	Material	Labor	Equip.	Total	Specification
3' 6" x 6' 8"							
	Demolish	Ea.		67		67	Cost includes material and labor to
	Install	Ea.	1050	37		1087	install 3' 6" x 6' 8" pre-hung 1-3/4"
	Demolish and Install	Ea.	1050	104		1154	18 gauge steel door with steel frame,
	Reinstall	Ea.		37.23		37.23	hinges, aluminum sill,
	Clean	Ea.	1.52	15.60		17.12	weather-stripped.
	Paint	Ea.	9.15	33.50		42.65	
	Minimum Charge	Job		242		242	
4' x 6' 8"							
	Demolish	Ea.		67		67	Cost includes material and labor to
	Install	Ea.	1150	48.50		1198.50	install 4' x 6' 8" pre-hung 1-3/4" 18
	Demolish and Install	Ea.	1150	115.50		1265.50	gauge steel door with steel frame,
	Reinstall	Ea.		48.40		48.40	hinges, aluminum sill,
	Clean	Ea.	1.52	15.60		17.12	weather-stripped.
	Paint	Ea.	9.15	33.50		42.65	
	Minimum Charge	Job		242		242	
Metal Jamb							
6' 8" High							
	Demolish	Ea.		27		27	Cost includes material and labor to
	Install	Ea.	175	30.50		205.50	install 18 gauge hollow metal door
	Demolish and Install	Ea.	175	57.50		232.50	frame to fit a 4-1/2" jamb 6' 8" high
	Reinstall	Ea.		30.25		30.25	and up to 3' 6" opening.
	Clean	Ea.	.32	7.80		8.12	
	Paint	Ea.	2.52	8.05		10.57	
	Minimum Charge	Job		242		242	
7' High							
	Demolish	Ea.		27		27	Cost includes material and labor to
	Install	Ea.	188	32.50		220.50	install 18 gauge hollow metal door
	Demolish and Install	Ea.	188	59.50		247.50	frame to fit a 4-1/2" jamb 7' high and
	Reinstall	Ea.		32.27		32.27	up to 3' 6" opening.
	Clean	Ea.	.32	7.80		8.12	
	Paint	Ea.	2.52	8.05		10.57	
	Minimum Charge	Job		242		242	
8' High							
	Demolish	Ea.		27		27	Cost includes material and labor to
	Install	Ea.	211	40.50		251.50	install 18 gauge hollow metal door
	Demolish and Install	Ea.	211	67.50		278.50	frame to fit a 4-1/2" jamb 8' high and
	Reinstall	Ea.		40.33		40.33	up to 3' 6" opening.
	Clean	Ea.	.45	9.35		9.80	
	Paint	Ea.	2.82	8.95		11.77	
	Minimum Charge	Job		242		242	
9' High							
	Demolish	Ea.		27		27	Cost includes material and labor to
	Install	Ea.	264	48.50		312.50	install 18 gauge hollow metal door
	Demolish and Install	Ea.	264	75.50		339.50	frame to fit a 4-1/2" jamb 9' high and
	Reinstall	Ea.		48.40		48.40	up to 3' 6" opening.
	Clean	Ea.	.45	9.35		9.80	
	Paint	Ea.	2.82	8.95		11.77	
	Minimum Charge	Job		242		242	
Hardware							
Doorknob w / Lock							
	Demolish	Ea.		19.20		19.20	Includes material and labor to install
	Install	Ea.	227	40.50		267.50	keyed entry lock set, bored type
	Demolish and Install	Ea.	227	59.70		286.70	including knob trim, strike and strike
	Reinstall	Ea.		32.27		32.27	box for standard commercial
	Clean	Ea.	.15	11.70		11.85	application.
	Minimum Charge	Job		242		242	

Exterior Doors and Windows

Commercial Metal Door		Unit	Material	Labor	Equip.	Total	Specification
Lever Handle							
	Demolish	Ea.		19.20		19.20	Includes material and labor to install
	Install	Ea.	145	48.50		193.50	passage lockset, keyless bored type
	Demolish and Install	Ea.	145	67.70		212.70	with lever handle, non-locking
	Reinstall	Ea.		38.72		38.72	standard commercial latchset.
	Clean	Ea.	.15	11.70		11.85	
	Minimum Charge	Job		242		242	
Deadbolt							
	Demolish	Ea.		16.80		16.80	Includes material and labor to install a
	Install	Ea.	90	34.50		124.50	deadbolt.
	Demolish and Install	Ea.	90	51.30		141.30	
	Reinstall	Ea.		27.66		27.66	
	Clean	Ea.	.15	11.70		11.85	
	Minimum Charge	Job		242		242	
Panic Device							
	Demolish	Ea.		41.50		41.50	Includes material and labor to install
	Install	Ea.	430	80.50		510.50	touch bar, low profile, exit only.
	Demolish and Install	Ea.	430	122		552	
	Reinstall	Ea.		64.53		64.53	
	Clean	Ea.	.08	21		21.08	
	Minimum Charge	Job		242		242	
Closer							
	Demolish	Ea.		10.10		10.10	Includes material and labor to install
	Install	Ea.	390	80.50		470.50	pneumatic heavy duty model for
	Demolish and Install	Ea.	390	90.60		480.60	exterior type doors.
	Reinstall	Ea.		64.53		64.53	
	Clean	Ea.	.08	15.60		15.68	
	Minimum Charge	Job		242		242	
Peep-Hole							
	Demolish	Ea.		9.25		9.25	Includes material and labor to install
	Install	Ea.	14.25	15.15		29.40	peep-hole for door.
	Demolish and Install	Ea.	14.25	24.40		38.65	
	Reinstall	Ea.		12.10		12.10	
	Clean	Ea.	.01	4.59		4.60	
	Minimum Charge	Job		242		242	
Exit Sign							
	Demolish	Ea.		28.50		28.50	Includes material and labor to install a
	Install	Ea.	116	71		187	wall mounted interior electric exit sign.
	Demolish and Install	Ea.	116	99.50		215.50	
	Reinstall	Ea.		56.84		56.84	
	Clean	Ea.	.01	9.35		9.36	
	Minimum Charge	Job		242		242	
Kickplate							
	Demolish	Ea.		14.95		14.95	Includes material and labor to install
	Install	Ea.	99	32.50		131.50	stainless steel kickplate, 10" x 36".
	Demolish and Install	Ea.	99	47.45		146.45	
	Reinstall	Ea.		25.81		25.81	
	Clean	Ea.	.15	11.70		11.85	
	Minimum Charge	Job		242		242	

Aluminum Window		Unit	Material	Labor	Equip.	Total	Specification
Single Hung Two Light							
2' x 2'							
	Demolish	Ea.		26		26	Cost includes material and labor to
	Install	Ea.	198	101		299	install 2' x 2' single hung, two light,
	Demolish and Install	Ea.	198	127		325	double glazed aluminum window
	Reinstall	Ea.		81.05		81.05	including screen.
	Clean	Ea.	.02	10.65		10.67	
	Minimum Charge	Job		279		279	

106

Exterior Doors and Windows

Aluminum Window	Unit	Material	Labor	Equip.	Total	Specification
2' x 2' 6"						
Demolish	Ea.		26		26	Cost includes material and labor to
Install	Ea.	70.50	101		171.50	install 2' x 2' 6" single hung, two
Demolish and Install	Ea.	70.50	127		197.50	light, double glazed aluminum
Reinstall	Ea.		81.05		81.05	window including screen.
Clean	Ea.	.02	10.65		10.67	
Minimum Charge	Job		279		279	
3' x 1' 6"						
Demolish	Ea.		26		26	Cost includes material and labor to
Install	Ea.	70.50	111		181.50	install 3' x 1' 6" single hung, two
Demolish and Install	Ea.	70.50	137		207.50	light, double glazed aluminum
Reinstall	Ea.		89.15		89.15	window including screen.
Clean	Ea.	.01	9.35		9.36	
Minimum Charge	Job		279		279	
3' x 2'						
Demolish	Ea.		26		26	Cost includes material and labor to
Install	Ea.	81	111		192	install 3' x 2' single hung, two light,
Demolish and Install	Ea.	81	137		218	double glazed aluminum window
Reinstall	Ea.		89.15		89.15	including screen.
Clean	Ea.	.01	9.35		9.36	
Minimum Charge	Job		279		279	
3' x 2' 6"						
Demolish	Ea.		26		26	Cost includes material and labor to
Install	Ea.	82.50	111		193.50	install 3' x 2' 6" single hung, two
Demolish and Install	Ea.	82.50	137		219.50	light, double glazed aluminum
Reinstall	Ea.		89.15		89.15	window including screen.
Clean	Ea.	.02	10.65		10.67	
Minimum Charge	Job		279		279	
3' x 3'						
Demolish	Ea.		26		26	Cost includes material and labor to
Install	Ea.	107	111		218	install 3' x 3' single hung, two light,
Demolish and Install	Ea.	107	137		244	double glazed aluminum window
Reinstall	Ea.		89.15		89.15	including screen.
Clean	Ea.	.02	10.65		10.67	
Minimum Charge	Job		279		279	
3' x 3' 6"						
Demolish	Ea.		26		26	Cost includes material and labor to
Install	Ea.	274	111		385	install 3' x 3' 6" single hung, two
Demolish and Install	Ea.	274	137		411	light, double glazed aluminum
Reinstall	Ea.		89.15		89.15	window including screen.
Clean	Ea.	.02	12.45		12.47	
Minimum Charge	Job		279		279	
3' x 4'						
Demolish	Ea.		26		26	Cost includes material and labor to
Install	Ea.	275	111		386	install 3' x 4' single hung, two light,
Demolish and Install	Ea.	275	137		412	double glazed aluminum window
Reinstall	Ea.		89.15		89.15	including screen.
Clean	Ea.	.02	12.45		12.47	
Minimum Charge	Job		279		279	
3' x 5'						
Demolish	Ea.		38		38	Cost includes material and labor to
Install	Ea.	149	111		260	install 3' x 5' single hung, two light,
Demolish and Install	Ea.	149	149		298	double glazed aluminum window
Reinstall	Ea.		89.15		89.15	including screen.
Clean	Ea.	.03	14.90		14.93	
Minimum Charge	Job		279		279	

For customer support on your Contractor's Pricing Guide: Residential Repair & Remodeling, call 800.448.8182.

Exterior Doors and Windows

Aluminum Window		Unit	Material	Labor	Equip.	Total	Specification
3' x 6'							
	Demolish	Ea.		38		38	Cost includes material and labor to
	Install	Ea.	530	124		654	install 3' x 6' single hung, two light,
	Demolish and Install	Ea.	530	162		692	double glazed aluminum window
	Reinstall	Ea.		99.06		99.06	including screen.
	Clean	Ea.	.03	17.55		17.58	
	Minimum Charge	Job		279		279	
Double Hung Two Light							
2' x 2'							
	Demolish	Ea.		26		26	Cost includes material and labor to
	Install	Ea.	130	101		231	install 2' x 2' double hung, two light,
	Demolish and Install	Ea.	130	127		257	double glazed aluminum window
	Reinstall	Ea.		81.05		81.05	including screen.
	Clean	Ea.	.01	9.35		9.36	
	Minimum Charge	Job		279		279	
2' x 2' 6"							
	Demolish	Ea.		26		26	Cost includes material and labor to
	Install	Ea.	146	101		247	install 2' x 2' 6" double hung, two
	Demolish and Install	Ea.	146	127		273	light, double glazed aluminum
	Reinstall	Ea.		81.05		81.05	window including screen.
	Clean	Ea.	.01	9.35		9.36	
	Minimum Charge	Job		279		279	
3' x 1' 6"							
	Demolish	Ea.		26		26	Cost includes material and labor to
	Install	Ea.	297	111		408	install 3' x 1' 6" double hung, two
	Demolish and Install	Ea.	297	137		434	light, double glazed aluminum
	Reinstall	Ea.		89.15		89.15	window including screen.
	Clean	Ea.	.01	9.35		9.36	
	Minimum Charge	Job		279		279	
3' x 2'							
	Demolish	Ea.		26		26	Cost includes material and labor to
	Install	Ea.	345	111		456	install 3' x 2' double hung, two light,
	Demolish and Install	Ea.	345	137		482	double glazed aluminum window
	Reinstall	Ea.		89.15		89.15	including screen.
	Clean	Ea.	.01	9.35		9.36	
	Minimum Charge	Job		279		279	
3' x 2' 6"							
	Demolish	Ea.		26		26	Cost includes material and labor to
	Install	Ea.	415	111		526	install 3' x 2' 6" double hung, two
	Demolish and Install	Ea.	415	137		552	light, double glazed aluminum
	Reinstall	Ea.		89.15		89.15	window including screen.
	Clean	Ea.	.02	10.65		10.67	
	Minimum Charge	Job		279		279	
3' x 3'							
	Demolish	Ea.		26		26	Cost includes material and labor to
	Install	Ea.	460	111		571	install 3' x 3' double hung, two light,
	Demolish and Install	Ea.	460	137		597	double glazed aluminum window
	Reinstall	Ea.		89.15		89.15	including screen.
	Clean	Ea.	.02	10.65		10.67	
	Minimum Charge	Job		279		279	
3' x 3' 6"							
	Demolish	Ea.		26		26	Cost includes material and labor to
	Install	Ea.	490	111		601	install 3' x 3' 6" double hung, two
	Demolish and Install	Ea.	490	137		627	light, double glazed aluminum
	Reinstall	Ea.		89.15		89.15	window including screen.
	Clean	Ea.	.02	12.45		12.47	
	Minimum Charge	Job		279		279	

For customer support on your Contractor's Pricing Guide: Residential Repair & Remodeling, call 800.448.8182.

Exterior Doors and Windows

Aluminum Window		Unit	Material	Labor	Equip.	Total	Specification
3' x 4'							
	Demolish	Ea.		26		26	Cost includes material and labor to
	Install	Ea.	530	111		641	install 3' x 4' double hung, two light,
	Demolish and Install	Ea.	530	137		667	double glazed aluminum window
	Reinstall	Ea.		89.15		89.15	including screen.
	Clean	Ea.	.02	12.45		12.47	
	Minimum Charge	Job		279		279	
3' x 5'							
	Demolish	Ea.		38		38	Cost includes material and labor to
	Install	Ea.	580	111		691	install 3' x 5' double hung, two light,
	Demolish and Install	Ea.	580	149		729	double glazed aluminum window
	Reinstall	Ea.		89.15		89.15	including screen.
	Clean	Ea.	.03	14.90		14.93	
	Minimum Charge	Job		279		279	
3' x 6'							
	Demolish	Ea.		38		38	Cost includes material and labor to
	Install	Ea.	1075	124		1199	install 3' x 6' double hung, two light,
	Demolish and Install	Ea.	1075	162		1237	double glazed aluminum window
	Reinstall	Ea.		99.06		99.06	including screen.
	Clean	Ea.	.03	17.55		17.58	
	Minimum Charge	Job		279		279	

Sliding Sash

		Unit	Material	Labor	Equip.	Total	Specification
1' 6" x 3'							
	Demolish	Ea.		26		26	Cost includes material and labor to
	Install	Ea.	248	101		349	install 1' 6" x 3' sliding sash, two
	Demolish and Install	Ea.	248	127		375	light, double glazed aluminum
	Reinstall	Ea.		81.05		81.05	window including screen.
	Clean	Ea.	.01	9.35		9.36	
	Minimum Charge	Job		279		279	
2' x 2'							
	Demolish	Ea.		26		26	Cost includes material and labor to
	Install	Ea.	71	111		182	install 2' x 2' sliding sash, two light,
	Demolish and Install	Ea.	71	137		208	double glazed aluminum window
	Reinstall	Ea.		89.15		89.15	including screen.
	Clean	Ea.	.01	9.35		9.36	
	Minimum Charge	Job		279		279	
3' x 2'							
	Demolish	Ea.		26		26	Cost includes material and labor to
	Install	Ea.	251	111		362	install 3' x 2' sliding sash, two light,
	Demolish and Install	Ea.	251	137		388	double glazed aluminum window
	Reinstall	Ea.		89.15		89.15	including screen.
	Clean	Ea.	.01	9.35		9.36	
	Minimum Charge	Job		279		279	
3' x 2' 6"							
	Demolish	Ea.		26		26	Cost includes material and labor to
	Install	Ea.	94	111		205	install 3' x 2' 6" sliding sash, two
	Demolish and Install	Ea.	94	137		231	light, double glazed aluminum
	Reinstall	Ea.		89.15		89.15	window including screen.
	Clean	Ea.	.02	10.65		10.67	
	Minimum Charge	Job		279		279	
3' x 3'							
	Demolish	Ea.		26		26	Cost includes material and labor to
	Install	Ea.	315	111		426	install 3' x 3' sliding sash, two light,
	Demolish and Install	Ea.	315	137		452	double glazed aluminum window
	Reinstall	Ea.		89.15		89.15	including screen.
	Clean	Ea.	.02	10.65		10.67	
	Minimum Charge	Job		279		279	

Aluminum Window		Unit	Material	Labor	Equip.	Total	Specification
3' x 3' 6"							
	Demolish	Ea.		26		26	Cost includes material and labor to
	Install	Ea.	87	111		198	install 3' x 3' 6" sliding sash, two
	Demolish and Install	Ea.	87	137		224	light, double glazed aluminum
	Reinstall	Ea.		89.15		89.15	window including screen.
	Clean	Ea.	.02	12.45		12.47	
	Minimum Charge	Job		279		279	
3' x 4'							
	Demolish	Ea.		26		26	Cost includes material and labor to
	Install	Ea.	125	124		249	install 3' x 4' sliding sash, two light,
	Demolish and Install	Ea.	125	150		275	double glazed aluminum window
	Reinstall	Ea.		99.06		99.06	including screen.
	Clean	Ea.	.02	12.45		12.47	
	Minimum Charge	Job		279		279	
3' x 5'							
	Demolish	Ea.		38		38	Cost includes material and labor to
	Install	Ea.	155	124		279	install 3' x 5' sliding sash, two light,
	Demolish and Install	Ea.	155	162		317	double glazed aluminum window
	Reinstall	Ea.		99.06		99.06	including screen.
	Clean	Ea.	.03	14.90		14.93	
	Minimum Charge	Job		279		279	
4' x 5'							
	Demolish	Ea.		38		38	Cost includes material and labor to
	Install	Ea.	157	159		316	install 4' x 5' sliding sash, two light,
	Demolish and Install	Ea.	157	197		354	double glazed aluminum window
	Reinstall	Ea.		127.36		127.36	including screen.
	Clean	Ea.	.04	21.50		21.54	
	Minimum Charge	Job		279		279	
5' x 5'							
	Demolish	Ea.		38		38	Cost includes material and labor to
	Install	Ea.	375	159		534	install 5' x 5' sliding sash, two light,
	Demolish and Install	Ea.	375	197		572	double glazed aluminum window
	Reinstall	Ea.		127.36		127.36	including screen.
	Clean	Ea.	.04	21.50		21.54	
	Minimum Charge	Job		279		279	
Awning Sash							
26" x 26", 2 Light							
	Demolish	Ea.		26		26	Cost includes material and labor to
	Install	Ea.	350	124		474	install 26" x 26" two light awning,
	Demolish and Install	Ea.	350	150		500	double glazed aluminum window
	Reinstall	Ea.		99.06		99.06	including screen.
	Clean	Ea.	.02	10.65		10.67	
	Minimum Charge	Job		279		279	
26" x 37", 2 Light							
	Demolish	Ea.		26		26	Cost includes material and labor to
	Install	Ea.	365	124		489	install 26" x 37" two light awning,
	Demolish and Install	Ea.	365	150		515	double glazed aluminum window
	Reinstall	Ea.		99.06		99.06	including screen.
	Clean	Ea.	.02	10.65		10.67	
	Minimum Charge	Job		279		279	
38" x 26", 3 Light							
	Demolish	Ea.		26		26	Cost includes material and labor to
	Install	Ea.	420	124		544	install 38" x 26" three light awning,
	Demolish and Install	Ea.	420	150		570	double glazed aluminum window
	Reinstall	Ea.		99.06		99.06	including screen.
	Clean	Ea.	.02	12.45		12.47	
	Minimum Charge	Job		279		279	

Exterior Doors and Windows

Aluminum Window		Unit	Material	Labor	Equip.	Total	Specification
38" x 37", 3 Light							
	Demolish	Ea.		26		26	Cost includes material and labor to
	Install	Ea.	405	139		544	install 38" x 37" three light awning,
	Demolish and Install	Ea.	405	165		570	double glazed aluminum window
	Reinstall	Ea.		111.44		111.44	including screen.
	Clean	Ea.	.02	12.45		12.47	
	Minimum Charge	Job		279		279	
38" x 53", 3 Light							
	Demolish	Ea.		38		38	Cost includes material and labor to
	Install	Ea.	800	139		939	install 38" x 53" three light awning,
	Demolish and Install	Ea.	800	177		977	double glazed aluminum window
	Reinstall	Ea.		111.44		111.44	including screen.
	Clean	Ea.	.03	14.90		14.93	
	Minimum Charge	Job		279		279	
50" x 37", 4 Light							
	Demolish	Ea.		38		38	Cost includes material and labor to
	Install	Ea.	720	139		859	install 50" x 37" four light awning,
	Demolish and Install	Ea.	720	177		897	double glazed aluminum window
	Reinstall	Ea.		111.44		111.44	including screen.
	Clean	Ea.	.03	14.90		14.93	
	Minimum Charge	Job		279		279	
50" x 53", 4 Light							
	Demolish	Ea.		38		38	Cost includes material and labor to
	Install	Ea.	1325	159		1484	install 50" x 53" four light awning,
	Demolish and Install	Ea.	1325	197		1522	double glazed aluminum window
	Reinstall	Ea.		127.36		127.36	including screen.
	Clean	Ea.	.03	17.55		17.58	
	Minimum Charge	Job		279		279	
63" x 37", 5 Light							
	Demolish	Ea.		38		38	Cost includes material and labor to
	Install	Ea.	930	159		1089	install 63" x 37" five light awning,
	Demolish and Install	Ea.	930	197		1127	double glazed aluminum window
	Reinstall	Ea.		127.36		127.36	including screen.
	Clean	Ea.	.03	17.55		17.58	
	Minimum Charge	Job		279		279	
63" x 53", 5 Light							
	Demolish	Ea.		38		38	Cost includes material and labor to
	Install	Ea.	1425	159		1584	install 63" x 53" five light awning,
	Demolish and Install	Ea.	1425	197		1622	double glazed aluminum window
	Reinstall	Ea.		127.36		127.36	including screen.
	Clean	Ea.	.04	21.50		21.54	
	Minimum Charge	Job		279		279	
Mullion Bars							
	Install	Ea.	14.40	44.50		58.90	Includes labor and material to install
	Minimum Charge	Job		279		279	mullions for sliding aluminum window.
Bay							
4' 8" x 6' 4"							
	Demolish	Ea.		27.50		27.50	Cost includes material and labor to
	Install	Ea.	1825	121		1946	install 4' 8" x 6' 4" angle bay double
	Demolish and Install	Ea.	1825	148.50		1973.50	hung unit, pine frame with aluminum
	Reinstall	Ea.		96.80		96.80	cladding, picture center sash,
	Clean	Ea.	.07	25		25.07	hardware, screens.
	Minimum Charge	Job		279		279	

111

Exterior Doors and Windows

Aluminum Window		Unit	Material	Labor	Equip.	Total	Specification
4' 8" x 7' 2"							
	Demolish	Ea.		27.50		27.50	Cost includes material and labor to
	Install	Ea.	1975	121		2096	install 4' 8" x 7' 2" angle bay double
	Demolish and Install	Ea.	1975	148.50		2123.50	hung unit, pine frame with aluminum
	Reinstall	Ea.		96.80		96.80	cladding, picture center sash,
	Clean	Ea.	.07	25		25.07	hardware, screens.
	Minimum Charge	Job		279		279	
4' 8" x 7' 10"							
	Demolish	Ea.		27.50		27.50	Cost includes material and labor to
	Install	Ea.	2025	121		2146	install 4' 8" x 7' 10" angle bay
	Demolish and Install	Ea.	2025	148.50		2173.50	double hung unit, pine frame with
	Reinstall	Ea.		96.80		96.80	aluminum cladding, picture center
	Clean	Ea.	.07	25		25.07	sash, hardware, screens.
	Minimum Charge	Job		279		279	
4' 8" x 8' 6"							
	Demolish	Ea.		32		32	Cost includes material and labor to
	Install	Ea.	2025	138		2163	install 4' 8" x 8' 6" angle bay double
	Demolish and Install	Ea.	2025	170		2195	hung unit, pine frame with aluminum
	Reinstall	Ea.		110.63		110.63	cladding, picture center sash,
	Clean	Ea.	.01	33		33.01	hardware, screens.
	Minimum Charge	Job		279		279	
4' 8" x 8' 10"							
	Demolish	Ea.		32		32	Cost includes material and labor to
	Install	Ea.	2025	138		2163	install 4' 8" x 8' 10" angle bay
	Demolish and Install	Ea.	2025	170		2195	double hung unit, pine frame with
	Reinstall	Ea.		110.63		110.63	aluminum cladding, picture center
	Clean	Ea.	.01	33		33.01	sash, hardware, screens.
	Minimum Charge	Job		279		279	
4' 8" x 9' 2"							
	Demolish	Ea.		32		32	Cost includes material and labor to
	Install	Ea.	1925	242		2167	install 4' 8" x 9' 2" angle bay double
	Demolish and Install	Ea.	1925	274		2199	hung unit, pine frame with aluminum
	Reinstall	Ea.		193.60		193.60	cladding, picture center sash,
	Clean	Ea.	.01	33		33.01	hardware, screens.
	Minimum Charge	Job		279		279	
4' 8" x 9' 10"							
	Demolish	Ea.		32		32	Cost includes material and labor to
	Install	Ea.	2075	242		2317	install 4' 8" x 9' 10" angle bay
	Demolish and Install	Ea.	2075	274		2349	double hung unit, pine frame with
	Reinstall	Ea.		193.60		193.60	aluminum cladding, picture center
	Clean	Ea.	.01	33		33.01	sash, hardware, screens.
	Minimum Charge	Job		279		279	
Casing							
	Demolish	Ea.		11.20		11.20	Cost includes material and labor to
	Install	Opng.	39.50	37		76.50	install 11/16" x 2-1/2" pine ranch
	Demolish and Install	Opng.	39.50	48.20		87.70	style window casing.
	Reinstall	Opng.		29.78		29.78	
	Clean	Opng.	.22	3.74		3.96	
	Paint	Ea.	7.45	8.05		15.50	
	Minimum Charge	Job		279		279	
Storm Window							
2' x 3' 6"							
	Demolish	Ea.		15.40		15.40	Includes material and labor to install
	Install	Ea.	124	32.50		156.50	residential aluminum storm window
	Demolish and Install	Ea.	124	47.90		171.90	with mill finish, 2' x 3' 6" high.
	Reinstall	Ea.		25.81		25.81	
	Clean	Ea.	.01	9.35		9.36	
	Minimum Charge	Job		279		279	

112

Exterior Doors and Windows

Aluminum Window

	Unit	Material	Labor	Equip.	Total	Specification
2' 6" x 5'						
Demolish	Ea.		15.40		15.40	Includes material and labor to install
Install	Ea.	156	34.50		190.50	residential aluminum storm window
Demolish and Install	Ea.	156	49.90		205.90	with mill finish, 2' 6" x 5' high.
Reinstall	Ea.		27.66		27.66	
Clean	Ea.	.02	12.45		12.47	
Minimum Charge	Job		279		279	
4' x 6'						
Demolish	Ea.		29.50		29.50	Includes material and labor to install
Install	Ea.	174	38.50		212.50	residential aluminum storm window, 4'
Demolish and Install	Ea.	174	68		242	x 6' high.
Reinstall	Ea.		30.98		30.98	
Clean	Ea.	.04	18.65		18.69	
Minimum Charge	Job		279		279	

Hardware
Weatherstripping

	Unit	Material	Labor	Equip.	Total	Specification
Install	L.F.	.34	4.03		4.37	Includes labor and material to install
Minimum Charge	Job		161		161	vinyl V strip weatherstripping for double-hung window.

Trim Set

	Unit	Material	Labor	Equip.	Total	Specification
Demolish	Ea.		11.20		11.20	Cost includes material and labor to
Install	Opng.	39.50	37		76.50	install 11/16" x 2-1/2" pine ranch
Demolish and Install	Opng.	39.50	48.20		87.70	style window casing.
Reinstall	Opng.		29.78		29.78	
Clean	Opng.	.22	3.74		3.96	
Paint	Ea.	7.45	8.05		15.50	
Minimum Charge	Job		242		242	

Vinyl Window

	Unit	Material	Labor	Equip.	Total	Specification
Double Hung						
2' x 2' 6"						
Demolish	Ea.		18.90		18.90	Cost includes material and labor to
Install	Ea.	246	64.50		310.50	install 2' x 2' 6" double hung, two
Demolish and Install	Ea.	246	83.40		329.40	light, vinyl window with screen.
Reinstall	Ea.		51.63		51.63	
Clean	Ea.	.01	9.35		9.36	
Minimum Charge	Job		161		161	
2' x 3' 6"						
Demolish	Ea.		18.90		18.90	Cost includes material and labor to
Install	Ea.	262	69		331	install 2' x 3' 6" double hung, two
Demolish and Install	Ea.	262	87.90		349.90	light, vinyl window with screen.
Reinstall	Ea.		55.31		55.31	
Clean	Ea.	.02	10.65		10.67	
Minimum Charge	Job		161		161	
2' 6" x 4' 6"						
Demolish	Ea.		18.90		18.90	Cost includes material and labor to
Install	Ea.	300	74.50		374.50	install 2' 6" x 4' 6" double hung, two
Demolish and Install	Ea.	300	93.40		393.40	light, vinyl window with screen.
Reinstall	Ea.		59.57		59.57	
Clean	Ea.	.02	12.45		12.47	
Minimum Charge	Job		161		161	
3' x 4'						
Demolish	Ea.		18.90		18.90	Includes material and labor to install
Install	Ea.	277	97		374	medium size double hung, 3' x 4',
Demolish and Install	Ea.	277	115.90		392.90	two light, vinyl window with screen.
Reinstall	Ea.		77.44		77.44	
Clean	Ea.	.02	12.45		12.47	
Minimum Charge	Job		161		161	

113

Exterior Doors and Windows

Vinyl Window

	Unit	Material	Labor	Equip.	Total	Specification
3' x 4' 6"						
Demolish	Ea.		23		23	Cost includes material and labor to
Install	Ea.	350	108		458	install 3' x 4' 6" double hung, two
Demolish and Install	Ea.	350	131		481	light, vinyl window with screen.
Reinstall	Ea.		86.04		86.04	
Clean	Ea.	.03	14.90		14.93	
Minimum Charge	Job		161		161	
4' x 4' 6"						
Demolish	Ea.		23		23	Cost includes material and labor to
Install	Ea.	395	121		516	install 4' x 4' 6" double hung, two
Demolish and Install	Ea.	395	144		539	light, vinyl window with screen.
Reinstall	Ea.		96.80		96.80	
Clean	Ea.	.03	17.55		17.58	
Minimum Charge	Job		161		161	
4' x 6'						
Demolish	Ea.		23		23	Includes material and labor to install
Install	Ea.	460	138		598	large double hung, 4' x 6', two light,
Demolish and Install	Ea.	460	161		621	vinyl window with screen.
Reinstall	Ea.		110.63		110.63	
Clean	Ea.	.04	21.50		21.54	
Minimum Charge	Job		161		161	
Trim Set						
Demolish	Ea.		11.20		11.20	Cost includes material and labor to
Install	Opng.	39.50	37		76.50	install 11/16" x 2-1/2" pine ranch
Demolish and Install	Opng.	39.50	48.20		87.70	style window casing.
Reinstall	Opng.		29.78		29.78	
Clean	Opng.	.22	3.74		3.96	
Paint	Ea.	7.45	8.05		15.50	
Minimum Charge	Job		279		279	

Wood Window

	Unit	Material	Labor	Equip.	Total	Specification
Double Hung						
2' 2" x 3' 4"						
Demolish	Ea.		18.90		18.90	Cost includes material and labor to
Install	Ea.	605	64.50		669.50	install 2' 2" x 3' 4" double hung, two
Demolish and Install	Ea.	605	83.40		688.40	light, double glazed wood window
Reinstall	Ea.		51.63		51.63	with screen.
Clean	Ea.	.02	10.65		10.67	
Paint	Ea.	4.64	16.80		21.44	
Minimum Charge	Job		242		242	
2' 2" x 4' 4"						
Demolish	Ea.		18.90		18.90	Cost includes material and labor to
Install	Ea.	278	69		347	install 2' 2" x 4' 4" double hung, two
Demolish and Install	Ea.	278	87.90		365.90	light, double glazed wood window
Reinstall	Ea.		55.31		55.31	with screen.
Clean	Ea.	.02	10.65		10.67	
Paint	Ea.	4.64	16.80		21.44	
Minimum Charge	Job		242		242	
2' 6" x 3' 4"						
Demolish	Ea.		18.90		18.90	Cost includes material and labor to
Install	Ea.	275	74.50		349.50	install 2' 6" x 3' 4" double hung, two
Demolish and Install	Ea.	275	93.40		368.40	light, double glazed wood window
Reinstall	Ea.		59.57		59.57	with screen.
Clean	Ea.	.02	12.45		12.47	
Paint	Ea.	5.95	20		25.95	
Minimum Charge	Job		242		242	

For customer support on your Contractor's Pricing Guide: Residential Repair & Remodeling, call 800.448.8182.

Exterior Doors and Windows

Wood Window		Unit	Material	Labor	Equip.	Total	Specification
2' 6" x 4'							
	Demolish	Ea.		18.90		18.90	Cost includes material and labor to
	Install	Ea.	282	80.50		362.50	install 2' 6" x 4' double hung, two
	Demolish and Install	Ea.	282	99.40		381.40	light, double glazed wood window
	Reinstall	Ea.		64.53		64.53	with screen.
	Clean	Ea.	.02	12.45		12.47	
	Paint	Ea.	5.95	20		25.95	
	Minimum Charge	Job		242		242	
2' 6" x 4' 8"							
	Demolish	Ea.		18.90		18.90	Cost includes material and labor to
	Install	Ea.	297	80.50		377.50	install 2' 6" x 4' 8" double hung, two
	Demolish and Install	Ea.	297	99.40		396.40	light, double glazed wood window
	Reinstall	Ea.		64.53		64.53	with screen.
	Clean	Ea.	.02	12.45		12.47	
	Paint	Ea.	5.95	20		25.95	
	Minimum Charge	Job		242		242	
2' 10" x 3' 4"							
	Demolish	Ea.		18.90		18.90	Cost includes material and labor to
	Install	Ea.	270	97		367	install 2' 10" x 3' 4" double hung, two
	Demolish and Install	Ea.	270	115.90		385.90	light, double glazed wood window
	Reinstall	Ea.		77.44		77.44	with screen.
	Clean	Ea.	.02	12.45		12.47	
	Paint	Ea.	5.95	20		25.95	
	Minimum Charge	Job		242		242	
2' 10" x 4'							
	Demolish	Ea.		18.90		18.90	Cost includes material and labor to
	Install	Ea.	300	97		397	install 2' 10" x 4' double hung, two
	Demolish and Install	Ea.	300	115.90		415.90	light, double glazed wood window
	Reinstall	Ea.		77.44		77.44	with screen.
	Clean	Ea.	.02	12.45		12.47	
	Paint	Ea.	5.95	20		25.95	
	Minimum Charge	Job		242		242	
3' 7" x 3' 4"							
	Demolish	Ea.		18.90		18.90	Cost includes material and labor to
	Install	Ea.	310	108		418	install 3' 7" x 3' 4" double hung, two
	Demolish and Install	Ea.	310	126.90		436.90	light, double glazed wood window
	Reinstall	Ea.		86.04		86.04	with screen.
	Clean	Ea.	.02	12.45		12.47	
	Paint	Ea.	5.95	20		25.95	
	Minimum Charge	Job		242		242	
3' 7" x 5' 4"							
	Demolish	Ea.		23		23	Cost includes material and labor to
	Install	Ea.	360	108		468	install 3' 7" x 5' 4" double hung, two
	Demolish and Install	Ea.	360	131		491	light, double glazed wood window
	Reinstall	Ea.		86.04		86.04	with screen.
	Clean	Ea.	.04	21.50		21.54	
	Paint	Ea.	9.90	33.50		43.40	
	Minimum Charge	Job		242		242	
4' 3" x 5' 4"							
	Demolish	Ea.		23		23	Cost includes material and labor to
	Install	Ea.	615	121		736	install 4' 3" x 5' 4" double hung, two
	Demolish and Install	Ea.	615	144		759	light, double glazed wood window
	Reinstall	Ea.		96.80		96.80	with screen.
	Clean	Ea.	.04	21.50		21.54	
	Paint	Ea.	7.80	40.50		48.30	
	Minimum Charge	Job		242		242	

115

Wood Window	Unit	Material	Labor	Equip.	Total	Specification
Casement						
2' x 3' 4"						
Demolish	Ea.		18.90		18.90	Cost includes material and labor to
Install	Ea.	365	48.50		413.50	install 2' x 3' 4" casement window
Demolish and Install	Ea.	365	67.40		432.40	with 3/4" insulated glass, screens,
Reinstall	Ea.		38.72		38.72	weatherstripping, hardware.
Clean	Ea.	.02	10.65		10.67	
Paint	Ea.	4.64	16.80		21.44	
Minimum Charge	Job		485		485	
2' x 4'						
Demolish	Ea.		18.90		18.90	Cost includes material and labor to
Install	Ea.	390	54		444	install 2' x 4' casement window with
Demolish and Install	Ea.	390	72.90		462.90	3/4" insulated glass, screens,
Reinstall	Ea.		43.02		43.02	weatherstripping, hardware.
Clean	Ea.	.02	10.65		10.67	
Paint	Ea.	4.64	16.80		21.44	
Minimum Charge	Job		485		485	
2' x 5'						
Demolish	Ea.		18.90		18.90	Cost includes material and labor to
Install	Ea.	470	57		527	install 2' x 5' casement window with
Demolish and Install	Ea.	470	75.90		545.90	3/4" insulated glass, screens,
Reinstall	Ea.		45.55		45.55	weatherstripping, hardware.
Clean	Ea.	.02	12.45		12.47	
Paint	Ea.	5.95	20		25.95	
Minimum Charge	Job		485		485	
2' x 6'						
Demolish	Ea.		18.90		18.90	Cost includes material and labor to
Install	Ea.	460	60.50		520.50	install 2' x 6' casement window with
Demolish and Install	Ea.	460	79.40		539.40	3/4" insulated glass, screens,
Reinstall	Ea.		48.40		48.40	weatherstripping, hardware.
Clean	Ea.	.02	12.45		12.47	
Paint	Ea.	5.95	20		25.95	
Minimum Charge	Job		485		485	
4' x 3' 4"						
Demolish	Ea.		23		23	Cost includes material and labor to
Install	Ea.	870	64.50		934.50	install 4' x 3' 4" casement windows
Demolish and Install	Ea.	870	87.50		957.50	with 3/4" insulated glass, screens,
Reinstall	Ea.		51.63		51.63	weatherstripping, hardware.
Clean	Ea.	.02	12.45		12.47	
Paint	Ea.	3.91	20		23.91	
Minimum Charge	Job		485		485	
4' x 4'						
Demolish	Ea.		23		23	Cost includes material and labor to
Install	Ea.	975	64.50		1039.50	install 4' x 4' casement windows with
Demolish and Install	Ea.	975	87.50		1062.50	3/4" insulated glass, screens,
Reinstall	Ea.		51.63		51.63	weatherstripping, hardware.
Clean	Ea.	.03	17.55		17.58	
Paint	Ea.	4.61	25		29.61	
Minimum Charge	Job		485		485	
4' x 5'						
Demolish	Ea.		23		23	Cost includes material and labor to
Install	Ea.	1100	69		1169	install 4' x 5' casement windows with
Demolish and Install	Ea.	1100	92		1192	3/4" insulated glass, screens,
Reinstall	Ea.		55.31		55.31	weatherstripping, hardware.
Clean	Ea.	.04	21.50		21.54	
Paint	Ea.	6.05	33.50		39.55	
Minimum Charge	Job		485		485	

For customer support on your Contractor's Pricing Guide: Residential Repair & Remodeling, call 800.448.8182.

Exterior Doors and Windows

Wood Window		Unit	Material	Labor	Equip.	Total	Specification
4' x 6'							
	Demolish	Ea.		23		23	Cost includes material and labor to
	Install	Ea.	1250	80.50		1330.50	install 4' x 6' casement windows with
	Demolish and Install	Ea.	1250	103.50		1353.50	3/4" insulated glass, screens,
	Reinstall	Ea.		64.53		64.53	weatherstripping, hardware.
	Clean	Ea.	.04	21.50		21.54	
	Paint	Ea.	7.10	40.50		47.60	
	Minimum Charge	Job		485		485	
Casement Bow							
4' 8" x 8' 1"							
	Demolish	Ea.		32		32	Includes material and labor to install
	Install	Ea.	2350	182		2532	casement bow window unit with 4
	Demolish and Install	Ea.	2350	214		2564	lights, pine frame, 1/2" insulated
	Reinstall	Ea.		145.20		145.20	glass, weatherstripping, hardware,
	Clean	Ea.	.07	25		25.07	screens.
	Paint	Ea.	13.15	80.50		93.65	
	Minimum Charge	Job		485		485	
4' 8" x 10'							
	Demolish	Ea.		32		32	Includes material and labor to install
	Install	Ea.	2825	242		3067	casement bow window unit with 5
	Demolish and Install	Ea.	2825	274		3099	lights, pine frame, 1/2" insulated
	Reinstall	Ea.		193.60		193.60	glass, weatherstripping, hardware,
	Clean	Ea.	.01	33		33.01	screens.
	Paint	Ea.	16.35	101		117.35	
	Minimum Charge	Job		485		485	
5' 4" x 8' 1"							
	Demolish	Ea.		32		32	Includes material and labor to install
	Install	Ea.	2500	182		2682	casement bow window unit with 4
	Demolish and Install	Ea.	2500	214		2714	lights, pine frame, 1/2" insulated
	Reinstall	Ea.		145.20		145.20	glass, weatherstripping, hardware,
	Clean	Ea.	.01	33		33.01	screens.
	Paint	Ea.	14.90	80.50		95.40	
	Minimum Charge	Job		485		485	
5' 4" x 10'							
	Demolish	Ea.		32		32	Includes material and labor to install
	Install	Ea.	2825	242		3067	casement bow window unit with 5
	Demolish and Install	Ea.	2825	274		3099	lights, pine frame, 1/2" insulated
	Reinstall	Ea.		193.60		193.60	glass, weatherstripping, hardware,
	Clean	Ea.	.01	33		33.01	screens.
	Paint	Ea.	18.80	101		119.80	
	Minimum Charge	Job		485		485	
6' x 8' 1"							
	Demolish	Ea.		32		32	Includes material and labor to install
	Install	Ea.	2825	242		3067	casement bow window unit with 5
	Demolish and Install	Ea.	2825	274		3099	lights, pine frame, 1/2" insulated
	Reinstall	Ea.		193.60		193.60	glass, weatherstripping, hardware,
	Clean	Ea.	.01	33		33.01	screens.
	Paint	Ea.	17	101		118	
	Minimum Charge	Job		485		485	
6' x 10'							
	Demolish	Ea.		32		32	Includes material and labor to install
	Install	Ea.	2975	242		3217	casement bow window unit with 5
	Demolish and Install	Ea.	2975	274		3249	lights, pine frame, 1/2" insulated
	Reinstall	Ea.		193.60		193.60	glass, weatherstripping, hardware,
	Clean	Ea.	.01	33		33.01	screens.
	Paint	Ea.	21.50	135		156.50	
	Minimum Charge	Job		485		485	

For customer support on your Contractor's Pricing Guide: Residential Repair & Remodeling, call 800.448.8182.

Exterior Doors and Windows

Wood Window		Unit	Material	Labor	Equip.	Total	Specification
Casing							
	Demolish	Ea.		11.20		11.20	Cost includes material and labor to
	Install	Opng.	39.50	37		76.50	install 11/16" x 2-1/2" pine ranch
	Demolish and Install	Opng.	39.50	48.20		87.70	style window casing.
	Reinstall	Opng.		29.78		29.78	
	Clean	Opng.	.22	3.74		3.96	
	Paint	Ea.	7.45	8.05		15.50	
	Minimum Charge	Job		279		279	
Trim Set							
	Demolish	Ea.		8.55		8.55	Cost includes material and labor to
	Install	Opng.	39.50	37		76.50	install 11/16" x 2-1/2" pine ranch
	Demolish and Install	Opng.	39.50	45.55		85.05	style window casing.
	Reinstall	Opng.		29.78		29.78	
	Clean	Opng.	.22	3.74		3.96	
	Paint	Ea.	7.45	8.05		15.50	
	Minimum Charge	Job		242		242	
Wood Picture Window							
3' 6" x 4'							
	Demolish	Ea.		23		23	Includes material and labor to install a
	Install	Ea.	650	80.50		730.50	wood framed picture window
	Demolish and Install	Ea.	650	103.50		753.50	including exterior trim.
	Clean	Ea.	.04	21.50		21.54	
	Paint	Ea.	5.70	31		36.70	
	Minimum Charge	Job		485		485	
4' x 4' 6"							
	Demolish	Ea.		23		23	Includes material and labor to install a
	Install	Ea.	880	88		968	wood framed picture window
	Demolish and Install	Ea.	880	111		991	including exterior trim.
	Clean	Ea.	.04	21.50		21.54	
	Paint	Ea.	5.70	31		36.70	
	Minimum Charge	Job		485		485	
5' x 4'							
	Demolish	Ea.		23		23	Includes material and labor to install a
	Install	Ea.	875	88		963	wood framed picture window
	Demolish and Install	Ea.	875	111		986	including exterior trim.
	Clean	Ea.	.04	21.50		21.54	
	Paint	Ea.	12.20	50.50		62.70	
	Minimum Charge	Job		485		485	
6' x 4' 6"							
	Demolish	Ea.		23		23	Includes material and labor to install a
	Install	Ea.	920	97		1017	wood framed picture window
	Demolish and Install	Ea.	920	120		1040	including exterior trim.
	Clean	Ea.	.04	21.50		21.54	
	Paint	Ea.	12.20	50.50		62.70	
	Minimum Charge	Job		485		485	

Window Screens		Unit	Material	Labor	Equip.	Total	Specification
Small							
	Demolish	Ea.		4.33		4.33	Includes material and labor to install
	Install	Ea.	21.50	14.65		36.15	small window screen, 4 S.F.
	Demolish and Install	Ea.	21.50	18.98		40.48	
	Reinstall	Ea.		11.73		11.73	
	Clean	Ea.	.15	7.80		7.95	
	Minimum Charge	Job		121		121	

For customer support on your Contractor's Pricing Guide: Residential Repair & Remodeling, call 800.448.8182.

Exterior Doors and Windows

Window Screens

Window Screens		Unit	Material	Labor	Equip.	Total	Specification
Standard Size							
	Demolish	Ea.		4.33		4.33	Includes material and labor to install
	Install	Ea.	61.50	14.65		76.15	medium window screen, 12 S.F.
	Demolish and Install	Ea.	61.50	18.98		80.48	
	Reinstall	Ea.		11.73		11.73	
	Clean	Ea.	.15	7.80		7.95	
	Minimum Charge	Job		121		121	
Large							
	Demolish	Ea.		4.33		4.33	Includes material and labor to install
	Install	Ea.	82.50	14.65		97.15	large window screen, 16 S.F.
	Demolish and Install	Ea.	82.50	18.98		101.48	
	Reinstall	Ea.		11.73		11.73	
	Clean	Ea.	.15	7.80		7.95	
	Minimum Charge	Job		121		121	

Window Glass

Window Glass		Unit	Material	Labor	Equip.	Total	Specification
Plate, Clear, 1/4"							
	Demolish	S.F.		2.08		2.08	Includes material and labor to install
	Install	S.F.	14.20	7.85		22.05	clear glass.
	Demolish and Install	S.F.	14.20	9.93		24.13	
	Reinstall	S.F.		6.27		6.27	
	Clean	S.F.	.01	.22		.23	
	Minimum Charge	Job		235		235	
Insulated							
	Demolish	S.F.		2.08		2.08	Includes material and labor to install
	Install	S.F.	31	13.45		44.45	insulated glass.
	Demolish and Install	S.F.	31	15.53		46.53	
	Reinstall	S.F.		10.75		10.75	
	Clean	S.F.	.01	.22		.23	
	Minimum Charge	Job		235		235	
Tinted							
	Demolish	S.F.		2.08		2.08	Includes material and labor to install
	Install	S.F.	8.45	7.25		15.70	tinted glass.
	Demolish and Install	S.F.	8.45	9.33		17.78	
	Reinstall	S.F.		5.79		5.79	
	Clean	S.F.	.01	.22		.23	
	Minimum Charge	Job		235		235	
Obscure / Florentine							
	Demolish	S.F.		2.08		2.08	Includes material and labor to install
	Install	S.F.	13.75	6.70		20.45	obscure glass.
	Demolish and Install	S.F.	13.75	8.78		22.53	
	Reinstall	S.F.		5.38		5.38	
	Clean	S.F.	.01	.22		.23	
	Minimum Charge	Job		235		235	
Laminated Safety							
	Demolish	S.F.		2.08		2.08	Includes material and labor to install
	Install	S.F.	14.70	10.45		25.15	clear laminated glass.
	Demolish and Install	S.F.	14.70	12.53		27.23	
	Reinstall	S.F.		8.36		8.36	
	Clean	S.F.	.01	.22		.23	
	Minimum Charge	Job		235		235	

For customer support on your Contractor's Pricing Guide: Residential Repair & Remodeling, call 800.448.8182.

Window Glass		Unit	Material	Labor	Equip.	Total	Specification
Wired							
	Demolish	S.F.		2.08		2.08	Includes material and labor to install
	Install	S.F.	28	6.95		34.95	rough obscure wire glass.
	Demolish and Install	S.F.	28	9.03		37.03	
	Reinstall	S.F.		5.58		5.58	
	Clean	S.F.	.01	.22		.23	
	Minimum Charge	Job		235		235	
Tempered							
	Demolish	S.F.		2.08		2.08	Includes material and labor to install
	Install	S.F.	14.20	7.85		22.05	clear glass.
	Demolish and Install	S.F.	14.20	9.93		24.13	
	Reinstall	S.F.		6.27		6.27	
	Clean	S.F.	.01	.22		.23	
	Minimum Charge	Job		235		235	
Stained							
	Demolish	S.F.		2.08		2.08	Includes material and labor to install
	Install	S.F.	72.50	7.85		80.35	stained glass.
	Demolish and Install	S.F.	72.50	9.93		82.43	
	Reinstall	S.F.		6.27		6.27	
	Clean	S.F.	.01	.22		.23	
	Minimum Charge	Job		235		235	
Etched							
	Demolish	S.F.		2.08		2.08	Includes material and labor to install
	Install	S.F.	60	7.85		67.85	etched glass.
	Demolish and Install	S.F.	60	9.93		69.93	
	Reinstall	S.F.		6.27		6.27	
	Clean	S.F.	.01	.22		.23	
	Minimum Charge	Job		235		235	
Wall Mirror							
	Demolish	S.F.		2.08		2.08	Includes material and labor to install
	Install	S.F.	10.50	7.55		18.05	distortion-free float glass mirror, 1/4"
	Demolish and Install	S.F.	10.50	9.63		20.13	thick, cut and polished edges.
	Reinstall	S.F.		6.02		6.02	
	Clean	S.F.	.01	.22		.23	
	Minimum Charge	Job		235		235	
Bullet Resistant							
	Demolish	S.F.		2.08		2.08	Includes material and labor to install
	Install	S.F.	146	94		240	bullet resistant clear laminated glass.
	Demolish and Install	S.F.	146	96.08		242.08	
	Reinstall	S.F.		75.26		75.26	
	Clean	S.F.	.01	.22		.23	
	Minimum Charge	Job		235		235	

Storefront System		Unit	Material	Labor	Equip.	Total	Specification
Clear Glass							
Stub Wall to 8'							
	Demolish	S.F.		3.25		3.25	Includes material and labor to install
	Install	S.F.	30.50	13.15		43.65	commercial grade 4-1/2" section,
	Demolish and Install	S.F.	30.50	16.40		46.90	anodized aluminum with clear glass in
	Reinstall	S.F.		10.52		10.52	upper section and safety glass in
	Clean	S.F.		.50		.50	lower section.
	Minimum Charge	Job		279		279	

For customer support on your Contractor's Pricing Guide: Residential Repair & Remodeling, call 800.448.8182.

Exterior Doors and Windows

Storefront System		Unit	Material	Labor	Equip.	Total	Specification
Floor to 10'							Includes material and labor to install
	Demolish	S.F.		3.25		3.25	commercial grade 4-1/2" section,
	Install	S.F.	31	13.15		44.15	anodized aluminum with clear glass in
	Demolish and Install	S.F.	31	16.40		47.40	upper section and safety glass in
	Reinstall	S.F.		10.52		10.52	lower section.
	Clean	S.F.		.50		.50	
	Minimum Charge	Job		279		279	
Polished Plate Glass							
Stub Wall to 9'							
	Demolish	S.F.		3.25		3.25	Cost includes material and labor to
	Install	S.F.	18.85	13.15		32	install 4' x 9' polished plate glass
	Demolish and Install	S.F.	18.85	16.40		35.25	window built on 8" stub wall.
	Reinstall	S.F.		10.52		10.52	
	Clean	S.F.		.50		.50	
	Minimum Charge	Job		279		279	
Stub Wall to 13'							
	Demolish	S.F.		3.25		3.25	Cost includes material and labor to
	Install	S.F.	21.50	13.15		34.65	install 4' x 13' polished plate glass
	Demolish and Install	S.F.	21.50	16.40		37.90	window built on 8" stub wall.
	Reinstall	S.F.		10.52		10.52	
	Clean	S.F.		.50		.50	
	Minimum Charge	Job		279		279	
Floor to 9'							
	Demolish	S.F.		3.25		3.25	Cost includes material and labor to
	Install	S.F.	22	13.15		35.15	install 4' x 9' polished plate glass
	Demolish and Install	S.F.	22	16.40		38.40	window built on floor.
	Reinstall	S.F.		10.52		10.52	
	Clean	S.F.		.50		.50	
	Minimum Charge	Job		279		279	
Floor to 13'							
	Demolish	S.F.		3.25		3.25	Cost includes material and labor to
	Install	S.F.	23.50	13.15		36.65	install 4' x 13' polished plate glass
	Demolish and Install	S.F.	23.50	16.40		39.90	window built on floor.
	Reinstall	S.F.		10.52		10.52	
	Clean	S.F.		.50		.50	
	Minimum Charge	Job		279		279	
Clear Plate Glass							
Floor to 9'							
	Demolish	S.F.		3.25		3.25	Cost includes material and labor to
	Install	S.F.	31.50	13.15		44.65	install 4' x 9' clear plate glass window
	Demolish and Install	S.F.	31.50	16.40		47.90	built on floor.
	Reinstall	S.F.		10.52		10.52	
	Clean	S.F.		.50		.50	
	Minimum Charge	Job		279		279	
Floor to 11'							
	Demolish	S.F.		3.25		3.25	Cost includes material and labor to
	Install	S.F.	31	13.15		44.15	install 4' x 11' clear plate glass
	Demolish and Install	S.F.	31	16.40		47.40	window built on floor.
	Reinstall	S.F.		10.52		10.52	
	Clean	S.F.		.50		.50	
	Minimum Charge	Job		279		279	
Floor to 13'							
	Demolish	S.F.		3.25		3.25	Cost includes material and labor to
	Install	S.F.	31	13.15		44.15	install 4' x 13' clear plate glass
	Demolish and Install	S.F.	31	16.40		47.40	window built on floor.
	Reinstall	S.F.		10.52		10.52	
	Clean	S.F.		.50		.50	
	Minimum Charge	Job		279		279	

For customer support on your Contractor's Pricing Guide: Residential Repair & Remodeling, call 800.448.8182.

Exterior Doors and Windows

Storefront System		Unit	Material	Labor	Equip.	Total	Specification
Entrance							
Narrow Stile 3' x 7"							
	Demolish	Ea.		27		27	Cost includes material and labor to
	Install	Ea.	1300	325		1625	install 3' x 7' narrow stile door, center
	Demolish and Install	Ea.	1300	352		1652	pivot concealed closer.
	Reinstall	Ea.		258.13		258.13	
	Clean	Ea.	1.90	47		48.90	
	Minimum Charge	Job		279		279	
Tempered Glass							
	Demolish	Ea.		27		27	Cost includes material and labor to
	Install	Ea.	2775	600		3375	install 3' x 7' tempered glass door,
	Demolish and Install	Ea.	2775	627		3402	center pivot concealed closer.
	Reinstall	Ea.		481.90		481.90	
	Clean	Ea.	1.90	47		48.90	
	Minimum Charge	Job		279		279	
Heavy Section							
	Demolish	Ea.		27		27	Includes material and labor to install
	Install	Ea.	1125	555		1680	wide stile door with 3' 6" x 7'
	Demolish and Install	Ea.	1125	582		1707	opening.
	Reinstall	Ea.		445.76		445.76	
	Clean	Ea.	.39	15.60		15.99	
	Minimum Charge	Job		940		940	

Curtain Wall System		Unit	Material	Labor	Equip.	Total	Specification
Regular Weight							
Bronze Anodized							
	Demolish	S.F.		3.25		3.25	Includes material and labor to install
	Install	S.F.	49	13.15		62.15	structural curtain wall system with
	Demolish and Install	S.F.	49	16.40		65.40	bronze anodized frame of regular
	Clean	S.F.		.50		.50	weight, float glass.
	Minimum Charge	Job		242		242	
Black Anodized							
	Demolish	S.F.		3.25		3.25	Includes material and labor to install
	Install	S.F.	53.50	13.15		66.65	structural curtain wall system with
	Demolish and Install	S.F.	53.50	16.40		69.90	black anodized frame of regular
	Clean	S.F.		.50		.50	weight, float glass.
	Minimum Charge	Job		242		242	
Heavy Weight							
Bronze Anodized							
	Demolish	S.F.		3.25		3.25	Includes material and labor to install
	Install	S.F.	49	13.15		62.15	structural curtain wall system with
	Demolish and Install	S.F.	49	16.40		65.40	bronze anodized frame of regular
	Clean	S.F.		.50		.50	weight, float glass.
	Minimum Charge	Job		242		242	
Black Anodized							
	Demolish	S.F.		3.25		3.25	Includes material and labor to install
	Install	S.F.	51	13.15		64.15	structural curtain wall system with
	Demolish and Install	S.F.	51	16.40		67.40	black anodized frame of heavy
	Clean	S.F.		.50		.50	weight, float glass.
	Minimum Charge	Job		242		242	
Glass							
1" Insulated							
	Install	S.F.	19.25	12.55		31.80	Cost includes labor and material to
	Minimum Charge	Job		235		235	install 1" insulated glass.

For customer support on your Contractor's Pricing Guide: Residential Repair & Remodeling, call 800.448.8182.

Exterior Doors and Windows

Skylight	Unit	Material	Labor	Equip.	Total	Specification
Dome						
22″ x 22″						
Demolish	Ea.		95.50		95.50	Cost includes material and labor to
Install	Ea.	135	151		286	install 22″ x 22″ fixed dome skylight.
Demolish and Install	Ea.	135	246.50		381.50	
Reinstall	Ea.		121.02		121.02	
Clean	Ea.	.18	13.35		13.53	
Minimum Charge	Job		910		910	
22″ x 46″						
Demolish	Ea.		95.50		95.50	Cost includes material and labor to
Install	Ea.	248	182		430	install 22″ x 46″ fixed dome skylight.
Demolish and Install	Ea.	248	277.50		525.50	
Reinstall	Ea.		145.23		145.23	
Clean	Ea.		21		21	
Minimum Charge	Job		910		910	
30″ x 30″						
Demolish	Ea.		95.50		95.50	Cost includes material and labor to
Install	Ea.	186	151		337	install 30″ x 30″ fixed dome skylight.
Demolish and Install	Ea.	186	246.50		432.50	
Reinstall	Ea.		121.02		121.02	
Clean	Ea.		21		21	
Minimum Charge	Job		910		910	
30″ x 46″						
Demolish	Ea.		95.50		95.50	Cost includes material and labor to
Install	Ea.	350	182		532	install 30″ x 46″ fixed dome skylight.
Demolish and Install	Ea.	350	277.50		627.50	
Reinstall	Ea.		145.23		145.23	
Clean	Ea.		21		21	
Minimum Charge	Job		910		910	
Fixed						
22″ x 27″						
Demolish	Ea.		95.50		95.50	Includes material and labor to install
Install	Ea.	259	151		410	double glazed skylight, 22″ x 27″,
Demolish and Install	Ea.	259	246.50		505.50	fixed.
Reinstall	Ea.		121.02		121.02	
Clean	Ea.	.18	13.35		13.53	
Minimum Charge	Job		910		910	
22″ x 46″						
Demolish	Ea.		95.50		95.50	Includes material and labor to install
Install	Ea.	248	182		430	double glazed skylight, 22″ x 46″,
Demolish and Install	Ea.	248	277.50		525.50	fixed.
Reinstall	Ea.		145.23		145.23	
Clean	Ea.		21		21	
Minimum Charge	Job		910		910	
44″ x 46″						
Demolish	Ea.		95.50		95.50	Includes material and labor to install
Install	Ea.	405	182		587	double glazed skylight, 44″ x 46″,
Demolish and Install	Ea.	405	277.50		682.50	fixed.
Reinstall	Ea.		145.23		145.23	
Clean	Ea.		23.50		23.50	
Minimum Charge	Job		910		910	
Operable						
22″ x 27″						
Demolish	Ea.		95.50		95.50	Includes material and labor to install
Install	Ea.	380	151		531	double glazed skylight, 22″ x 27″,
Demolish and Install	Ea.	380	246.50		626.50	operable.
Reinstall	Ea.		121.02		121.02	
Clean	Ea.	.18	13.35		13.53	
Minimum Charge	Job		910		910	

For customer support on your Contractor's Pricing Guide: Residential Repair & Remodeling, call 800.448.8182.

Exterior Doors and Windows

Skylight

Skylight	Unit	Material	Labor	Equip.	Total	Specification
22" x 46"						
Demolish	Ea.		95.50		95.50	Includes material and labor to install
Install	Ea.	880	182		1062	double glazed skylight, 22" x 46",
Demolish and Install	Ea.	880	277.50		1157.50	operable.
Reinstall	Ea.		145.23		145.23	
Clean	Ea.		21		21	
Minimum Charge	Job		910		910	
44" x 46"						
Demolish	Ea.		95.50		95.50	Includes material and labor to install
Install	Ea.	790	182		972	double glazed skylight, 44" x 46",
Demolish and Install	Ea.	790	277.50		1067.50	operable.
Reinstall	Ea.		145.23		145.23	
Clean	Ea.		23.50		23.50	
Minimum Charge	Job		910		910	

Industrial Steel Window

Industrial Steel Window	Unit	Material	Labor	Equip.	Total	Specification
100% Fixed						
Demolish	S.F.		2.66		2.66	Includes material and labor to install
Install	S.F.	32	5.55		37.55	commercial grade steel windows,
Demolish and Install	S.F.	32	8.21		40.21	including glazing.
Reinstall	S.F.		4.46		4.46	
Clean	Ea.	.07	1.55		1.62	
Paint	S.F.	.25	.90		1.15	
Minimum Charge	Job		279		279	
50% Vented						
Demolish	S.F.		2.66		2.66	Includes material and labor to install
Install	S.F.	62	5.55		67.55	commercial grade steel windows,
Demolish and Install	S.F.	62	8.21		70.21	including glazing.
Reinstall	S.F.		4.46		4.46	
Clean	Ea.	.07	1.55		1.62	
Paint	S.F.	.25	.90		1.15	
Minimum Charge	Job		279		279	

For customer support on your Contractor's Pricing Guide: Residential Repair & Remodeling, call 800.448.8182.

Brick Veneer

Concrete Masonry Units

Masonry Wall		Unit	Material	Labor	Equip.	Total	Specification
Unreinforced CMU							
4″ x 8″ x 16″							
	Demolish	S.F.		.14	.40	.54	Includes material and labor to install
	Install	S.F.	2.08	4.89		6.97	normal weight 4″ x 8″ x 16″ block.
	Demolish and Install	S.F.	2.08	5.03	.40	7.51	
	Clean	S.F.	.12	.58		.70	
	Paint	S.F.	.28	.92		1.20	
	Minimum Charge	Job		239		239	
6″ x 8″ x 16″							
	Demolish	S.F.		.19	.53	.72	Includes material and labor to install
	Install	S.F.	4.41	5.10		9.51	normal weight 6″ x 8″ x 16″ block.
	Demolish and Install	S.F.	4.41	5.29	.53	10.23	
	Clean	S.F.	.12	.58		.70	
	Paint	S.F.	.28	.92		1.20	
	Minimum Charge	Job		239		239	
8″ x 8″ x 16″							
	Demolish	S.F.		.39	1.06	1.45	Includes material and labor to install
	Install	S.F.	2.86	5.65		8.51	normal weight 8″ x 8″ x 16″ block.
	Demolish and Install	S.F.	2.86	6.04	1.06	9.96	
	Clean	S.F.	.12	.58		.70	
	Paint	S.F.	.28	.92		1.20	
	Minimum Charge	Job		239		239	
12″ x 8″ x 16″							
	Demolish	S.F.		.39	1.06	1.45	Includes material and labor to install
	Install	S.F.	4.93	8.60		13.53	normal weight 12″ x 8″ x 16″ block.
	Demolish and Install	S.F.	4.93	8.99	1.06	14.98	
	Clean	S.F.	.12	.58		.70	
	Paint	S.F.	.28	.92		1.20	
	Minimum Charge	Job		239		239	
Reinforced CMU							
4″ x 8″ x 16″							
	Demolish	S.F.		2.31	1.08	3.39	Includes material and labor to install
	Install	S.F.	2.26	5		7.26	normal weight 4″ x 8″ x 16″ block,
	Demolish and Install	S.F.	2.26	7.31	1.08	10.65	mortar and 1/2″ re-bar installed
	Clean	S.F.	.12	.58		.70	vertically every 24″ O.C.
	Paint	S.F.	.28	.92		1.20	
	Minimum Charge	Job		445		445	
6″ x 8″ x 16″							
	Demolish	S.F.		2.43	.79	3.22	Includes material and labor to install
	Install	S.F.	4.59	5.25		9.84	normal weight 6″ x 8″ x 16″ block,
	Demolish and Install	S.F.	4.59	7.68	.79	13.06	mortar and 1/2″ re-bar installed
	Clean	S.F.	.12	.58		.70	vertically every 24″ O.C.
	Paint	S.F.	.28	.92		1.20	
	Minimum Charge	Job		445		445	

125

Masonry

Masonry Wall		Unit	Material	Labor	Equip.	Total	Specification
8" x 8" x 16"							
	Demolish	S.F.		2.57	.83	3.40	Includes material and labor to install
	Install	S.F.	3.05	5.70		8.75	normal weight 8" x 8" x 16" block,
	Demolish and Install	S.F.	3.05	8.27	.83	12.15	mortar and 1/2" re-bar installed
	Clean	S.F.	.12	.58		.70	vertically every 24" O.C.
	Paint	S.F.	.28	.92		1.20	
	Minimum Charge	Job		445		445	
12" x 8" x 16"							
	Demolish	S.F.		2.65	.86	3.51	Includes material and labor to install
	Install	S.F.	5.15	8.85		14	normal weight 12" x 8" x 16" block,
	Demolish and Install	S.F.	5.15	11.50	.86	17.51	mortar and 1/2" re-bar installed
	Clean	S.F.	.12	.58		.70	vertically every 24" O.C.
	Paint	S.F.	.28	.92		1.20	
	Minimum Charge	Job		445		445	
Structural Brick							
4" Single Row							
	Demolish	S.F.		3.32		3.32	Cost includes material and labor to
	Install	S.F.	5.50	10.45		15.95	install 4" red brick wall, running bond
	Demolish and Install	S.F.	5.50	13.77		19.27	with 3/8" concave joints, mortar and
	Clean	S.F.	.12	.49		.61	ties.
	Paint	S.F.	.28	.92		1.20	
	Minimum Charge	Job		445		445	
8" Double Row							
	Demolish	S.F.		3.32		3.32	Cost includes material and labor to
	Install	S.F.	16.50	16.70		33.20	install 8" red brick wall, running bond,
	Demolish and Install	S.F.	16.50	20.02		36.52	with 3/8" concave joints, mortar and
	Clean	S.F.	.12	.49		.61	wall ties.
	Paint	S.F.	.28	.92		1.20	
	Minimum Charge	Job		445		445	
12" Triple Row							
	Demolish	S.F.		4.62		4.62	Cost includes material and labor to
	Install	S.F.	25	23.50		48.50	install 12" red brick wall, running
	Demolish and Install	S.F.	25	28.12		53.12	bond, with 3/8" concave joints,
	Clean	S.F.	.12	.49		.61	mortar and wall ties.
	Paint	S.F.	.28	.92		1.20	
	Minimum Charge	Job		445		445	
Brick Veneer							
4" Red Common							
	Demolish	S.F.		3.32		3.32	Cost includes material and labor to
	Install	S.F.	7.85	9.80		17.65	install 4" common red face brick,
	Demolish and Install	S.F.	7.85	13.12		20.97	running bond with 3/8" concave
	Clean	S.F.	.12	.49		.61	joints, mortar and wall ties.
	Paint	S.F.	.28	.92		1.20	
	Minimum Charge	Job		239		239	
4" Used							
	Demolish	S.F.		3.32		3.32	Cost includes material and labor to
	Install	S.F.	5.50	10.45		15.95	install 4" red brick wall, running bond
	Demolish and Install	S.F.	5.50	13.77		19.27	with 3/8" concave joints, mortar and
	Clean	S.F.	.12	.49		.61	ties.
	Paint	S.F.	.28	.92		1.20	
	Minimum Charge	Job		239		239	

For customer support on your Contractor's Pricing Guide: Residential Repair & Remodeling, call 800.448.8182.

Masonry Wall		Unit	Material	Labor	Equip.	Total	Specification
6" Jumbo							
	Demolish	S.F.		3.32		3.32	Cost includes material and labor to
	Install	S.F.	5.95	5.20		11.15	install 6" red jumbo brick veneer,
	Demolish and Install	S.F.	5.95	8.52		14.47	running bond with 3/8" concave
	Clean	S.F.	.12	.49		.61	joints, mortar and ties.
	Paint	S.F.	.28	.92		1.20	
	Minimum Charge	Job		239		239	
Norman (11-1/2")							
	Demolish	S.F.		3.32		3.32	Cost includes material and labor to
	Install	S.F.	7.45	7.05		14.50	install 4" norman red brick wall,
	Demolish and Install	S.F.	7.45	10.37		17.82	running bond with 3/8" concave
	Clean	S.F.	.12	.49		.61	joints, mortar and ties.
	Paint	S.F.	.28	.92		1.20	
	Minimum Charge	Job		239		239	
Glazed							
	Demolish	S.F.		3.32		3.32	Includes material, labor and
	Install	S.F.	23.50	10.70		34.20	equipment to install glazed brick.
	Demolish and Install	S.F.	23.50	14.02		37.52	
	Clean	S.F.	.12	.49		.61	
	Paint	S.F.	.28	.92		1.20	
	Minimum Charge	Job		239		239	
Slumpstone							
6" x 4"							
	Demolish	S.F.		.14	.40	.54	Cost includes material and labor to
	Install	S.F.	7.60	5.55		13.15	install 6" x 4" x 16" concrete slump
	Demolish and Install	S.F.	7.60	5.69	.40	13.69	block.
	Clean	S.F.	.12	.58		.70	
	Paint	S.F.	.28	.92		1.20	
	Minimum Charge	Job		239		239	
6" x 6"							
	Demolish	S.F.		.19	.53	.72	Cost includes material and labor to
	Install	S.F.	5.80	5.70		11.50	install 6" x 6" x 16" concrete slump
	Demolish and Install	S.F.	5.80	5.89	.53	12.22	block.
	Clean	S.F.	.12	.58		.70	
	Paint	S.F.	.28	.92		1.20	
	Minimum Charge	Job		239		239	
8" x 6"							
	Demolish	S.F.		.19	.53	.72	Cost includes material and labor to
	Install	S.F.	8.75	5.90		14.65	install 8" x 6" x 16" concrete slump
	Demolish and Install	S.F.	8.75	6.09	.53	15.37	block.
	Clean	S.F.	.12	.58		.70	
	Paint	S.F.	.28	.92		1.20	
	Minimum Charge	Job		239		239	
12" x 4"							
	Demolish	S.F.		.14	.40	.54	Cost includes material and labor to
	Install	S.F.	15.30	6.80		22.10	install 12" x 4" x 16" concrete slump
	Demolish and Install	S.F.	15.30	6.94	.40	22.64	block.
	Clean	S.F.	.12	.58		.70	
	Paint	S.F.	.28	.92		1.20	
	Minimum Charge	Job		239		239	
12" x 6"							
	Demolish	S.F.		.19	.53	.72	Cost includes material and labor to
	Install	S.F.	16.60	7.40		24	install 12" x 6" x 16" concrete slump
	Demolish and Install	S.F.	16.60	7.59	.53	24.72	block.
	Clean	S.F.	.12	.58		.70	
	Paint	S.F.	.28	.92		1.20	
	Minimum Charge	Job		239		239	

For customer support on your Contractor's Pricing Guide: Residential Repair & Remodeling, call 800.448.8182.

Masonry

Masonry Wall	Unit	Material	Labor	Equip.	Total	Specification
Stucco						
Three Coat						
Demolish	S.Y.		2.87		2.87	Includes material and labor to install
Install	S.Y.	4.04	11.15	.62	15.81	stucco on exterior masonry walls.
Demolish and Install	S.Y.	4.04	14.02	.62	18.68	
Clean	S.Y.		5.20		5.20	
Paint	S.F.	.35	.68		1.03	
Minimum Charge	Job		234		234	
Three Coat w / Wire Mesh						
Demolish	S.Y.		5.20		5.20	Includes material and labor to install
Install	S.Y.	9.80	43	1.97	54.77	stucco on wire mesh over wood
Demolish and Install	S.Y.	9.80	48.20	1.97	59.97	framing.
Clean	S.Y.		5.20		5.20	
Paint	S.F.	.35	.68		1.03	
Minimum Charge	Job		234		234	
Fill Cracks						
Install	S.F.	.48	3.75		4.23	Includes labor and material to fill in
Minimum Charge	Job		234		234	hairline cracks up to 1/8" with filler, including sand and fill per S.F.
Furring						
1" x 2"						
Demolish	L.F.		.17		.17	Cost includes material and labor to
Install	L.F.	.36	.98		1.34	install 1" x 2" furring on masonry wall.
Demolish and Install	L.F.	.36	1.15		1.51	
Minimum Charge	Job		242		242	
2" x 2"						
Demolish	L.F.		.18		.18	Cost includes material and labor to
Install	L.F.	.37	1.08		1.45	install 2" x 2" furring on masonry wall.
Demolish and Install	L.F.	.37	1.26		1.63	
Minimum Charge	Job		242		242	
2" x 4"						
Demolish	L.F.		.20		.20	Cost includes material and labor to
Install	L.F.	.86	1.23		2.09	install 2" x 4" furring on masonry wall.
Demolish and Install	L.F.	.86	1.43		2.29	
Reinstall	L.F.		.98		.98	
Minimum Charge	Job		242		242	
Maintenance / Cleaning						
Repoint						
Install	S.F.	.91	6		6.91	Includes labor and material to cut and
Minimum Charge	Hr.		60		60	repoint brick, hard mortar, running bond.
Acid Etch						
Install	S.F.	.08	.86		.94	Includes labor and material to acid
Minimum Charge	Job		187	60.50	247.50	wash smooth brick.
Pressure Wash						
Clean	S.F.		1	.30	1.30	Includes labor and material to high
Minimum Charge	Job		187	60.50	247.50	pressure water wash masonry, average.
Waterproof						
Install	S.F.	.50	.91		1.41	Includes labor and material to install
Minimum Charge	Job		227		227	masonry waterproofing.

Masonry

Chimney

Chimney		Unit	Material	Labor	Equip.	Total	Specification
Demo & Haul Exterior	Demolish	C.F.		6.80		6.80	Includes minimum labor and
	Minimum Charge	Job		208		208	equipment to remove masonry and haul debris to truck or dumpster.
Demo & Haul Interior	Demolish	C.F.		9.35		9.35	Includes minimum labor and
	Minimum Charge	Job		208		208	equipment to remove masonry and haul debris to truck or dumpster.
Install New (w / flue)							
Brick	Install	V.L.F.	67.50	55.50		123	Includes material and labor to install
	Minimum Charge	Job		239		239	brick chimney with flue. Foundation not included.
Natural Stone	Install	V.L.F.	97.50	88.50		186	Includes labor and material to install
	Minimum Charge	Job		239		239	natural stone chimney.
Install From Roof Up							
Brick	Install	V.L.F.	67.50	55.50		123	Includes material and labor to install
	Minimum Charge	Job		239		239	brick chimney with flue. Foundation not included.
Natural Stone	Install	V.L.F.	97.50	74		171.50	Includes labor and material to install
	Minimum Charge	Job		239		239	natural stone chimney.
Install Base Pad	Install	Ea.	420	269	1.26	690.26	Includes labor and material to replace
	Minimum Charge	Job		1175	30.50	1205.50	spread footing, 5' square x 16" deep.
Metal Cap	Install	Ea.	64	30		94	Includes labor and materials to install
	Minimum Charge	Job		242		242	bird and squirrel screens in a chimney flue.

Fireplace

Fireplace		Unit	Material	Labor	Equip.	Total	Specification
Common Brick							
30"	Demolish	Ea.		1475		1475	Includes material and labor to install a
	Install	Ea.	2725	4425		7150	complete unit which includes common
	Demolish and Install	Ea.	2725	5900		8625	brick, foundation, damper, flue lining
	Minimum Charge	Job		239		239	stack to 15' high and a 30" fire box.
36"	Demolish	Ea.		1475		1475	Includes material and labor to install a
	Install	Ea.	2850	4925		7775	complete unit which includes common
	Demolish and Install	Ea.	2850	6400		9250	brick, foundation, damper, flue lining
	Minimum Charge	Job		239		239	stack to 15' high and a 36" fire box.
42"	Demolish	Ea.		1475		1475	Includes material and labor to install a
	Install	Ea.	3025	5550		8575	complete unit which includes common
	Demolish and Install	Ea.	3025	7025		10050	brick, foundation, damper, flue lining
	Minimum Charge	Job		239		239	stack to 15' high and a 42" fire box.

For customer support on your Contractor's Pricing Guide: Residential Repair & Remodeling, call 800.448.8182.

Fireplace	Unit	Material	Labor	Equip.	Total	Specification
48"						
Demolish	Ea.		1475		1475	Includes material and labor to install a complete unit which includes common brick, foundation, damper, flue lining stack to 15' high and a 48" fire box.
Install	Ea.	3175	6325		9500	
Demolish and Install	Ea.	3175	7800		10975	
Minimum Charge	Job		239		239	
Raised Hearth						
Brick Hearth						
Demolish	S.F.		6.50		6.50	Includes material and labor to install a brick fireplace hearth based on 30" x 29" opening.
Install	S.F.	18.85	37		55.85	
Demolish and Install	S.F.	18.85	43.50		62.35	
Clean	S.F.	.12	.49		.61	
Minimum Charge	Job		239		239	
Marble Hearth / Facing						
Demolish	Ea.		41.50		41.50	Includes material and labor to install marble fireplace hearth.
Install	Ea.	335	148		483	
Demolish and Install	Ea.	335	189.50		524.50	
Reinstall	Ea.		118.63		118.63	
Clean	Ea.	42.50	7.50		50	
Minimum Charge	Job		239		239	
Breast (face) Brick						
Demolish	S.F.		6.50		6.50	Includes material and labor to install standard red brick, running bond.
Install	S.F.	5.55	10.25		15.80	
Demolish and Install	S.F.	5.55	16.75		22.30	
Clean	S.F.		1	.30	1.30	
Minimum Charge	Hr.		60		60	
Fire Box						
Demolish	Ea.		330		330	Includes material and labor to install fireplace box only (110 brick).
Install	Ea.	172	445		617	
Demolish and Install	Ea.	172	775		947	
Clean	Ea.	.06	18.65		18.71	
Minimum Charge	Hr.		60		60	
Natural Stone						
Demolish	S.F.		6.50		6.50	Includes material and labor to install high priced stone, split or rock face.
Install	S.F.	19.40	18.75		38.15	
Demolish and Install	S.F.	19.40	25.25		44.65	
Clean	S.F.	.10	1.11		1.21	
Minimum Charge	Job		445		445	
Slumpstone Veneer						
Demolish	S.F.		.19	.53	.72	Cost includes material and labor to install 8" x 6" x 16" concrete slump block.
Install	S.F.	8.75	5.90		14.65	
Demolish and Install	S.F.	8.75	6.09	.53	15.37	
Clean	S.F.	.12	.58		.70	
Paint	S.F.	.28	.92		1.20	
Minimum Charge	Job		239		239	
Prefabricated						
Firebox						
Demolish	Ea.		59.50		59.50	Includes material and labor to install up to 43" radiant heat, zero clearance, prefabricated fireplace. Flues, doors or blowers not included.
Install	Ea.	2400	440		2840	
Demolish and Install	Ea.	2400	499.50		2899.50	
Reinstall	Ea.		352		352	
Clean	Ea.	.76	23.50		24.26	
Minimum Charge	Job		242		242	

For customer support on your Contractor's Pricing Guide: Residential Repair & Remodeling, call 800.448.8182.

Fireplace	Unit	Material	Labor	Equip.	Total	Specification
Flue						
Demolish	V.L.F.		6.95		6.95	Includes material and labor to install
Install	V.L.F.	32	18.45		50.45	prefabricated chimney vent, 8"
Demolish and Install	V.L.F.	32	25.40		57.40	exposed metal flue.
Reinstall	V.L.F.		14.78		14.78	
Clean	L.F.	.31	4.68		4.99	
Minimum Charge	Job		350		350	
Blower						
Demolish	Ea.		5.55		5.55	Includes material and labor to install
Install	Ea.	180	33.50		213.50	blower for metal fireplace flue.
Demolish and Install	Ea.	180	39.05		219.05	
Reinstall	Ea.		26.68		26.68	
Clean	Ea.	1.52	47		48.52	
Minimum Charge	Job		350		350	
Accessories						
Wood Mantle						
Demolish	L.F.		6.95		6.95	Includes material and labor to install
Install	L.F.	14.15	13.85		28	wood fireplace mantel.
Demolish and Install	L.F.	14.15	20.80		34.95	
Reinstall	L.F.		13.83		13.83	
Clean	S.F.		.47		.47	
Paint	L.F.	.17	10.95		11.12	
Minimum Charge	Job		242		242	
Colonial Style						
Demolish	Ea.		60		60	Includes material and labor to install
Install	Opng.	2450	242		2692	wood fireplace mantel.
Demolish and Install	Opng.	2450	302		2752	
Reinstall	Opng.		242		242	
Clean	Ea.		.83		.83	
Paint	Ea.	27.50	151		178.50	
Minimum Charge	Job		242		242	
Modern Design						
Demolish	Ea.		60		60	Includes material and labor to install
Install	Opng.	730	97		827	wood fireplace mantel.
Demolish and Install	Opng.	730	157		887	
Reinstall	Opng.		96.80		96.80	
Clean	Ea.		.83		.83	
Paint	Ea.	27.50	151		178.50	
Minimum Charge	Job		242		242	
Glass Door Unit						
Demolish	Ea.		8.65		8.65	Includes material and labor to install a
Install	Ea.	305	120		425	fireplace glass door unit.
Demolish and Install	Ea.	305	128.65		433.65	
Reinstall	Ea.		119.70		119.70	
Clean	Ea.	.19	14.40		14.59	
Minimum Charge	Job		242		242	
Fire Screen						
Demolish	Ea.		4.33		4.33	Includes material and labor to install a
Install	Ea.	134	4.99		138.99	fireplace fire screen.
Demolish and Install	Ea.	134	9.32		143.32	
Reinstall	Ea.		4.99		4.99	
Clean	Ea.	.09	12.45		12.54	
Minimum Charge	Job		242		242	

For customer support on your Contractor's Pricing Guide: Residential Repair & Remodeling, call 800.448.8182.

Shingles

Board & Batten

Texture 1-11

Exterior Siding		Unit	Material	Labor	Equip.	Total	Specification
Trim							
1″ x 2″							
	Demolish	L.F.		.35		.35	Cost includes material and labor to
	Install	L.F.	.36	1.47		1.83	install 1″ x 2″ pine trim.
	Demolish and Install	L.F.	.36	1.82		2.18	
	Clean	L.F.		.25		.25	
	Paint	L.F.	.04	.63		.67	
	Minimum Charge	Job		242		242	
1″ x 3″							
	Demolish	L.F.		.35		.35	Cost includes material and labor to
	Install	L.F.	.37	1.67		2.04	install 1″ x 3″ pine trim.
	Demolish and Install	L.F.	.37	2.02		2.39	
	Clean	L.F.		.25		.25	
	Paint	L.F.	.04	.63		.67	
	Minimum Charge	Job		242		242	
1″ x 4″							
	Demolish	L.F.		.35		.35	Cost includes material and labor to
	Install	L.F.	.76	1.94		2.70	install 1″ x 4″ pine trim.
	Demolish and Install	L.F.	.76	2.29		3.05	
	Clean	L.F.		.25		.25	
	Paint	L.F.	.04	.63		.67	
	Minimum Charge	Job		242		242	
1″ x 6″							
	Demolish	L.F.		.35		.35	Cost includes material and labor to
	Install	L.F.	1.13	1.94		3.07	install 1″ x 6″ pine trim.
	Demolish and Install	L.F.	1.13	2.29		3.42	
	Clean	L.F.	.01	.25		.26	
	Paint	L.F.	.14	.63		.77	
	Minimum Charge	Job		242		242	
1″ x 8″							
	Demolish	L.F.		.38		.38	Cost includes material and labor to
	Install	L.F.	1.55	2.42		3.97	install 1″ x 8″ pine trim.
	Demolish and Install	L.F.	1.55	2.80		4.35	
	Clean	L.F.	.01	.27		.28	
	Paint	L.F.	.14	.63		.77	
	Minimum Charge	Job		242		242	
Aluminum							
8″ Non-Insulated							
	Demolish	S.F.		.94		.94	Includes material and labor to install
	Install	S.F.	3.53	1.97		5.50	horizontal, colored clapboard, 8″ to
	Demolish and Install	S.F.	3.53	2.91		6.44	10″ wide, plain.
	Reinstall	S.F.		1.97		1.97	
	Clean	S.F.		.37		.37	
	Minimum Charge	Job		242		242	

For customer support on your Contractor's Pricing Guide: Residential Repair & Remodeling, call 800.448.8182.

Siding

Exterior Siding	Unit	Material	Labor	Equip.	Total	Specification
8" Insulated						
Demolish	S.F.		.94		.94	Includes material and labor to install
Install	S.F.	3.41	1.97		5.38	horizontal, colored insulated
Demolish and Install	S.F.	3.41	2.91		6.32	clapboard, 8" wide, plain.
Reinstall	S.F.		1.97		1.97	
Clean	S.F.		.37		.37	
Minimum Charge	Job		242		242	
12" Non-Insulated						
Demolish	S.F.		.94		.94	Includes material and labor to install
Install	S.F.	4.02	1.70		5.72	horizontal, colored clapboard, plain.
Demolish and Install	S.F.	4.02	2.64		6.66	
Reinstall	S.F.		1.70		1.70	
Clean	S.F.		.37		.37	
Minimum Charge	Job		242		242	
12" Insulated						
Demolish	S.F.		.94		.94	Includes material and labor to install
Install	S.F.	3.64	1.70		5.34	horizontal, colored insulated
Demolish and Install	S.F.	3.64	2.64		6.28	clapboard, plain.
Reinstall	S.F.		1.70		1.70	
Clean	S.F.		.37		.37	
Minimum Charge	Job		242		242	
Clapboard 8" to 10"						
Demolish	S.F.		.94		.94	Includes material and labor to install
Install	S.F.	3.53	1.97		5.50	horizontal, colored clapboard, 8" to
Demolish and Install	S.F.	3.53	2.91		6.44	10" wide, plain.
Reinstall	S.F.		1.97		1.97	
Clean	S.F.		.37		.37	
Minimum Charge	Job		242		242	
Corner Strips						
Demolish	L.F.		.49		.49	Includes material and labor to install
Install	V.L.F.	4.54	1.97		6.51	corners for horizontal lapped
Demolish and Install	V.L.F.	4.54	2.46		7	aluminum siding.
Reinstall	V.L.F.		1.97		1.97	
Clean	L.F.		.25		.25	
Minimum Charge	Job		242		242	
Vinyl						
Non-Insulated						
Demolish	S.F.		.73		.73	Cost includes material and labor to
Install	S.F.	2.59	1.76		4.35	install 10" non-insulated solid vinyl
Demolish and Install	S.F.	2.59	2.49		5.08	siding.
Reinstall	S.F.		1.41		1.41	
Clean	S.F.		.37		.37	
Minimum Charge	Job		242		242	
Insulated						
Demolish	S.F.		.73		.73	Cost includes material and labor to
Install	S.F.	4.07	2.42		6.49	install insulated solid vinyl siding
Demolish and Install	S.F.	4.07	3.15		7.22	panels.
Reinstall	S.F.		1.94		1.94	
Clean	S.F.		.37		.37	
Minimum Charge	Job		242		242	
Corner Strips						
Demolish	L.F.		.46		.46	Cost includes material and labor to
Install	L.F.	3.06	1.38		4.44	install corner posts for vinyl siding
Demolish and Install	L.F.	3.06	1.84		4.90	panels.
Reinstall	L.F.		1.11		1.11	
Clean	L.F.		.25		.25	
Minimum Charge	Job		242		242	

For customer support on your Contractor's Pricing Guide: Residential Repair & Remodeling, call 800.448.8182.

Siding

Exterior Siding	Unit	Material	Labor	Equip.	Total	Specification
Fiber Cement Siding						
4' x 8' Panel						
Demolish	S.F.		.73		.73	Cost includes material and labor to
Install	S.F.	1.19	1.06		2.25	install 4' x 8' fiber cement 7/16"
Demolish and Install	S.F.	1.19	1.79		2.98	thick.
Clean	S.F.		.37		.37	
Paint	S.F.	.75	1.41		2.16	
Minimum Charge	Job		242		242	
Lap						
Demolish	S.F.		.72		.72	Cost includes material and labor to
Install	S.F.	1.73	1.92		3.65	install 8" or 12" fiber cement lap
Demolish and Install	S.F.	1.73	2.64		4.37	siding 7/16" thick with wood grain
Clean	S.F.		.37		.37	finish.
Paint	S.F.	.75	1.41		2.16	
Minimum Charge	Job		242		242	
Plywood Panel						
Douglas Fir (T-1-11)						
Demolish	S.F.		.57		.57	Includes material and labor to install
Install	S.F.	1.62	1.51		3.13	Douglas fir 4' x 8' plywood exterior
Demolish and Install	S.F.	1.62	2.08		3.70	siding plain or patterns.
Reinstall	S.F.		1.51		1.51	
Clean	S.F.		.37		.37	
Paint	S.F.	.75	1.41		2.16	
Minimum Charge	Job		242		242	
Redwood						
Demolish	S.F.		.57		.57	Includes material and labor to install
Install	S.F.	2.35	1.51		3.86	redwood 4' x 8' plywood exterior
Demolish and Install	S.F.	2.35	2.08		4.43	siding plain or patterns.
Reinstall	S.F.		1.51		1.51	
Clean	S.F.		.37		.37	
Paint	S.F.	.75	1.41		2.16	
Minimum Charge	Job		242		242	
Cedar						
Demolish	S.F.		.57		.57	Cost includes material and labor to
Install	S.F.	2.96	1.51		4.47	install 4' x 8' cedar plywood exterior
Demolish and Install	S.F.	2.96	2.08		5.04	siding plain or patterns.
Reinstall	S.F.		1.51		1.51	
Clean	S.F.		.37		.37	
Paint	S.F.	.75	1.41		2.16	
Minimum Charge	Job		242		242	
Tongue and Groove						
Fir, Tongue and Groove						
Demolish	S.F.		1.04		1.04	Includes material and labor to install
Install	S.F.	6.50	1.47		7.97	fir tongue and groove exterior siding
Demolish and Install	S.F.	6.50	2.51		9.01	1" x 8".
Reinstall	S.F.		1.17		1.17	
Clean	S.F.		.37		.37	
Paint	S.F.	.75	1.41		2.16	
Minimum Charge	Job		242		242	
Pine, Tongue and Groove						
Demolish	S.F.		1.04		1.04	Includes labor and material to install
Install	S.F.	2.84	1.47		4.31	Pine tongue and groove exterior
Demolish and Install	S.F.	2.84	2.51		5.35	siding 1" x 8".
Reinstall	S.F.		1.47		1.47	
Clean	S.F.		.37		.37	
Paint	S.F.	.75	1.41		2.16	
Minimum Charge	Job		242		242	

135

Exterior Siding	Unit	Material	Labor	Equip.	Total	Specification
Cedar, Tongue and Groove						
Demolish	S.F.		1.04		1.04	Includes labor and material to install
Install	S.F.	10.05	2.20		12.25	cedar tongue and groove exterior
Demolish and Install	S.F.	10.05	3.24		13.29	siding 1" x 8".
Reinstall	S.F.		1.76		1.76	
Clean	S.F.		.37		.37	
Paint	S.F.	.75	1.41		2.16	
Minimum Charge	Job		242		242	
Redwood, Tongue and Groove						
Demolish	S.F.		1.04		1.04	Includes labor and material to install
Install	S.F.	8.05	2.20		10.25	redwood tongue and groove exterior
Demolish and Install	S.F.	8.05	3.24		11.29	siding 1" x 4".
Reinstall	S.F.		2.20		2.20	
Clean	S.F.		.37		.37	
Paint	S.F.	.75	1.41		2.16	
Minimum Charge	Job		242		242	
Board and Batten						
Fir						
Demolish	S.F.		1.04		1.04	Cost includes material and labor to
Install	S.F.	6.75	1.47		8.22	install 1" x 8" fir boards with 1" x 3"
Demolish and Install	S.F.	6.75	2.51		9.26	battens.
Reinstall	S.F.		1.17		1.17	
Clean	S.F.		.37		.37	
Paint	S.F.	.75	1.41		2.16	
Minimum Charge	Job		242		242	
Pine						
Demolish	S.F.		1.04		1.04	Cost includes material and labor to
Install	S.F.	3.73	1.29		5.02	install 1" x 10" pine boards with 1" x
Demolish and Install	S.F.	3.73	2.33		6.06	3" battens.
Reinstall	S.F.		1.03		1.03	
Clean	S.F.		.37		.37	
Paint	S.F.	.75	1.41		2.16	
Minimum Charge	Job		242		242	
Cedar						
Demolish	S.F.		1.04		1.04	Cost includes material and labor to
Install	S.F.	8.40	1.29		9.69	install 1" x 10" cedar boards with 1" x
Demolish and Install	S.F.	8.40	2.33		10.73	3" battens.
Reinstall	S.F.		1.03		1.03	
Clean	S.F.		.37		.37	
Paint	S.F.	.75	1.41		2.16	
Minimum Charge	Job		242		242	
Redwood						
Demolish	S.F.		1.04		1.04	Cost includes material and labor to
Install	S.F.	6.65	1.29		7.94	install 1" x 10" redwood boards with
Demolish and Install	S.F.	6.65	2.33		8.98	1" x 3" battens.
Reinstall	S.F.		1.03		1.03	
Clean	S.F.		.37		.37	
Paint	S.F.	.75	1.41		2.16	
Minimum Charge	Job		242		242	
Wood Shingle						
Demolish	Sq.		96		96	Includes material and labor to install
Install	Sq.	375	215		590	wood shingles/shakes.
Demolish and Install	Sq.	375	311		686	
Clean	Sq.		39.50		39.50	
Paint	Sq.	49.50	126		175.50	
Minimum Charge	Job		242		242	

For customer support on your Contractor's Pricing Guide: Residential Repair & Remodeling, call 800.448.8182.

Siding

Exterior Siding	Unit	Material	Labor	Equip.	Total	Specification
Synthetic Stucco System						
Demolish	S.F.		3.46		3.46	Cost includes material and labor to
Install	S.F.	1.24	8.30	.46	10	install 2 coats of adhesive mixture,
Demolish and Install	S.F.	1.24	11.76	.46	13.46	glass fiber mesh, 1" insulation board,
Clean	S.F.		.34		.34	and textured finish coat.
Paint	S.F.	.35	.68		1.03	
Minimum Charge	Job		1125	62	1187	

For customer support on your Contractor's Pricing Guide: Residential Repair & Remodeling, call 800.448.8182.

Front-end Loader

Bulldozer

Grading	Unit	Material	Labor	Equip.	Total	Specification
Hand (fine)						
Install	S.F.		.75		.75	Includes labor and material to install fine grade for slab on grade, hand grade.
Minimum Charge	Job		187		187	
Machine (fine)						
Install	S.Y.		.65	.62	1.27	Includes labor and material to install fine grade for roadway, base or leveling course.
Minimum Charge	Job		445	425	870	

Asphalt Paving	Unit	Material	Labor	Equip.	Total	Specification
Seal Coat						
Install	S.F.	.11	.14		.25	Includes labor and material to install sealcoat for a small area.
Minimum Charge	Job		187		187	
Wearing Course						
Install	S.F.	75	.48	.34	75.82	Cost includes labor and material to install 1″ thick wearing course.
Minimum Charge	Job		187		187	

Masonry Pavers	Unit	Material	Labor	Equip.	Total	Specification
Natural Concrete						
Sand Base						
Demolish	S.F.		4.16		4.16	Includes material and labor to install concrete pavers on sand base.
Install	S.F.	4.74	8		12.74	
Demolish and Install	S.F.	4.74	12.16		16.90	
Reinstall	S.F.		6.38		6.38	
Clean	S.F.		1.50		1.50	
Minimum Charge	Job		239		239	
Mortar Base						
Demolish	S.F.		6.85	2.01	8.86	Includes material and labor to install concrete pavers on mortar base.
Install	S.F.	3.78	6.65		10.43	
Demolish and Install	S.F.	3.78	13.50	2.01	19.29	
Reinstall	S.F.		5.32		5.32	
Clean	S.F.		1.36		1.36	
Minimum Charge	Job		239		239	

Masonry Pavers

Masonry Pavers	Unit	Material	Labor	Equip.	Total	Specification
Adobe						
Sand Base						
Demolish	S.F.		4.16		4.16	Includes material and labor to install
Install	S.F.	2.09	5.30		7.39	adobe pavers on sand base.
Demolish and Install	S.F.	2.09	9.46		11.55	
Reinstall	S.F.		4.26		4.26	
Clean	S.F.		1.50		1.50	
Minimum Charge	Job		239		239	
Mortar Base						
Demolish	S.F.		6.85	2.01	8.86	Includes material and labor to install
Install	S.F.	3.15	6.65		9.80	adobe pavers on mortar base.
Demolish and Install	S.F.	3.15	13.50	2.01	18.66	
Reinstall	S.F.		5.32		5.32	
Clean	S.F.		1.36		1.36	
Minimum Charge	Job		239		239	
Standard Brick						
Sand Base						
Demolish	S.F.		4.16		4.16	Includes material and labor to install
Install	S.F.	4.19	8.85		13.04	standard brick pavers on sand base.
Demolish and Install	S.F.	4.19	13.01		17.20	
Reinstall	S.F.		7.09		7.09	
Clean	S.F.		1.50		1.50	
Minimum Charge	Job		239		239	
Mortar Base						
Demolish	S.F.		6.85	2.01	8.86	Includes material and labor to install
Install	S.F.	4.61	3.19		7.80	standard brick pavers on mortar base.
Demolish and Install	S.F.	4.61	10.04	2.01	16.66	
Reinstall	S.F.		2.55		2.55	
Clean	S.F.		1.36		1.36	
Minimum Charge	Job		239		239	
Deluxe Brick						
Sand Base						
Demolish	S.F.		4.16		4.16	Includes material and labor to install
Install	S.F.	5.40	6.65		12.05	deluxe brick pavers on sand base.
Demolish and Install	S.F.	5.40	10.81		16.21	
Reinstall	S.F.		5.32		5.32	
Clean	S.F.		1.50		1.50	
Minimum Charge	Job		239		239	
Mortar Base						
Demolish	S.F.		6.85	2.01	8.86	Includes material and labor to install
Install	S.F.	5.20	11.95		17.15	deluxe brick pavers on mortar base.
Demolish and Install	S.F.	5.20	18.80	2.01	26.01	
Reinstall	S.F.		9.58		9.58	
Clean	S.F.		1.36		1.36	
Minimum Charge	Job		239		239	

Fencing

Fencing	Unit	Material	Labor	Equip.	Total	Specification
Block w / Footing						
4"						
Demolish	S.F.		.14	.40	.54	Cost includes material and labor to
Install	S.F.	1.97	5.25		7.22	install 4" non-reinforced concrete
Demolish and Install	S.F.	1.97	5.39	.40	7.76	block.
Clean	S.F.	.25	1.25	.37	1.87	
Paint	S.F.	.41	.77		1.18	
Minimum Charge	Job		239		239	

Fencing	Unit	Material	Labor	Equip.	Total	Specification
4″ Reinforced						
Demolish	S.F.		2.31	1.08	3.39	Cost includes material and labor to
Install	S.F.	2.15	5.30		7.45	install 4″ reinforced concrete block.
Demolish and Install	S.F.	2.15	7.61	1.08	10.84	
Clean	S.F.	.25	1.25	.37	1.87	
Paint	S.F.	.41	.77		1.18	
Minimum Charge	Job		239		239	
6″						
Demolish	S.F.		.19	.53	.72	Cost includes material and labor to
Install	S.F.	4.30	5.65		9.95	install 6″ non-reinforced concrete
Demolish and Install	S.F.	4.30	5.84	.53	10.67	block.
Clean	S.F.	.25	1.25	.37	1.87	
Paint	S.F.	.41	.77		1.18	
Minimum Charge	Job		239		239	
6″ Reinforced						
Demolish	S.F.		2.43	.79	3.22	Cost includes material and labor to
Install	S.F.	4.48	5.70		10.18	install 6″ reinforced concrete block.
Demolish and Install	S.F.	4.48	8.13	.79	13.40	
Clean	S.F.	.25	1.25	.37	1.87	
Paint	S.F.	.41	.77		1.18	
Minimum Charge	Job		239		239	
Cap for Block Wall						
Demolish	L.F.		2.60		2.60	Includes labor and equipment to install
Install	L.F.	7.90	11.80		19.70	precast concrete coping.
Demolish and Install	L.F.	7.90	14.40		22.30	
Clean	L.F.	.03	3.90		3.93	
Paint	S.F.	.41	.77		1.18	
Minimum Charge	Job		239		239	
Block Pilaster						
Demolish	Ea.		178		178	Includes material and labor to install
Install	Ea.	112	177		289	block pilaster for fence.
Demolish and Install	Ea.	112	355		467	
Reinstall	Ea.		177.28		177.28	
Paint	Ea.	9.35	17.95		27.30	
Minimum Charge	Job		239		239	
Block w / o Footing						
4″						
Demolish	S.F.		.14	.40	.54	Cost includes material and labor to
Install	S.F.	2.13	4.03		6.16	install 4″ non-reinforced hollow
Demolish and Install	S.F.	2.13	4.17	.40	6.70	lightweight block.
Clean	S.F.	.25	1.25	.37	1.87	
Paint	S.F.	.41	.77		1.18	
Minimum Charge	Job		239		239	
4″ Reinforced						
Demolish	S.F.		2.31	1.08	3.39	Cost includes material and labor to
Install	S.F.	2.27	5.20		7.47	install 4″ reinforced hollow lightweight
Demolish and Install	S.F.	2.27	7.51	1.08	10.86	block.
Clean	S.F.	.25	1.25	.37	1.87	
Paint	S.F.	.41	.77		1.18	
Minimum Charge	Job		239		239	
6″						
Demolish	S.F.		.19	.53	.72	Cost includes material and labor to
Install	S.F.	1.93	4.43		6.36	install 6″ non-reinforced hollow
Demolish and Install	S.F.	1.93	4.62	.53	7.08	lightweight block.
Clean	S.F.	.25	1.25	.37	1.87	
Paint	S.F.	.41	.77		1.18	
Minimum Charge	Job		239		239	

141

Fencing

Fencing	Unit	Material	Labor	Equip.	Total	Specification
6" Reinforced						
Demolish	S.F.		2.43	.79	3.22	Cost includes material and labor to
Install	S.F.	2.11	5.55		7.66	install 6" reinforced hollow lightweight
Demolish and Install	S.F.	2.11	7.98	.79	10.88	block.
Clean	S.F.	.25	1.25	.37	1.87	
Paint	S.F.	.41	.77		1.18	
Minimum Charge	Job		239		239	
Cap for Block Wall						
Demolish	L.F.		2.60		2.60	Includes labor and equipment to install
Install	L.F.	7.90	11.80		19.70	precast concrete coping.
Demolish and Install	L.F.	7.90	14.40		22.30	
Clean	L.F.	.03	3.90		3.93	
Paint	S.F.	.41	.77		1.18	
Minimum Charge	Job		239		239	
Block Pilaster						
Demolish	Ea.		178		178	Includes material and labor to install
Install	Ea.	112	177		289	block pilaster for fence.
Demolish and Install	Ea.	112	355		467	
Reinstall	Ea.		177.28		177.28	
Paint	Ea.	9.35	17.95		27.30	
Minimum Charge	Job		239		239	

Chain Link

	Unit	Material	Labor	Equip.	Total	Specification
3' High w / o Top Rail						
Demolish	L.F.		3.22	.95	4.17	Includes material and labor to install
Install	L.F.	6.10	2.28	.81	9.19	residential galvanized fence 3' high
Demolish and Install	L.F.	6.10	5.50	1.76	13.36	with 1-3/8" steel post 10' O.C. set
Reinstall	L.F.		1.82	.65	2.47	without concrete.
Clean	L.F.	.04	1.36		1.40	
Minimum Charge	Job		575		575	
4' High w / o Top Rail						
Demolish	L.F.		3.22	.95	4.17	Includes material and labor to install
Install	L.F.	7.10	2.81	1.01	10.92	residential galvanized fence 4' high
Demolish and Install	L.F.	7.10	6.03	1.96	15.09	with 1-3/8" steel post 10' O.C. set
Reinstall	L.F.		2.25	.81	3.06	without concrete.
Clean	L.F.	.07	1.50		1.57	
Minimum Charge	Job		575		575	
5' High w / o Top Rail						
Demolish	L.F.		3.22	.95	4.17	Includes material and labor to install
Install	L.F.	7.55	3.68	1.32	12.55	residential galvanized fence 5' high
Demolish and Install	L.F.	7.55	6.90	2.27	16.72	with 1-3/8" steel post 10' O.C. set
Reinstall	L.F.		2.94	1.05	3.99	without concrete.
Clean	L.F.	.08	1.66		1.74	
Minimum Charge	Job		575		575	
3' High w / Top Rail						
Demolish	L.F.		3.42	1	4.42	Includes material and labor to install
Install	L.F.	8.20	2.39	.86	11.45	residential galvanized fence 3' high
Demolish and Install	L.F.	8.20	5.81	1.86	15.87	with 1-3/8" steel post 10' O.C. set
Reinstall	L.F.		1.91	.68	2.59	without concrete.
Clean	L.F.	.04	1.39		1.43	
Minimum Charge	Job		575		575	
4' High w / Top Rail						
Demolish	L.F.		3.42	1	4.42	Includes material and labor to install
Install	L.F.	9.25	2.99	1.07	13.31	residential galvanized fence 4' high
Demolish and Install	L.F.	9.25	6.41	2.07	17.73	with 1-5/8" steel post 10' O.C. set
Reinstall	L.F.		2.39	.86	3.25	without concrete and 1-5/8" top rail
Clean	L.F.	.07	1.53		1.60	with sleeves.
Minimum Charge	Job		575		575	

For customer support on your Contractor's Pricing Guide: Residential Repair & Remodeling, call 800.448.8182.

Fencing	Unit	Material	Labor	Equip.	Total	Specification
5' High w / Top Rail						Includes material and labor to install
Demolish	L.F.		3.42	1	4.42	residential galvanized fence 5' high
Install	L.F.	9.70	3.99	1.43	15.12	with 1-5/8" steel post 10' O.C. set
Demolish and Install	L.F.	9.70	7.41	2.43	19.54	without concrete and 1-5/8" top rail
Reinstall	L.F.		3.19	1.14	4.33	with sleeves.
Clean	L.F.	.08	1.70		1.78	
Minimum Charge	Job		575		575	
6' High w / Top Rail						Includes material and labor to install
Demolish	L.F.		3.42	1	4.42	residential galvanized fence 6' high
Install	L.F.	11.95	6	2.14	20.09	with 2" steel post 10' O.C. set without
Demolish and Install	L.F.	11.95	9.42	3.14	24.51	concrete and 1-3/8" top rail with
Reinstall	L.F.		4.78	1.71	6.49	sleeves.
Clean	L.F.	.09	1.92		2.01	
Minimum Charge	Job		575		575	
Barbed Wire						Cost includes material and labor to
Demolish	L.F.		1.93		1.93	install 3 strand galvanized barbed
Install	L.F.	.56	.75	.28	1.59	wire fence.
Demolish and Install	L.F.	.56	2.68	.28	3.52	
Reinstall	L.F.		.60	.22	.82	
Minimum Charge	Job		575	298	873	
6' Redwood						Cost includes material and labor to
Demolish	L.F.		12.75		12.75	install 6' high redwood fencing.
Install	L.F.	32	9.55	3.42	44.97	
Demolish and Install	L.F.	32	22.30	3.42	57.72	
Reinstall	L.F.		7.65	2.74	10.39	
Clean	L.F.	.09	1.74		1.83	
Paint	L.F.	2.04	4.25		6.29	
Minimum Charge	Job		575		575	
6' Red Cedar						Cost includes material and labor to
Demolish	L.F.		12.75		12.75	install 6' cedar fence.
Install	L.F.	26	9.55	3.42	38.97	
Demolish and Install	L.F.	26	22.30	3.42	51.72	
Reinstall	L.F.		7.65	2.74	10.39	
Clean	L.F.	.09	1.74		1.83	
Paint	L.F.	2.04	4.25		6.29	
Minimum Charge	Job		575		575	
6' Shadowbox						Cost includes material and labor to
Demolish	L.F.		12.75		12.75	install 6' pressure treated pine fence.
Install	L.F.	39	7.95	2.85	49.80	
Demolish and Install	L.F.	39	20.70	2.85	62.55	
Reinstall	L.F.		6.38	2.28	8.66	
Clean	L.F.	.09	1.74		1.83	
Paint	L.F.	2.04	4.25		6.29	
Minimum Charge	Job		575		575	
3' Picket						Includes material, labor and
Demolish	L.F.		12.75		12.75	equipment to install pressure treated
Install	L.F.	3.98	8.55	3.05	15.58	pine picket fence.
Demolish and Install	L.F.	3.98	21.30	3.05	28.33	
Reinstall	L.F.		6.83	2.44	9.27	
Clean	L.F.	.04	1.50		1.54	
Paint	L.F.	1.01	2.31		3.32	
Minimum Charge	Job		575		575	

Fencing	Unit	Material	Labor	Equip.	Total	Specification
Wrought Iron						
Demolish	L.F.		27.50		27.50	Cost includes material and labor to
Install	L.F.	43	15.30		58.30	install 5' high wrought iron fence with
Demolish and Install	L.F.	43	42.80		85.80	pickets at 4-1/2" O.C.
Reinstall	L.F.		12.24		12.24	
Clean	L.F.		1.50		1.50	
Paint	L.F.	8.40	4.75		13.15	
Minimum Charge	Job		575		575	
Post						
Dirt Based Wood						
Demolish	Ea.		32		32	Includes material and labor to install
Install	Ea.	44	115		159	wood post set in earth.
Demolish and Install	Ea.	44	147		191	
Reinstall	Ea.		91.83		91.83	
Clean	Ea.		4.68		4.68	
Paint	Ea.	2.23	1.61		3.84	
Minimum Charge	Job		575		575	
Concrete Based Wood						
Demolish	Ea.		45.50	13.40	58.90	Includes material and labor to install
Install	Ea.	45	143		188	wood post set in concrete.
Demolish and Install	Ea.	45	188.50	13.40	246.90	
Clean	Ea.		4.68		4.68	
Paint	Ea.	2.23	1.61		3.84	
Minimum Charge	Job		575		575	
Block Pilaster						
Demolish	Ea.		178		178	Includes material and labor to install
Install	Ea.	112	177		289	block pilaster for fence.
Demolish and Install	Ea.	112	355		467	
Reinstall	Ea.		177.28		177.28	
Paint	Ea.	9.35	17.95		27.30	
Minimum Charge	Job		239		239	
Gate						
Chain Link						
Demolish	Ea.		42.50		42.50	Cost includes material and labor to
Install	Ea.	138	64		202	install 3' wide chain link pass gate
Demolish and Install	Ea.	138	106.50		244.50	and hardware.
Reinstall	Ea.		51.02		51.02	
Clean	Ea.		5.20		5.20	
Minimum Charge	Job		575		575	
Redwood						
Demolish	Ea.		36.50		36.50	Includes material, labor and
Install	Ea.	510	64		574	equipment to install redwood pass
Demolish and Install	Ea.	510	100.50		610.50	gate and hardware.
Reinstall	Ea.		51.02		51.02	
Clean	Ea.		5.20		5.20	
Paint	Ea.	3.36	8.05		11.41	
Minimum Charge	Job		575		575	
Red Cedar						
Demolish	Ea.		36.50		36.50	Cost includes material and labor to
Install	Ea.	425	64		489	install 3' wide red cedar pass gate
Demolish and Install	Ea.	425	100.50		525.50	and hardware.
Reinstall	Ea.		51.02		51.02	
Clean	Ea.		5.20		5.20	
Paint	Ea.	3.36	8.05		11.41	
Minimum Charge	Job		575		575	

Fencing	Unit	Material	Labor	Equip.	Total	Specification
Wrought Iron						
Demolish	Ea.		27.50		27.50	Includes material, labor and
Install	Ea.	350	143		493	equipment to install wrought iron gate
Demolish and Install	Ea.	350	170.50		520.50	with hardware.
Reinstall	Ea.		114.79		114.79	
Clean	Ea.		5.20		5.20	
Paint	Ea.	16.85	8.05		24.90	
Minimum Charge	Job		575		575	

145

For customer support on your Contractor's Pricing Guide: Residential Repair & Remodeling, call 800.448.8182.

Fire-extinguishing System	Space Heater	Hot-air Furnace

Plumbing	Unit	Material	Labor	Equip.	Total	Specification
Components						
3/4″ Supply						
Demolish	L.F.		3.04		3.04	Includes material and labor to install
Install	L.F.	7.10	7.20		14.30	type L copper tubing including
Demolish and Install	L.F.	7.10	10.24		17.34	couplings and hangers at 10′ O.C.
Reinstall	L.F.		7.19		7.19	
Minimum Charge	Job		273		273	
1/2″ Fixture Supply						
Install	Ea.	78	182		260	Includes labor and materials to install
						supply pipe only to a plumbing fixture.
2″ Single Vent Stack						
Demolish	Ea.		50.50		50.50	Includes material and labor to install
Install	Ea.	93.50	273		366.50	PVC vent stack pipe with couplings at
Demolish and Install	Ea.	93.50	323.50		417	10′ O.C.
Reinstall	Ea.		273.40		273.40	
Minimum Charge	Job		273		273	
4″ Drain						
Demolish	L.F.		4.05		4.05	Includes material and labor to install
Install	L.F.	13.90	8.95		22.85	DVW type PVC pipe with couplings at
Demolish and Install	L.F.	13.90	13		26.90	10′ O.C. and 3 strap hangers per 10′
Reinstall	L.F.		8.95		8.95	run.
Minimum Charge	Job		273		273	
1-1/2″ Drain						
Demolish	L.F.		3.04		3.04	Includes material and labor to install
Install	L.F.	5.35	5.80		11.15	DVW type PVC pipe with couplings at
Demolish and Install	L.F.	5.35	8.84		14.19	10′ O.C. and 3 strap hangers per 10′
Reinstall	L.F.		5.82		5.82	run.
Minimum Charge	Job		273		273	
Maintenance						
Flush Out / Unclog						
Clean	Job		273		273	Includes 4 hour minimum labor charge
Minimum Charge	Job		273		273	for a plumber.
Vent Cap						
Large						
Demolish	Ea.		13.80		13.80	Includes labor and material to install a
Install	Ea.	55.50	25		80.50	PVC plumbing pipe vent cap.
Demolish and Install	Ea.	55.50	38.80		94.30	
Clean	Ea.	.21	8.50		8.71	
Paint	Ea.	.33	9.15		9.48	
Minimum Charge	Job		137		137	

Plumbing

Plumbing	Unit	Material	Labor	Equip.	Total	Specification
Medium						
Demolish	Ea.		13.80		13.80	Includes labor and material to install a
Install	Ea.	48	24		72	PVC plumbing pipe vent cap.
Demolish and Install	Ea.	48	37.80		85.80	
Clean	Ea.	.21	8.50		8.71	
Paint	Ea.	.33	9.15		9.48	
Minimum Charge	Job		137		137	
Small						
Demolish	Ea.		13.80		13.80	Includes material and labor to install a
Install	Ea.	16.30	23		39.30	PVC plumbing pipe vent cap.
Demolish and Install	Ea.	16.30	36.80		53.10	
Clean	Ea.	.21	8.50		8.71	
Paint	Ea.	.33	9.15		9.48	
Minimum Charge	Job		137		137	

Plumbing	Unit	Material	Labor	Equip.	Total	Specification
Sump Pump						
1/4 HP						
Demolish	Ea.		54.50		54.50	Includes material and labor to install a
Install	Ea.	190	85.50		275.50	submersible sump pump.
Demolish and Install	Ea.	190	140		330	
Reinstall	Ea.		68.35		68.35	
Clean	Ea.	.53	23.50		24.03	
Minimum Charge	Job		273		273	
1/3 HP						
Demolish	Ea.		54.50		54.50	Includes material and labor to install a
Install	Ea.	252	91		343	submersible sump pump.
Demolish and Install	Ea.	252	145.50		397.50	
Reinstall	Ea.		72.91		72.91	
Clean	Ea.	.53	23.50		24.03	
Minimum Charge	Job		273		273	
1/2 HP						
Demolish	Ea.		54.50		54.50	Includes material and labor to install a
Install	Ea.	274	101		375	submersible sump pump.
Demolish and Install	Ea.	274	155.50		429.50	
Reinstall	Ea.		81.01		81.01	
Clean	Ea.	.53	23.50		24.03	
Minimum Charge	Job		273		273	
Clean and Service						
Clean	Ea.		34		34	Includes labor to clean and service a
Minimum Charge	Job		273		273	sump pump.

Water Heater	Unit	Material	Labor	Equip.	Total	Specification
Electric						
30 Gallon						
Demolish	Ea.		171		171	Includes material and labor to install a
Install	Ea.	880	249		1129	30 gallon electric water heater.
Demolish and Install	Ea.	880	420		1300	
Reinstall	Ea.		198.84		198.84	
Clean	Ea.	.53	23.50		24.03	
Minimum Charge	Job		273		273	

For customer support on your Contractor's Pricing Guide: Residential Repair & Remodeling, call 800.448.8182.

Water Heater		Unit	Material	Labor	Equip.	Total	Specification
40 Gallon							
	Demolish	Ea.		171		171	Includes material and labor to install a
	Install	Ea.	915	273		1188	40 gallon electric water heater.
	Demolish and Install	Ea.	915	444		1359	
	Reinstall	Ea.		218.72		218.72	
	Clean	Ea.	.53	23.50		24.03	
	Minimum Charge	Job		273		273	
52 Gallon							
	Demolish	Ea.		171		171	Includes material and labor to install a
	Install	Ea.	1600	273		1873	52 gallon electric water heater.
	Demolish and Install	Ea.	1600	444		2044	
	Reinstall	Ea.		218.72		218.72	
	Clean	Ea.	.53	23.50		24.03	
	Minimum Charge	Job		273		273	
82 Gallon							
	Demolish	Ea.		171		171	Includes material and labor to install
	Install	Ea.	2450	340		2790	an 82 gallon electric water heater.
	Demolish and Install	Ea.	2450	511		2961	
	Reinstall	Ea.		273.40		273.40	
	Clean	Ea.	.53	23.50		24.03	
	Minimum Charge	Job		273		273	

Gas-fired

		Unit	Material	Labor	Equip.	Total	Specification
30 Gallon Gas-fired							
	Demolish	Ea.		210		210	Includes material and labor to install a
	Install	Ea.	2225	273		2498	30 gallon gas water heater.
	Demolish and Install	Ea.	2225	483		2708	
	Reinstall	Ea.		218.72		218.72	
	Clean	Ea.	.53	23.50		24.03	
	Minimum Charge	Job		273		273	
30 Gallon Gas-fired, Flue							
	Install	Ea.	57	78		135	Includes labor and materials to install
	Minimum Charge	Job		273		273	a flue for a water heater (up to 50 gallon size).
40 Gallon Gas-fired							
	Demolish	Ea.		210		210	Includes material and labor to install a
	Install	Ea.	765	288		1053	40 gallon gas water heater.
	Demolish and Install	Ea.	765	498		1263	
	Reinstall	Ea.		230.23		230.23	
	Clean	Ea.	.53	23.50		24.03	
	Minimum Charge	Job		273		273	
40 Gallon Gas-fired, Flue							
	Install	Ea.	57	78		135	Includes labor and materials to install
	Minimum Charge	Job		273		273	a flue for a water heater (up to 50 gallon size).
50 Gallon Gas-fired							
	Demolish	Ea.		210		210	Includes material and labor to install a
	Install	Ea.	865	305		1170	50 gallon gas water heater.
	Demolish and Install	Ea.	865	515		1380	
	Reinstall	Ea.		243.02		243.02	
	Clean	Ea.	.53	23.50		24.03	
	Minimum Charge	Job		273		273	
50 Gallon Gas-fired, Flue							
	Install	Ea.	57	78		135	Includes labor and materials to install
	Minimum Charge	Job		273		273	a flue for a water heater (up to 50 gallon size).

Water Heater

	Unit	Material	Labor	Equip.	Total	Specification
Solar Hot Water System						
Water Heater						
Demolish	Ea.		137		137	Includes material and labor to install a
Install	Ea.	1600	365		1965	solar hot water heat exchanger
Demolish and Install	Ea.	1600	502		2102	storage tank, 100 gallon.
Reinstall	Ea.		291.63		291.63	
Minimum Charge	Job		284		284	
Solar Panel						
Demolish	S.F.		2		2	Includes material and labor to install
Install	S.F.	36.50	3.60		40.10	solar panel.
Demolish and Install	S.F.	36.50	5.60		42.10	
Reinstall	S.F.		2.88		2.88	
Minimum Charge	Job		284		284	
Repair Plumbing						
Install	Job		284		284	Includes 4 hour minimum labor charge
Minimum Charge	Job		284		284	for a steamfitter or pipefitter.
Insulating Wrap						
Demolish	Ea.		6.35		6.35	Includes material and labor to install
Install	Ea.	20.50	17.10		37.60	insulating wrap around a 100 gallon
Demolish and Install	Ea.	20.50	23.45		43.95	solar hot water heat exchanger
Reinstall	Ea.		13.67		13.67	storage tank.
Minimum Charge	Job		273		273	
Water Softener						
Demolish	Ea.		101		101	Includes material and labor to install a
Install	Ea.	3175	246		3421	14 GPM water softener.
Demolish and Install	Ea.	3175	347		3522	
Reinstall	Ea.		196.80		196.80	
Clean	Ea.	.53	23.50		24.03	
Minimum Charge	Ea.		273		273	

Ductwork

	Unit	Material	Labor	Equip.	Total	Specification
Galvanized, Square or Rectangular						
6"						
Demolish	L.F.		2.47		2.47	Includes material and labor to install
Install	L.F.	1.61	11.85		13.46	rectangular ductwork with a greatest
Demolish and Install	L.F.	1.61	14.32		15.93	dimension of 6".
Reinstall	L.F.		9.49		9.49	
Clean	L.F.		2.85		2.85	
Minimum Charge	Job		267		267	
8"						
Demolish	L.F.		2.47		2.47	Includes material and labor to install
Install	L.F.	2.13	15.25		17.38	rectangular ductwork with a greatest
Demolish and Install	L.F.	2.13	17.72		19.85	dimension of 8".
Reinstall	L.F.		12.20		12.20	
Clean	L.F.		3.79		3.79	
Minimum Charge	Job		267		267	
10"						
Demolish	L.F.		2.47		2.47	Includes material and labor to install
Install	L.F.	2.67	19.75		22.42	rectangular ductwork with a greatest
Demolish and Install	L.F.	2.67	22.22		24.89	dimension of 10".
Reinstall	L.F.		15.81		15.81	
Clean	L.F.		4.74		4.74	
Minimum Charge	Job		267		267	

For customer support on your Contractor's Pricing Guide: Residential Repair & Remodeling, call 800.448.8182.

Rough Mechanical

Ductwork

Ductwork	Unit	Material	Labor	Equip.	Total	Specification
Galvanized, Spiral						
4"						
Demolish	L.F.		2.96		2.96	Cost includes material and labor to
Install	L.F.	7.70	2.67		10.37	install 4" diameter spiral pre-formed
Demolish and Install	L.F.	7.70	5.63		13.33	steel galvanized ductwork.
Reinstall	L.F.		2.13		2.13	
Clean	L.F.		1.49		1.49	
Minimum Charge	Job		267		267	
6"						
Demolish	L.F.		2.96		2.96	Cost includes material and labor to
Install	L.F.	8.60	3.43		12.03	install 6" diameter spiral pre-formed
Demolish and Install	L.F.	8.60	6.39		14.99	steel galvanized ductwork.
Reinstall	L.F.		2.74		2.74	
Clean	L.F.		2.24		2.24	
Minimum Charge	Job		267		267	
8"						
Demolish	L.F.		2.96		2.96	Cost includes material and labor to
Install	L.F.	9.90	4.80		14.70	install 8" diameter spiral pre-formed
Demolish and Install	L.F.	9.90	7.76		17.66	steel galvanized ductwork.
Reinstall	L.F.		3.84		3.84	
Clean	L.F.		2.98		2.98	
Minimum Charge	Job		267		267	
10"						
Demolish	L.F.		3.71		3.71	Cost includes material and labor to
Install	L.F.	12.10	6		18.10	install 10" diameter spiral pre-formed
Demolish and Install	L.F.	12.10	9.71		21.81	steel galvanized ductwork.
Reinstall	L.F.		4.80		4.80	
Clean	L.F.		3.73		3.73	
Minimum Charge	Job		267		267	
12"						
Demolish	L.F.		4.94		4.94	Cost includes material and labor to
Install	L.F.	12	8		20	install 12" diameter spiral pre-formed
Demolish and Install	L.F.	12	12.94		24.94	steel galvanized ductwork.
Reinstall	L.F.		6.40		6.40	
Clean	L.F.		4.47		4.47	
Minimum Charge	Job		267		267	
16"						
Demolish	L.F.		9.90		9.90	Cost includes material and labor to
Install	L.F.	24	16		40	install 16" diameter spiral pre-formed
Demolish and Install	L.F.	24	25.90		49.90	steel galvanized ductwork.
Reinstall	L.F.		12.81		12.81	
Clean	L.F.		5.95		5.95	
Minimum Charge	Job		267		267	
Flexible Fiberglass						
6" Non-insulated						
Demolish	L.F.		2.28		2.28	Cost includes material and labor to
Install	L.F.	2.13	3.43		5.56	install 6" diameter non-insulated
Demolish and Install	L.F.	2.13	5.71		7.84	flexible fiberglass fabric on corrosion
Reinstall	L.F.		2.74		2.74	resistant metal ductwork.
Clean	L.F.		2.24		2.24	
Minimum Charge	Job		267		267	
8" Non-insulated						
Demolish	L.F.		3.29		3.29	Cost includes material and labor to
Install	L.F.	2.70	4.80		7.50	install 8" diameter non-insulated
Demolish and Install	L.F.	2.70	8.09		10.79	flexible fiberglass fabric on corrosion
Reinstall	L.F.		3.84		3.84	resistant metal ductwork.
Clean	L.F.		2.98		2.98	
Minimum Charge	Job		267		267	

Rough Mechanical

Ductwork	Unit	Material	Labor	Equip.	Total	Specification
10" Non-insulated						
Demolish	L.F.		4.23		4.23	Cost includes material and labor to
Install	L.F.	3.43	6		9.43	install 10" diameter non-insulated
Demolish and Install	L.F.	3.43	10.23		13.66	flexible fiberglass fabric on corrosion
Reinstall	L.F.		4.80		4.80	resistant metal ductwork.
Clean	L.F.		3.73		3.73	
Minimum Charge	Job		267		267	
12" Non-insulated						
Demolish	L.F.		5.95		5.95	Cost includes material and labor to
Install	L.F.	4.04	8		12.04	install 12" diameter non-insulated
Demolish and Install	L.F.	4.04	13.95		17.99	flexible fiberglass fabric on corrosion
Reinstall	L.F.		6.40		6.40	resistant metal ductwork.
Clean	L.F.		4.47		4.47	
Minimum Charge	Job		267		267	
4" Insulated						
Demolish	L.F.		1.74		1.74	Cost includes material and labor to
Install	L.F.	3.78	2.82		6.60	install 4" diameter insulated flexible
Demolish and Install	L.F.	3.78	4.56		8.34	fiberglass fabric on corrosion resistant
Reinstall	L.F.		2.26		2.26	metal ductwork.
Clean	L.F.		1.49		1.49	
Minimum Charge	Job		267		267	
6" Insulated						
Demolish	L.F.		2.28		2.28	Cost includes material and labor to
Install	L.F.	4.19	3.69		7.88	install 6" diameter insulated flexible
Demolish and Install	L.F.	4.19	5.97		10.16	fiberglass fabric on corrosion resistant
Reinstall	L.F.		2.96		2.96	metal ductwork.
Clean	L.F.		2.24		2.24	
Minimum Charge	Job		267		267	
8" Insulated						
Demolish	L.F.		3.29		3.29	Cost includes material and labor to
Install	L.F.	4.93	5.35		10.28	install 8" diameter insulated flexible
Demolish and Install	L.F.	4.93	8.64		13.57	fiberglass fabric on corrosion resistant
Reinstall	L.F.		4.27		4.27	metal ductwork.
Clean	L.F.		2.98		2.98	
Minimum Charge	Job		267		267	
10" Insulated						
Demolish	L.F.		4.23		4.23	Cost includes material and labor to
Install	L.F.	5.95	6.85		12.80	install 10" diameter insulated flexible
Demolish and Install	L.F.	5.95	11.08		17.03	fiberglass fabric on corrosion resistant
Reinstall	L.F.		5.49		5.49	metal ductwork.
Clean	L.F.		3.73		3.73	
Minimum Charge	Job		267		267	
12" Insulated						
Demolish	L.F.		5.95		5.95	Cost includes material and labor to
Install	L.F.	6.95	9.60		16.55	install 12" diameter insulated flexible
Demolish and Install	L.F.	6.95	15.55		22.50	fiberglass fabric on corrosion resistant
Reinstall	L.F.		7.68		7.68	metal ductwork.
Clean	L.F.		4.47		4.47	
Minimum Charge	Job		267		267	

For customer support on your Contractor's Pricing Guide: Residential Repair & Remodeling, call 800.448.8182.

Heating

Heating	Unit	Material	Labor	Equip.	Total	Specification
Hot Water Baseboard						
Baseboard						
Demolish	L.F.		9.75		9.75	Includes material and labor to install
Install	L.F.	12.70	17.05		29.75	baseboard radiation.
Demolish and Install	L.F.	12.70	26.80		39.50	
Reinstall	L.F.		13.66		13.66	
Clean	L.F.	.53	5.85		6.38	
Minimum Charge	Job		273		273	
3'						
Demolish	Ea.		30.50		30.50	Includes material and labor to install
Install	Ea.	25.50	57		82.50	fin tube baseboard radiation in 3'
Demolish and Install	Ea.	25.50	87.50		113	sections.
Reinstall	Ea.		45.53		45.53	
Clean	Ea.	.15	17.80		17.95	
Minimum Charge	Job		273		273	
4'						
Demolish	Ea.		40.50		40.50	Includes material and labor to install
Install	Ea.	34	64		98	fin tube baseboard radiation in 4'
Demolish and Install	Ea.	34	104.50		138.50	sections.
Reinstall	Ea.		51.22		51.22	
Clean	Ea.	.31	23.50		23.81	
Minimum Charge	Job		273		273	
5'						
Demolish	Ea.		50.50		50.50	Includes material and labor to install
Install	Ea.	42.50	73		115.50	fin tube baseboard radiation in 5'
Demolish and Install	Ea.	42.50	123.50		166	sections.
Reinstall	Ea.		58.54		58.54	
Clean	Ea.	.31	29		29.31	
Minimum Charge	Job		273		273	
6'						
Demolish	Ea.		61		61	Includes material and labor to install
Install	Ea.	51	85.50		136.50	fin tube baseboard radiation in 6'
Demolish and Install	Ea.	51	146.50		197.50	sections.
Reinstall	Ea.		68.30		68.30	
Clean	Ea.	.31	34		34.31	
Minimum Charge	Job		273		273	
7'						
Demolish	Ea.		67.50		67.50	Includes material and labor to install
Install	Ea.	59.50	109		168.50	fin tube baseboard radiation in 7'
Demolish and Install	Ea.	59.50	176.50		236	sections.
Reinstall	Ea.		87.49		87.49	
Clean	Ea.	.39	41.50		41.89	
Minimum Charge	Job		273		273	
9'						
Demolish	Ea.		101		101	Includes material and labor to install
Install	Ea.	76.50	156		232.50	fin tube baseboard radiation in 9'
Demolish and Install	Ea.	76.50	257		333.50	sections.
Reinstall	Ea.		124.98		124.98	
Clean	Ea.	.61	53.50		54.11	
Minimum Charge	Job		273		273	

Heating

Heating	Unit	Material	Labor	Equip.	Total	Specification
Cabinet Unit Heater						
50 MBH						
Demolish	Ea.		51		51	Includes material and labor to install a
Install	Ea.	2025	146		2171	cabinet space heater with an output of
Demolish and Install	Ea.	2025	197		2222	50 MBH.
Reinstall	Ea.		117.08		117.08	
Clean	Ea.	.39	47		47.39	
Minimum Charge	Job		284		284	
75 MBH						
Demolish	Ea.		51		51	Includes material and labor to install a
Install	Ea.	1150	171		1321	cabinet space heater with an output of
Demolish and Install	Ea.	1150	222		1372	75 MBH.
Reinstall	Ea.		136.60		136.60	
Clean	Ea.	.39	47		47.39	
Minimum Charge	Job		284		284	
125 MBH						
Demolish	Ea.		61.50		61.50	Includes material and labor to install a
Install	Ea.	1425	205		1630	cabinet space heater with an output of
Demolish and Install	Ea.	1425	266.50		1691.50	125 MBH.
Reinstall	Ea.		163.92		163.92	
Clean	Ea.	.39	47		47.39	
Minimum Charge	Job		284		284	
175 MBH						
Demolish	Ea.		77		77	Includes material and labor to install a
Install	Ea.	4125	293		4418	cabinet space heater with an output of
Demolish and Install	Ea.	4125	370		4495	175 MBH.
Reinstall	Ea.		234.17		234.17	
Clean	Ea.	.39	47		47.39	
Minimum Charge	Job		284		284	

Furnace Unit

Furnace Unit	Unit	Material	Labor	Equip.	Total	Specification
Wall (gas-fired)						
25 MBH w / Thermostat						
Demolish	Ea.		120		120	Includes material and labor to install a
Install	Ea.	890	192		1082	gas fired hot air furnace rated at 25
Demolish and Install	Ea.	890	312		1202	MBH.
Reinstall	Ea.		153.68		153.68	
Clean	Ea.	.76	104		104.76	
Minimum Charge	Job		284		284	
35 MBH w / Thermostat						
Demolish	Ea.		120		120	Includes material and labor to install a
Install	Ea.	950	213		1163	gas fired hot air furnace rated at 35
Demolish and Install	Ea.	950	333		1283	MBH.
Reinstall	Ea.		170.75		170.75	
Clean	Ea.	.76	104		104.76	
Minimum Charge	Job		284		284	
50 MBH w / Thermostat						
Demolish	Ea.		120		120	Includes material and labor to install a
Install	Ea.	1850	240		2090	gas fired hot air furnace rated at 50
Demolish and Install	Ea.	1850	360		2210	MBH.
Reinstall	Ea.		192.10		192.10	
Clean	Ea.	.76	104		104.76	
Minimum Charge	Job		284		284	

Rough Mechanical

Furnace Unit		Unit	Material	Labor	Equip.	Total	Specification
60 MBH w / Thermostat							
	Demolish	Ea.		120		120	Includes material and labor to install a
	Install	Ea.	1650	267		1917	gas fired hot air furnace rated at 60
	Demolish and Install	Ea.	1650	387		2037	MBH.
	Reinstall	Ea.		213.44		213.44	
	Clean	Ea.	.76	104		104.76	
	Minimum Charge	Job		284		284	
Floor (gas-fired)							
30 MBH w / Thermostat							
	Demolish	Ea.		120		120	Includes material and labor to install a
	Install	Ea.	1525	192		1717	gas fired hot air floor furnace rated at
	Demolish and Install	Ea.	1525	312		1837	30 MBH.
	Reinstall	Ea.		153.68		153.68	
	Clean	Ea.	.76	104		104.76	
	Minimum Charge	Job		284		284	
50 MBH w / Thermostat							
	Demolish	Ea.		120		120	Includes material and labor to install a
	Install	Ea.	1750	240		1990	gas fired hot air floor furnace rated at
	Demolish and Install	Ea.	1750	360		2110	50 MBH.
	Reinstall	Ea.		192.10		192.10	
	Clean	Ea.	.76	104		104.76	
	Minimum Charge	Job		284		284	
Gas Forced Air							
50 MBH							
	Demolish	Ea.		120		120	Includes material and labor to install
	Install	Ea.	1250	240		1490	an up flow gas fired hot air furnace
	Demolish and Install	Ea.	1250	360		1610	rated at 50 MBH.
	Reinstall	Ea.		192.10		192.10	
	Clean	Ea.	.76	104		104.76	
	Minimum Charge	Job		284		284	
80 MBH							
	Demolish	Ea.		120		120	Includes material and labor to install
	Install	Ea.	1350	267		1617	an up flow gas fired hot air furnace
	Demolish and Install	Ea.	1350	387		1737	rated at 80 MBH.
	Reinstall	Ea.		213.44		213.44	
	Clean	Ea.	.76	104		104.76	
	Minimum Charge	Job		284		284	
100 MBH							
	Demolish	Ea.		120		120	Includes material and labor to install
	Install	Ea.	1525	300		1825	an up flow gas fired hot air furnace
	Demolish and Install	Ea.	1525	420		1945	rated at 100 MBH.
	Reinstall	Ea.		240.12		240.12	
	Clean	Ea.	.76	104		104.76	
	Minimum Charge	Job		284		284	
120 MBH							
	Demolish	Ea.		120		120	Includes material and labor to install
	Install	Ea.	1550	320		1870	an up flow gas fired hot air furnace
	Demolish and Install	Ea.	1550	440		1990	rated at 120 MBH.
	Reinstall	Ea.		256.13		256.13	
	Clean	Ea.	.76	104		104.76	
	Minimum Charge	Job		284		284	

For customer support on your Contractor's Pricing Guide: Residential Repair & Remodeling, call 800.448.8182.

Furnace Unit

Furnace Unit	Unit	Material	Labor	Equip.	Total	Specification
Electric Forced Air						
31.4 MBH						
Demolish	Ea.		120		120	Includes material and labor to install
Install	Ea.	1850	277		2127	an electric hot air furnace rated at
Demolish and Install	Ea.	1850	397		2247	31.4 MBH.
Reinstall	Ea.		221.26		221.26	
Clean	Ea.	.76	104		104.76	
Minimum Charge	Job		284		284	
47.1 MBH						
Demolish	Ea.		120		120	Includes material and labor to install
Install	Ea.	2150	289		2439	an electric hot air furnace rated at
Demolish and Install	Ea.	2150	409		2559	47.1 MBH.
Reinstall	Ea.		231.55		231.55	
Clean	Ea.	.76	104		104.76	
Minimum Charge	Job		284		284	
62.9 MBH						
Demolish	Ea.		120		120	Includes material and labor to install
Install	Ea.	2350	310		2660	an electric hot air furnace rated at
Demolish and Install	Ea.	2350	430		2780	62.9 MBH.
Reinstall	Ea.		248.92		248.92	
Clean	Ea.	.76	104		104.76	
Minimum Charge	Job		284		284	
78.5 MBH						
Demolish	Ea.		120		120	Includes material and labor to install
Install	Ea.	3300	320		3620	an electric hot air furnace rated at
Demolish and Install	Ea.	3300	440		3740	78.5 MBH.
Reinstall	Ea.		255.30		255.30	
Clean	Ea.	.76	104		104.76	
Minimum Charge	Job		284		284	
110 MBH						
Demolish	Ea.		120		120	Includes material and labor to install
Install	Ea.	4175	335		4510	an electric hot air furnace rated at
Demolish and Install	Ea.	4175	455		4630	110 MBH.
Reinstall	Ea.		269.10		269.10	
Clean	Ea.	.76	104		104.76	
Minimum Charge	Job		284		284	
A/C Junction Box						
Install	Ea.	74.50	142		216.50	Includes labor and materials to install
Minimum Charge	Job		284		284	a heating system emergency shut-off switch. Cost includes up to 40 feet of non-metallic cable.

Boiler

Boiler	Unit	Material	Labor	Equip.	Total	Specification
Gas-Fired						
60 MBH						
Demolish	Ea.		1025		1025	Includes material and labor to install a
Install	Ea.	3325	1875		5200	gas fired cast iron boiler rated at 60
Demolish and Install	Ea.	3325	2900		6225	MBH.
Reinstall	Ea.		1490.15		1490.15	
Clean	Ea.	1.52	62.50		64.02	
Minimum Charge	Job		284		284	

For customer support on your Contractor's Pricing Guide: Residential Repair & Remodeling, call 800.448.8182.

Boiler		Unit	Material	Labor	Equip.	Total	Specification
Gas-Fired							
125 MBH							
	Demolish	Ea.		2850		2850	Includes material and labor to install a
	Install	Ea.	4225	2275		6500	gas fired cast iron boiler rated at 125
	Demolish and Install	Ea.	4225	5125		9350	MBH.
	Reinstall	Ea.		1821.30		1821.30	
	Clean	Ea.	1.52	62.50		64.02	
	Minimum Charge	Job		284		284	
Gas-Fired							
400 MBH							
	Demolish	Ea.		2850		2850	Includes material and labor to install a
	Install	Ea.	5675	3200		8875	gas fired cast iron boiler rated at 400
	Demolish and Install	Ea.	5675	6050		11725	MBH.
	Reinstall	Ea.		3201.50		3201.50	
	Clean	Ea.	1.52	62.50		64.02	
	Minimum Charge	Job		284		284	

Heat Pump		Unit	Material	Labor	Equip.	Total	Specification
2 Ton							
	Demolish	Ea.		284		284	Includes material and labor to install a
	Install	Ea.	1800	510		2310	2 ton heat pump.
	Demolish and Install	Ea.	1800	794		2594	
	Reinstall	Ea.		409.79		409.79	
	Clean	Ea.	1.52	47		48.52	
	Minimum Charge	Job		284		284	
3 Ton							
	Demolish	Ea.		475		475	Includes material and labor to install a
	Install	Ea.	2200	855		3055	3 ton heat pump.
	Demolish and Install	Ea.	2200	1330		3530	
	Reinstall	Ea.		682.99		682.99	
	Clean	Ea.	1.52	47		48.52	
	Minimum Charge	Job		284		284	
4 Ton							
	Demolish	Ea.		710		710	Includes material and labor to install a
	Install	Ea.	2525	1275		3800	4 ton heat pump.
	Demolish and Install	Ea.	2525	1985		4510	
	Reinstall	Ea.		1024.48		1024.48	
	Clean	Ea.	1.52	47		48.52	
	Minimum Charge	Job		284		284	
5 Ton							
	Demolish	Ea.		1150		1150	Includes material and labor to install a
	Install	Ea.	2750	2050		4800	5 ton heat pump.
	Demolish and Install	Ea.	2750	3200		5950	
	Reinstall	Ea.		1639.17		1639.17	
	Clean	Ea.	1.52	47		48.52	
	Minimum Charge	Job		284		284	
Thermostat							
	Demolish	Ea.		15.80		15.80	Includes material and labor to install
	Install	Ea.	276	71		347	an electric thermostat, 2 set back.
	Demolish and Install	Ea.	276	86.80		362.80	
	Reinstall	Ea.		56.88		56.88	
	Clean	Ea.	.03	11.70		11.73	
	Minimum Charge	Job		284		284	

Heat Pump	Unit	Material	Labor	Equip.	Total	Specification
A/C Junction Box						
Install	Ea.	550	435		985	Includes labor and materials to install
Minimum Charge	Job		284		284	a heat pump circuit including up to 40 feet of non-metallic cable.

Heating	Unit	Material	Labor	Equip.	Total	Specification
Electric Baseboard						
Demolish	L.F.		7.90		7.90	Includes material and labor to install
Install	L.F.	12	18.95		30.95	electric baseboard heating elements.
Demolish and Install	L.F.	12	26.85		38.85	
Reinstall	L.F.		15.16		15.16	
Clean	L.F.	.53	5.85		6.38	
Minimum Charge	Job		284		284	
2' 6"						
Demolish	Ea.		26.50		26.50	Includes material and labor to install a
Install	Ea.	36	71		107	2' 6" long section of electric
Demolish and Install	Ea.	36	97.50		133.50	baseboard heating elements.
Reinstall	Ea.		56.84		56.84	
Clean	Ea.	.15	14.40		14.55	
Minimum Charge	Job		284		284	
3'						
Demolish	Ea.		28.50		28.50	Includes material and labor to install a
Install	Ea.	41.50	71		112.50	3' long section of electric baseboard
Demolish and Install	Ea.	41.50	99.50		141	heating elements.
Reinstall	Ea.		56.84		56.84	
Clean	Ea.	.15	17.80		17.95	
Minimum Charge	Job		284		284	
4'						
Demolish	Ea.		31.50		31.50	Includes material and labor to install a
Install	Ea.	50	85		135	4' long section of electric baseboard
Demolish and Install	Ea.	50	116.50		166.50	heating elements.
Reinstall	Ea.		67.87		67.87	
Clean	Ea.	.07	3.90		3.97	
Minimum Charge	Job		284		284	
5'						
Demolish	Ea.		35		35	Includes material and labor to install a
Install	Ea.	60	99.50		159.50	5' long section of electric baseboard
Demolish and Install	Ea.	60	134.50		194.50	heating elements.
Reinstall	Ea.		79.78		79.78	
Clean	Ea.	.31	29		29.31	
Minimum Charge	Job		284		284	
6'						
Demolish	Ea.		39.50		39.50	Includes material and labor to install a
Install	Ea.	69	114		183	6' long section of electric baseboard
Demolish and Install	Ea.	69	153.50		222.50	heating elements.
Reinstall	Ea.		90.94		90.94	
Clean	Ea.	.07	4.68		4.75	
Minimum Charge	Job		284		284	
8'						
Demolish	Ea.		52.50		52.50	Includes material and labor to install
Install	Ea.	83.50	142		225.50	an 8' long section of electric
Demolish and Install	Ea.	83.50	194.50		278	baseboard heating elements.
Reinstall	Ea.		113.68		113.68	
Clean	Ea.	.07	5.85		5.92	
Minimum Charge	Job		284		284	

Rough Mechanical

Heating

Heating	Unit	Material	Labor	Equip.	Total	Specification
10'						
Demolish	Ea.		63		63	Includes material and labor to install a
Install	Ea.	108	172		280	10' long section of electric baseboard
Demolish and Install	Ea.	108	235		343	heating elements.
Reinstall	Ea.		137.79		137.79	
Clean	Ea.	.61	58.50		59.11	
Minimum Charge	Job		284		284	
Electric Radiant Wall Panel						
Demolish	Ea.		42		42	Includes material and labor to install
Install	Ea.	485	114		599	an 18" x 23" electric radiant heat wall
Demolish and Install	Ea.	485	156		641	panel.
Reinstall	Ea.		90.94		90.94	
Clean	Ea.	.08	11.70		11.78	
Minimum Charge	Job		284		284	

Heating Control

Heating Control	Unit	Material	Labor	Equip.	Total	Specification
Thermostat						
Demolish	Ea.		15.80		15.80	Includes material and labor to install
Install	Ea.	276	71		347	an electric thermostat, 2 set back.
Demolish and Install	Ea.	276	86.80		362.80	
Reinstall	Ea.		56.88		56.88	
Clean	Ea.	.03	11.70		11.73	
Minimum Charge	Job		284		284	

Air Conditioning

Air Conditioning	Unit	Material	Labor	Equip.	Total	Specification
Self-contained						
2 Ton						
Demolish	Ea.		505		505	Includes material and labor to install a
Install	Ea.	5100	885		5985	rooftop air conditioning unit, 2 ton.
Demolish and Install	Ea.	5100	1390		6490	No hoisting is included.
Reinstall	Ea.		883.93		883.93	
Clean	Ea.	1.52	47		48.52	
Minimum Charge	Job		284		284	
3 Ton						
Demolish	Ea.		505		505	Includes material and labor to install a
Install	Ea.	4450	1175		5625	rooftop air conditioning unit, 3 ton.
Demolish and Install	Ea.	4450	1680		6130	No hoisting is included.
Reinstall	Ea.		1165.51		1165.51	
Clean	Ea.	1.52	47		48.52	
Minimum Charge	Job		284		284	
4 Ton						
Demolish	Ea.		505		505	Includes material and labor to install a
Install	Ea.	5425	1350		6775	rooftop air conditioning unit, 4 ton.
Demolish and Install	Ea.	5425	1855		7280	No hoisting is included.
Reinstall	Ea.		1351.56		1351.56	
Clean	Ea.	1.52	47		48.52	
Minimum Charge	Job		284		284	
5 Ton						
Demolish	Ea.		505		505	Includes material and labor to install a
Install	Ea.	5475	1450		6925	rooftop air conditioning unit, 5 ton.
Demolish and Install	Ea.	5475	1955		7430	No hoisting is included.
Reinstall	Ea.		1459.37		1459.37	
Clean	Ea.	1.52	47		48.52	
Minimum Charge	Job		284		284	

For customer support on your Contractor's Pricing Guide: Residential Repair & Remodeling, call 800.448.8182.

Air Conditioning	Unit	Material	Labor	Equip.	Total	Specification
Compressor						
2 Ton						
Demolish	Ea.		475		475	Includes material and labor to install a
Install	Ea.	330	284		614	refrigeration compressor.
Demolish and Install	Ea.	330	759		1089	
Reinstall	Ea.		227.52		227.52	
Clean	Ea.	1.52	47		48.52	
Minimum Charge	Job		284		284	
2.5 Ton						
Demolish	Ea.		475		475	Includes material and labor to install a
Install	Ea.	390	310		700	refrigeration compressor.
Demolish and Install	Ea.	390	785		1175	
Reinstall	Ea.		246.50		246.50	
Clean	Ea.	1.52	47		48.52	
Minimum Charge	Job		284		284	
3 Ton						
Demolish	Ea.		475		475	Includes material and labor to install a
Install	Ea.	430	320		750	refrigeration compressor.
Demolish and Install	Ea.	430	795		1225	
Reinstall	Ea.		257.23		257.23	
Clean	Ea.	1.52	47		48.52	
Minimum Charge	Job		284		284	
4 Ton						
Demolish	Ea.		475		475	Includes material and labor to install a
Install	Ea.	550	259		809	refrigeration compressor.
Demolish and Install	Ea.	550	734		1284	
Reinstall	Ea.		206.84		206.84	
Clean	Ea.	1.52	47		48.52	
Minimum Charge	Job		284		284	
5 Ton						
Demolish	Ea.		475		475	Includes material and labor to install a
Install	Ea.	605	350		955	refrigeration compressor.
Demolish and Install	Ea.	605	825		1430	
Reinstall	Ea.		281.76		281.76	
Clean	Ea.	1.52	47		48.52	
Minimum Charge	Job		284		284	
Condensing Unit						
2 Ton Air-cooled						
Demolish	Ea.		135		135	Includes material and labor to install a
Install	Ea.	1475	490		1965	2 ton condensing unit.
Demolish and Install	Ea.	1475	625		2100	
Reinstall	Ea.		390.28		390.28	
Clean	Ea.	1.52	47		48.52	
Minimum Charge	Job		284		284	
2.5 Ton Air-cooled						
Demolish	Ea.		158		158	Includes material and labor to install a
Install	Ea.	1600	605		2205	2.5 ton condensing unit.
Demolish and Install	Ea.	1600	763		2363	
Reinstall	Ea.		482.11		482.11	
Clean	Ea.	1.52	47		48.52	
Minimum Charge	Job		284		284	
3 Ton Air-cooled						
Demolish	Ea.		190		190	Includes material and labor to install a
Install	Ea.	1800	790		2590	3 ton condensing unit.
Demolish and Install	Ea.	1800	980		2780	
Reinstall	Ea.		630.45		630.45	
Clean	Ea.	1.52	47		48.52	
Minimum Charge	Job		284		284	

Rough Mechanical

Air Conditioning	Unit	Material	Labor	Equip.	Total	Specification
4 Ton Air-cooled						
Demolish	Ea.		285		285	Includes material and labor to install a
Install	Ea.	2200	1150		3350	4 ton condensing unit.
Demolish and Install	Ea.	2200	1435		3635	
Reinstall	Ea.		910.65		910.65	
Clean	Ea.	1.52	47		48.52	
Minimum Charge	Job		284		284	
5 Ton Air-cooled						
Demolish	Ea.		425		425	Includes material and labor to install a
Install	Ea.	2550	1700		4250	5 ton condensing unit.
Demolish and Install	Ea.	2550	2125		4675	
Reinstall	Ea.		1365.97		1365.97	
Clean	Ea.	1.52	47		48.52	
Minimum Charge	Job		284		284	
Air Handler						
1,200 CFM						
Demolish	Ea.		340		340	Includes material and labor to install a
Install	Ea.	5150	790		5940	1200 CFM capacity air handling unit.
Demolish and Install	Ea.	5150	1130		6280	
Reinstall	Ea.		630.45		630.45	
Clean	Ea.	1.52	47		48.52	
Minimum Charge	Job		284		284	
2,000 CFM						
Demolish	Ea.		340		340	Includes material and labor to install a
Install	Ea.	10900	930		11830	2000 CFM capacity air handling unit.
Demolish and Install	Ea.	10900	1270		12170	
Reinstall	Ea.		745.08		745.08	
Clean	Ea.	1.52	47		48.52	
Minimum Charge	Job		284		284	
3,000 CFM						
Demolish	Ea.		380		380	Includes material and labor to install a
Install	Ea.	11300	1025		12325	3000 CFM capacity air handling unit.
Demolish and Install	Ea.	11300	1405		12705	
Reinstall	Ea.		819.58		819.58	
Clean	Ea.	1.52	47		48.52	
Minimum Charge	Job		284		284	
Evaporative Cooler						
4,300 CFM, 1/3 HP						
Demolish	Ea.		135		135	Includes material and labor to install
Install	Ea.	1175	300		1475	an evaporative cooler.
Demolish and Install	Ea.	1175	435		1610	
Reinstall	Ea.		240.12		240.12	
Clean	Ea.	.61	47		47.61	
Minimum Charge	Job		284		284	
4,700 CFM, 1/2 HP						
Demolish	Ea.		135		135	Includes material and labor to install
Install	Ea.	1125	310		1435	an evaporative cooler.
Demolish and Install	Ea.	1125	445		1570	
Reinstall	Ea.		247.87		247.87	
Clean	Ea.	.61	47		47.61	
Minimum Charge	Job		284		284	
5,600 CFM, 3/4 HP						
Demolish	Ea.		135		135	Includes material and labor to install
Install	Ea.	2275	320		2595	an evaporative cooler.
Demolish and Install	Ea.	2275	455		2730	
Reinstall	Ea.		256.13		256.13	
Clean	Ea.	.61	47		47.61	
Minimum Charge	Job		284		284	

For customer support on your Contractor's Pricing Guide: Residential Repair & Remodeling, call 800.448.8182.

Air Conditioning

Air Conditioning	Unit	Material	Labor	Equip.	Total	Specification
Master Cool Brand						
Demolish	Ea.		135		135	Includes material and labor to install
Install	Ea.	1150	330		1480	an evaporative cooler.
Demolish and Install	Ea.	1150	465		1615	
Reinstall	Ea.		264.96		264.96	
Clean	Ea.	.61	47		47.61	
Minimum Charge	Job		284		284	
Motor						
Demolish	Ea.		76		76	Includes material and labor to install
Install	Ea.	276	233		509	an evaporative cooler motor.
Demolish and Install	Ea.	276	309		585	
Reinstall	Ea.		186.74		186.74	
Clean	Ea.	.61	47		47.61	
Minimum Charge	Job		284		284	
Water Pump						
Demolish	Ea.		76		76	Includes material and labor to install
Install	Ea.	72	133		205	an evaporative cooler water pump.
Demolish and Install	Ea.	72	209		281	
Reinstall	Ea.		106.72		106.72	
Clean	Ea.	.61	47		47.61	
Minimum Charge	Job		284		284	
Fan Belt						
Demolish	Ea.		15.20		15.20	Includes material and labor to install
Install	Ea.	51.50	26.50		78	an evaporative cooler fan belt.
Demolish and Install	Ea.	51.50	41.70		93.20	
Reinstall	Ea.		21.34		21.34	
Minimum Charge	Job		284		284	
Pads (4 ea.)						
Demolish	Ea.		15.20		15.20	Includes material and labor to install
Install	Ea.	26	66.50		92.50	pre-fabricated roof pads for an
Demolish and Install	Ea.	26	81.70		107.70	evaporative cooler.
Reinstall	Ea.		53.36		53.36	
Clean	Ea.	.61	47		47.61	
Minimum Charge	Job		284		284	
Service call						
Install	Job		267		267	Includes 4 hour minimum labor charge for a sheetmetal worker.
Window Mounted						
Low Output BTU						
Install	Ea.	340	60.50		400.50	Includes material and labor to install a
Reinstall	Ea.		48.40		48.40	window air conditioning unit.
Clean	Ea.	.23	11.70		11.93	
Minimum Charge	Job		284		284	
Mid-range Output BTU						
Demolish	Ea.		42.50		42.50	Includes material and labor to install a
Install	Ea.	430	80.50		510.50	window air conditioning unit.
Demolish and Install	Ea.	430	123		553	
Reinstall	Ea.		64.53		64.53	
Clean	Ea.	.23	11.70		11.93	
Minimum Charge	Job		284		284	
High Output BTU						
Demolish	Ea.		54		54	Includes material and labor to install a
Install	Ea.	515	80.50		595.50	window air conditioning unit.
Demolish and Install	Ea.	515	134.50		649.50	
Reinstall	Ea.		64.53		64.53	
Clean	Ea.	.23	11.70		11.93	
Minimum Charge	Job		284		284	

Rough Mechanical

Air Conditioning

Air Conditioning	Unit	Material	Labor	Equip.	Total	Specification
Hotel-type Hot / Cold						
Demolish	Ea.		81.50		81.50	Includes material and labor to install
Install	Ea.	3275	340		3615	packaged air conditioning unit,
Demolish and Install	Ea.	3275	421.50		3696.50	15000 BTU cooling, 13900 BTU
Reinstall	Ea.		273.19		273.19	heating.
Clean	Ea.	.31	23.50		23.81	
Minimum Charge	Job		284		284	
Thermostat						
Demolish	Ea.		15.80		15.80	Includes material and labor to install
Install	Ea.	276	71		347	an electric thermostat, 2 set back.
Demolish and Install	Ea.	276	86.80		362.80	
Reinstall	Ea.		56.88		56.88	
Clean	Ea.	.03	11.70		11.73	
Minimum Charge	Job		284		284	
Concrete Pad						
Demolish	S.F.		2.60		2.60	Includes material and labor to install
Install	S.F.	4	5.95		9.95	air conditioner pad.
Demolish and Install	S.F.	4	8.55		12.55	
Reinstall	S.F.		4.77		4.77	
Clean	S.F.	.03	.75		.78	
Paint	S.F.	.17	.41		.58	
Minimum Charge	Job		284		284	
A/C Junction Box						
Install	Ea.	174	162		336	Includes labor and materials to install
Minimum Charge	Job		284		284	an air conditioning circuit including up to 40 feet of non-metallic cable.

Sprinkler System

Sprinkler System	Unit	Material	Labor	Equip.	Total	Specification
Exposed						
Wet Type						
Demolish	SF Flr.		.24		.24	Includes material and labor to install
Install	SF Flr.	3.26	2.40		5.66	an exposed wet pipe fire sprinkler
Demolish and Install	SF Flr.	3.26	2.64		5.90	system.
Reinstall	SF Flr.		2.40		2.40	
Minimum Charge	Job		273		273	
Dry Type						
Demolish	SF Flr.		.24		.24	Includes material and labor to install
Install	SF Flr.	3.74	2.40		6.14	an exposed dry pipe fire sprinkler
Demolish and Install	SF Flr.	3.74	2.64		6.38	system.
Reinstall	SF Flr.		2.40		2.40	
Minimum Charge	Job		273		273	
Concealed						
Wet Type						
Demolish	SF Flr.		.49		.49	Includes material and labor to install a
Install	SF Flr.	3.26	3.28		6.54	concealed wet pipe fire sprinkler
Demolish and Install	SF Flr.	3.26	3.77		7.03	system.
Reinstall	SF Flr.		3.28		3.28	
Minimum Charge	Job		273		273	
Dry Type						
Demolish	SF Flr.		.49		.49	Includes material and labor to install a
Install	SF Flr.	3.74	3.28		7.02	concealed dry pipe fire sprinkler
Demolish and Install	SF Flr.	3.74	3.77		7.51	system.
Reinstall	SF Flr.		3.28		3.28	
Minimum Charge	Job		273		273	

| Junction Boxes | Cable | Switch Box |

Electrical Per S.F.	Unit	Material	Labor	Equip.	Total	Specification
Residential						
Wiring, Lighting and Fixtures						Includes labor and material to replace house wiring, lighting fixtures, and built-in appliances. Apply S.F. cost to floor area including garages but not basements.
Install	S.F.	1.62	2.37		3.99	
Minimum Charge	Job		284		284	
Wiring Only						Includes labor and material to replace house wiring only. Apply S.F. cost to floor area including garages but not basements.
Install	S.F.	.19	.93		1.12	

Main Service	Unit	Material	Labor	Equip.	Total	Specification
Complete Panel						
100 Amp.						Cost includes material and labor to install 100 Amp. residential service panel including 12 breaker interior panel, 10 single pole breakers and hook-up. Wiring is not included.
Demolish	Ea.		234		234	
Install	Ea.	455	480		935	
Demolish and Install	Ea.	455	714		1169	
Reinstall	Ea.		382.12		382.12	
Clean	Ea.	.15	93.50		93.65	
Minimum Charge	Job		284		284	
150 Amp.						Cost includes material and labor to install 150 Amp. residential service, incl. 20 breaker exterior panel, main switch, 1 GFI and 14 single pole breakers, meter socket including hookup. Wiring is not included.
Demolish	Ea.		234		234	
Install	Ea.	645	550		1195	
Demolish and Install	Ea.	645	784		1429	
Reinstall	Ea.		441.48		441.48	
Clean	Ea.	.15	93.50		93.65	
Minimum Charge	Job		284		284	
200 Amp.						Cost includes material and labor to install 200 Amp. residential service, incl. 20 breaker exterior panel, main switch, 1 GFI and 18 single pole breakers, meter socket including hookup. Wiring is not included.
Demolish	Ea.		234		234	
Install	Ea.	1375	630		2005	
Demolish and Install	Ea.	1375	864		2239	
Reinstall	Ea.		505.24		505.24	
Clean	Ea.	.15	93.50		93.65	
Minimum Charge	Job		284		284	

For customer support on your Contractor's Pricing Guide: Residential Repair & Remodeling, call 800.448.8182.

Main Service	Unit	Material	Labor	Equip.	Total	Specification
Masthead						
Demolish	Ea.		39.50		39.50	Includes material and labor to install
Install	Ea.	50	237		287	galvanized steel conduit including
Demolish and Install	Ea.	50	276.50		326.50	masthead.
Reinstall	Ea.		189.47		189.47	
Minimum Charge	Job		284		284	
Allowance to Repair						
Install	Job		284		284	Minimum labor charge for residential
Minimum Charge	Job		284		284	wiring.

For customer support on your Contractor's Pricing Guide: Residential Repair & Remodeling, call 800.448.8182.

Insulation

Wall Insulation

Floor Insulation		Unit	Material	Labor	Equip.	Total	Specification
Kraft Faced Batt							
3-1/2″ (R-13)							
	Demolish	S.F.		.17		.17	Cost includes material and labor to
	Install	S.F.	.80	.69		1.49	install 3-1/2″ (R-13) kraft paper or foil
	Demolish and Install	S.F.	.80	.86		1.66	faced roll and batt fiberglass
	Minimum Charge	Job		242		242	insulation, joists 16″ O.C.
6″ (R-19)							
	Demolish	S.F.		.17		.17	Cost includes material and labor to
	Install	S.F.	1.09	.81		1.90	install 6″ (R-19) kraft paper or foil
	Demolish and Install	S.F.	1.09	.98		2.07	faced roll and batt fiberglass
	Minimum Charge	Job		242		242	insulation, joists 16″ O.C.
10″ (R-30)							
	Demolish	S.F.		.17		.17	Cost includes material and labor to
	Install	S.F.	1.25	.97		2.22	install 10″ (R-30) kraft paper-faced roll
	Demolish and Install	S.F.	1.25	1.14		2.39	and batt fiberglass insulation, joists
	Minimum Charge	Job		242		242	16″ O.C.
12″ (R-38)							
	Demolish	S.F.		.17		.17	Cost includes material and labor to
	Install	S.F.	1.58	1.02		2.60	install 12″ (R-38) kraft paper-faced roll
	Demolish and Install	S.F.	1.58	1.19		2.77	and batt fiberglass insulation, joists
	Minimum Charge	Job		242		242	16″ O.C.
Foil Faced Batt							
3-1/2″ (R-13)							
	Demolish	S.F.		.17		.17	Cost includes material and labor to
	Install	S.F.	.80	.69		1.49	install 3-1/2″ (R-13) kraft paper or foil
	Demolish and Install	S.F.	.80	.86		1.66	faced roll and batt fiberglass
	Minimum Charge	Job		242		242	insulation, joists 16″ O.C.
6″ (R-19)							
	Demolish	S.F.		.17		.17	Cost includes material and labor to
	Install	S.F.	1.09	.81		1.90	install 6″ (R-19) kraft paper or foil
	Demolish and Install	S.F.	1.09	.98		2.07	faced roll and batt fiberglass
	Minimum Charge	Job		242		242	insulation, joists 16″ O.C.
Unfaced Batt							
3-1/2″ (R-13)							
	Demolish	S.F.		.17		.17	Cost includes material and labor to
	Install	S.F.	.59	.81		1.40	install 3-1/2″ (R-13) unfaced roll and
	Demolish and Install	S.F.	.59	.98		1.57	batt fiberglass insulation, joists 16″
	Minimum Charge	Job		242		242	O.C.

For customer support on your Contractor's Pricing Guide: Residential Repair & Remodeling, call 800.448.8182.

Insulation

Floor Insulation

Floor Insulation		Unit	Material	Labor	Equip.	Total	Specification
6″ (R-19)							
	Demolish	S.F.		.17		.17	Cost includes material and labor to
	Install	S.F.	.86	.97		1.83	install 6″ (R-19) unfaced roll and batt
	Demolish and Install	S.F.	.86	1.14		2	fiberglass insulation, joists 16″ O.C.
	Minimum Charge	Job		242		242	
10″ (R-30)							
	Demolish	S.F.		.17		.17	Cost includes material and labor to
	Install	S.F.	1.24	1.08		2.32	install 10″ (R-30) unfaced roll and batt
	Demolish and Install	S.F.	1.24	1.25		2.49	fiberglass insulation, joists 16″ O.C.
	Minimum Charge	Job		242		242	
12″ (R-38)							
	Demolish	S.F.		.17		.17	Cost includes material and labor to
	Install	S.F.	1.83	1.14		2.97	install 12″ (R-38) unfaced roll and batt
	Demolish and Install	S.F.	1.83	1.31		3.14	fiberglass insulation, joists 16″ O.C.
	Minimum Charge	Job		242		242	

Rigid Board

Rigid Board		Unit	Material	Labor	Equip.	Total	Specification
1/2″							
	Demolish	S.F.		.17		.17	Includes material and labor to install
	Install	S.F.	.41	.61		1.02	rigid foam insulation board foil faced
	Demolish and Install	S.F.	.41	.78		1.19	two sides.
	Minimum Charge	Job		242		242	
3/4″							
	Demolish	S.F.		.17		.17	Includes material and labor to install
	Install	S.F.	.62	.61		1.23	rigid foam insulation board foil faced
	Demolish and Install	S.F.	.62	.78		1.40	two sides.
	Minimum Charge	Job		242		242	
1-1/2″							
	Demolish	S.F.		.17		.17	Includes material and labor to install
	Install	S.F.	.91	.65		1.56	rigid foam insulation board foil faced
	Demolish and Install	S.F.	.91	.82		1.73	two sides.
	Minimum Charge	Job		242		242	

Support Wire

Support Wire		Unit	Material	Labor	Equip.	Total	Specification
	Demolish	S.F.		.07		.07	Includes material and labor to install
	Install	S.F.	.03	.13		.16	friction fit support wire.
	Demolish and Install	S.F.	.03	.20		.23	
	Minimum Charge	Job		242		242	

Vapor Barrier

Vapor Barrier		Unit	Material	Labor	Equip.	Total	Specification
	Demolish	S.F.		.09		.09	Cost includes material and labor to
	Install	S.F.	.04	.13		.17	install 4 mil polyethylene vapor barrier
	Demolish and Install	S.F.	.04	.22		.26	10′ wide sheets with 6″ overlaps.
	Minimum Charge	Job		242		242	

Wall Insulation

Wall Insulation		Unit	Material	Labor	Equip.	Total	Specification
Kraft Faced Batt							
3-1/2″ (R-13)							
	Demolish	S.F.		.17		.17	Cost includes material and labor to
	Install	S.F.	.66	.36		1.02	install 3-1/2″ (R-13) kraft paper-faced
	Demolish and Install	S.F.	.66	.53		1.19	roll and batt fiberglass insulation, joist
	Minimum Charge	Job		242		242	or studs 16″ O.C.
6″ (R-19)							
	Demolish	S.F.		.17		.17	Cost includes material and labor to
	Install	S.F.	.95	.36		1.31	install 6″ (R-19) kraft paper-faced roll
	Demolish and Install	S.F.	.95	.53		1.48	and batt fiberglass insulation, joist or
	Minimum Charge	Job		242		242	studs 16″ O.C.

For customer support on your Contractor's Pricing Guide: Residential Repair & Remodeling, call 800.448.8182.

Insulation

Wall Insulation		Unit	Material	Labor	Equip.	Total	Specification
10″ (R-30)							
	Demolish	S.F.		.17		.17	Cost includes material and labor to
	Install	S.F.	1.25	.42		1.67	install 10″ (R-30) kraft paper-faced roll
	Demolish and Install	S.F.	1.25	.59		1.84	and batt fiberglass insulation, joist or
	Minimum Charge	Job		242		242	studs 16″ O.C.
12″ (R-38)							
	Demolish	S.F.		.17		.17	Cost includes material and labor to
	Install	S.F.	1.58	.42		2	install 12″ (R-38) kraft paper-faced roll
	Demolish and Install	S.F.	1.58	.59		2.17	and batt fiberglass insulation, joist or
	Minimum Charge	Job		242		242	studs 16″ O.C.
Foil Faced Batt							
3-1/2″ (R-13)							
	Demolish	S.F.		.17		.17	Cost includes material and labor to
	Install	S.F.	.77	.36		1.13	install 3-1/2″ (R-13) foil faced roll and
	Demolish and Install	S.F.	.77	.53		1.30	batt fiberglass insulation, joist or studs
	Minimum Charge	Job		242		242	16″ O.C.
6″ (R-19)							
	Demolish	S.F.		.17		.17	Cost includes material and labor to
	Install	S.F.	1.07	.36		1.43	install 6″ (R-19) foil faced roll and batt
	Demolish and Install	S.F.	1.07	.53		1.60	fiberglass insulation, joist or studs 16″
	Minimum Charge	Job		242		242	O.C.
Unfaced Batt							
3-1/2″ (R-13)							
	Demolish	S.F.		.17		.17	Cost includes material and labor to
	Install	S.F.	.59	.36		.95	install 3-1/2″ (R-13) unfaced roll and
	Demolish and Install	S.F.	.59	.53		1.12	batt fiberglass insulation, joist or studs
	Minimum Charge	Job		242		242	16″ O.C.
6″ (R-19)							
	Demolish	S.F.		.17		.17	Cost includes material and labor to
	Install	S.F.	.86	.42		1.28	install 6″ (R-19) unfaced roll and batt
	Demolish and Install	S.F.	.86	.59		1.45	fiberglass insulation, joist or studs 16″
	Minimum Charge	Job		242		242	O.C.
10″ (R-30)							
	Demolish	S.F.		.17		.17	Cost includes material and labor to
	Install	S.F.	1.24	.42		1.66	install 10″ (R-30) unfaced roll and batt
	Demolish and Install	S.F.	1.24	.59		1.83	fiberglass insulation, joist or studs 16″
	Minimum Charge	Job		242		242	O.C.
12″ (R-38)							
	Demolish	S.F.		.17		.17	Cost includes material and labor to
	Install	S.F.	1.83	.48		2.31	install 12″ (R-38) unfaced roll and batt
	Demolish and Install	S.F.	1.83	.65		2.48	fiberglass insulation, joist or studs 16″
	Minimum Charge	Job		242		242	O.C.
Blown							
R-13							
	Demolish	S.F.		.17		.17	Includes material and labor to install
	Install	S.F.	.36	1.66	.87	2.89	fiberglass blown-in insulation in wall
	Demolish and Install	S.F.	.36	1.83	.87	3.06	with wood siding.
	Minimum Charge	Job		575	300	875	
R-19							
	Demolish	S.F.		.17		.17	Includes material and labor to install
	Install	S.F.	.57	1.91	1	3.48	fiberglass blown-in insulation in wall
	Demolish and Install	S.F.	.57	2.08	1	3.65	with wood siding.
	Minimum Charge	Job		575	300	875	

For customer support on your Contractor's Pricing Guide: Residential Repair & Remodeling, call 800.448.8182.

Insulation

Wall Insulation

	Unit	Material	Labor	Equip.	Total	Specification
Rigid Board						
1/2″						
Demolish	S.F.		.21		.21	Includes material and labor to install
Install	S.F.	.41	.61		1.02	rigid foam insulation board, foil faced,
Demolish and Install	S.F.	.41	.82		1.23	two sides.
Minimum Charge	Job		242		242	
3/4″						
Demolish	S.F.		.21		.21	Includes material and labor to install
Install	S.F.	.62	.61		1.23	rigid foam insulation board, foil faced,
Demolish and Install	S.F.	.62	.82		1.44	two sides.
Minimum Charge	Job		242		242	
1-1/2″						
Demolish	S.F.		.21		.21	Includes material and labor to install
Install	S.F.	.91	.66		1.57	rigid foam insulation board, foil faced,
Demolish and Install	S.F.	.91	.87		1.78	two sides.
Minimum Charge	Job		242		242	
Vapor Barrier						
Demolish	S.F.		.06		.06	Cost includes material and labor to
Install	S.F.	.04	.13		.17	install 4 mil polyethylene vapor barrier
Demolish and Install	S.F.	.04	.19		.23	10′ wide sheets with 6″ overlaps.
Minimum Charge	Job		242		242	

Ceiling Insulation

	Unit	Material	Labor	Equip.	Total	Specification
Kraft Faced Batt						
3-1/2″ (R-13)						
Demolish	S.F.		.17		.17	Cost includes material and labor to
Install	S.F.	.66	.36		1.02	install 3-1/2″ (R-13) kraft paper-faced
Demolish and Install	S.F.	.66	.53		1.19	roll and batt fiberglass insulation, joist
Minimum Charge	Job		242		242	or studs 16″ O.C.
6″ (R-19)						
Demolish	S.F.		.17		.17	Cost includes material and labor to
Install	S.F.	.95	.36		1.31	install 6″ (R-19) kraft paper-faced roll
Demolish and Install	S.F.	.95	.53		1.48	and batt fiberglass insulation, joist or
Minimum Charge	Job		242		242	studs 16″ O.C.
10″ (R-30)						
Demolish	S.F.		.17		.17	Cost includes material and labor to
Install	S.F.	1.25	.42		1.67	install 10″ (R-30) kraft paper-faced roll
Demolish and Install	S.F.	1.25	.59		1.84	and batt fiberglass insulation, joist or
Minimum Charge	Job		242		242	studs 16″ O.C.
12″ (R-38)						
Demolish	S.F.		.17		.17	Cost includes material and labor to
Install	S.F.	1.58	.42		2	install 12″ (R-38) kraft paper-faced roll
Demolish and Install	S.F.	1.58	.59		2.17	and batt fiberglass insulation, joist or
Minimum Charge	Job		242		242	studs 16″ O.C.
Foil Faced Batt						
3-1/2″ (R-13)						
Demolish	S.F.		.17		.17	Cost includes material and labor to
Install	S.F.	.77	.36		1.13	install 3-1/2″ (R-13) foil faced roll and
Demolish and Install	S.F.	.77	.53		1.30	batt fiberglass insulation, joist or studs
Minimum Charge	Job		242		242	16″ O.C.

For customer support on your Contractor's Pricing Guide: Residential Repair & Remodeling, call 800.448.8182.

Insulation

Ceiling Insulation		Unit	Material	Labor	Equip.	Total	Specification
6″ (R-19)							
	Demolish	S.F.		.17		.17	Cost includes material and labor to
	Install	S.F.	1.07	.36		1.43	install 6″ (R-19) foil faced roll and batt
	Demolish and Install	S.F.	1.07	.53		1.60	fiberglass insulation, joist or studs 16″
	Minimum Charge	Job		242		242	O.C.
Unfaced Batt							
3-1/2″ (R-13)							
	Demolish	S.F.		.17		.17	Cost includes material and labor to
	Install	S.F.	.59	.36		.95	install 3-1/2″ (R-13) unfaced roll and
	Demolish and Install	S.F.	.59	.53		1.12	batt fiberglass insulation, joist or studs
	Minimum Charge	Job		242		242	16″ O.C.
6″ (R-19)							
	Demolish	S.F.		.17		.17	Cost includes material and labor to
	Install	S.F.	.86	.42		1.28	install 6″ (R-19) unfaced roll and batt
	Demolish and Install	S.F.	.86	.59		1.45	fiberglass insulation, joist or studs 16″
	Minimum Charge	Job		242		242	O.C.
10″ (R-30)							
	Demolish	S.F.		.17		.17	Cost includes material and labor to
	Install	S.F.	1.24	.42		1.66	install 10″ (R-30) unfaced roll and batt
	Demolish and Install	S.F.	1.24	.59		1.83	fiberglass insulation, joist or studs 16″
	Minimum Charge	Job		242		242	O.C.
12″ (R-38)							
	Demolish	S.F.		.17		.17	Cost includes material and labor to
	Install	S.F.	1.83	.48		2.31	install 12″ (R-38) unfaced roll and batt
	Demolish and Install	S.F.	1.83	.65		2.48	fiberglass insulation, joist or studs 16″
	Minimum Charge	Job		242		242	O.C.
Blown							
5″							
	Demolish	S.F.		.52		.52	Includes material and labor to install
	Install	S.F.	.26	.30	.16	.72	fiberglass blown-in insulation.
	Demolish and Install	S.F.	.26	.82	.16	1.24	
	Minimum Charge	Job		575	300	875	
6″							
	Demolish	S.F.		.52		.52	Includes material and labor to install
	Install	S.F.	.39	.38	.20	.97	fiberglass blown-in insulation.
	Demolish and Install	S.F.	.39	.90	.20	1.49	
	Minimum Charge	Job		575	300	875	
9″							
	Demolish	S.F.		.52		.52	Includes material and labor to install
	Install	S.F.	.47	.52	.27	1.26	fiberglass blown-in insulation.
	Demolish and Install	S.F.	.47	1.04	.27	1.78	
	Minimum Charge	Job		575	300	875	
12″							
	Demolish	S.F.		.52		.52	Includes material and labor to install
	Install	S.F.	.66	.77	.40	1.83	fiberglass blown-in insulation.
	Demolish and Install	S.F.	.66	1.29	.40	2.35	
	Minimum Charge	Job		575	300	875	
Rigid Board							
1/2″							
	Demolish	S.F.		.29		.29	Includes material and labor to install
	Install	S.F.	.41	.61		1.02	rigid foam insulation board, foil faced,
	Demolish and Install	S.F.	.41	.90		1.31	two sides.
	Minimum Charge	Job		242		242	

For customer support on your Contractor's Pricing Guide: Residential Repair & Remodeling, call 800.448.8182.

Insulation

Ceiling Insulation

	Unit	Material	Labor	Equip.	Total	Specification
3/4″						
Demolish	S.F.		.29		.29	Includes material and labor to install
Install	S.F.	.62	.61		1.23	rigid foam insulation board, foil faced,
Demolish and Install	S.F.	.62	.90		1.52	two sides.
Minimum Charge	Job		242		242	
1-1/2″						
Demolish	S.F.		.29		.29	Includes material and labor to install
Install	S.F.	.91	.66		1.57	rigid foam insulation board, foil faced,
Demolish and Install	S.F.	.91	.95		1.86	two sides.
Minimum Charge	Job		242		242	

Exterior Insulation

	Unit	Material	Labor	Equip.	Total	Specification
Rigid Board						
1/2″						
Demolish	S.F.		.16	•	.16	Includes material and labor to install
Install	S.F.	.41	.61		1.02	rigid foam insulation board foil faced
Demolish and Install	S.F.	.41	.77		1.18	two sides.
Minimum Charge	Job		242		242	
3/4″						
Demolish	S.F.		.16		.16	Includes material and labor to install
Install	S.F.	.62	.61		1.23	rigid foam insulation board foil faced
Demolish and Install	S.F.	.62	.77		1.39	two sides.
Minimum Charge	Job		242		242	
1-1/2″						
Demolish	S.F.		.16		.16	Includes material and labor to install
Install	S.F.	.91	.65		1.56	rigid foam insulation board foil faced
Demolish and Install	S.F.	.91	.81		1.72	two sides.
Minimum Charge	Job		242		242	
Vapor Barrier						
Demolish	S.F.		.09		.09	Cost includes material and labor to
Install	S.F.	.04	.13		.17	install 4 mil polyethylene vapor barrier
Demolish and Install	S.F.	.04	.22		.26	10′ wide sheets with 6″ overlaps.
Minimum Charge	Job		242		242	

Roof Insulation Board

	Unit	Material	Labor	Equip.	Total	Specification
Fiberglass						
3/4″ Fiberglass						
Demolish	Sq.		54		54	Includes material and labor to install
Install	Sq.	70.50	27.50		98	fiberglass board insulation 3/4″ thick,
Demolish and Install	Sq.	70.50	81.50		152	R-2.80 and C-0.36 values.
Minimum Charge	Job		242		242	
1″ Fiberglass						
Demolish	Sq.		54		54	Includes material and labor to install
Install	Sq.	103	31		134	fiberglass board insulation 1″ thick,
Demolish and Install	Sq.	103	85		188	R-4.20 and C-0.24 values.
Minimum Charge	Job		242		242	
1-3/8″ Fiberglass						
Demolish	Sq.		54		54	Includes material and labor to install
Install	Sq.	168	34		202	fiberglass board insulation 1-3/8″
Demolish and Install	Sq.	168	88		256	thick, R-5.30 and C-0.19 values.
Minimum Charge	Job		242		242	

For customer support on your Contractor's Pricing Guide: Residential Repair & Remodeling, call 800.448.8182.

Insulation

Roof Insulation Board		Unit	Material	Labor	Equip.	Total	Specification
1-5/8″ Fiberglass							Includes material and labor to install
	Demolish	Sq.		54		54	fiberglass board insulation 1-5/8″
	Install	Sq.	169	35.50		204.50	thick, R-6.70 and C-0.15 values.
	Demolish and Install	Sq.	169	89.50		258.50	
	Minimum Charge	Job		242		242	
2-1/4″ Fiberglass							Includes material and labor to install
	Demolish	Sq.		54		54	fiberglass board insulation 2-1/4″
	Install	Sq.	202	39		241	thick, R-8.30 and C-0.12 values.
	Demolish and Install	Sq.	202	93		295	
	Minimum Charge	Job		242		242	
Perlite							
1″ Perlite							Includes material and labor to install
	Demolish	Sq.		54		54	perlite board insulation 1″ thick,
	Install	Sq.	68	31		99	R-2.80 and C-0.36 values.
	Demolish and Install	Sq.	68	85		153	
	Minimum Charge	Job		242		242	
4″ Perlite							Includes material and labor to install
	Demolish	Sq.		54		54	perlite board insulation 4″ thick,
	Install	Sq.	273	59.50		332.50	R-10.0 and C-0.10 values.
	Demolish and Install	Sq.	273	113.50		386.50	
	Minimum Charge	Job		242		242	
1-1/2″ Perlite							Includes material and labor to install
	Demolish	Sq.		54		54	perlite board insulation 1-1/2″ thick,
	Install	Sq.	97	34		131	R-4.20 and C-0.24 values.
	Demolish and Install	Sq.	97	88		185	
	Minimum Charge	Job		242		242	
2″ Perlite							Includes material and labor to install
	Demolish	Sq.		54		54	perlite board insulation 2″ thick,
	Install	Sq.	121	41.50		162.50	R-5.30 and C-0.19 values.
	Demolish and Install	Sq.	121	95.50		216.50	
	Minimum Charge	Job		242		242	
2-1/2″ Perlite							Includes material and labor to install
	Demolish	Sq.		54		54	perlite board insulation 2-1/2″ thick,
	Install	Sq.	155	46.50		201.50	R-6.70 and C-0.15 values.
	Demolish and Install	Sq.	155	100.50		255.50	
	Minimum Charge	Job		242		242	
3″ Perlite							Includes material and labor to install
	Demolish	Sq.		54		54	perlite board insulation 3″ thick,
	Install	Sq.	205	49.50		254.50	R-8.30 and C-0.12 values.
	Demolish and Install	Sq.	205	103.50		308.50	
	Minimum Charge	Job		242		242	
Polystyrene							
2″ Polystyrene							Includes material and labor to install
	Demolish	Sq.		54		54	extruded polystyrene board, R5.0/ 2″
	Install	Sq.	61.50	39		100.50	in thickness @25PSI compressive
	Demolish and Install	Sq.	61.50	93		154.50	strength.
	Minimum Charge	Job		242		242	
1″ Polystyrene							Includes material and labor to install
	Demolish	Sq.		54		54	extruded polystyrene board, R5.0/ 1″
	Install	Sq.	31	26.50		57.50	in thickness @25PSI compressive
	Demolish and Install	Sq.	31	80.50		111.50	strength.
	Minimum Charge	Job		242		242	
1-1/2″ Polystyrene							Includes material and labor to install
	Demolish	Sq.		54		54	extruded polystyrene board, R5.0/
	Install	Sq.	46	31		77	1-1/2″ in thickness @25PSI
	Demolish and Install	Sq.	46	85		131	compressive strength.
	Minimum Charge	Job		242		242	

For customer support on your Contractor's Pricing Guide: Residential Repair & Remodeling, call 800.448.8182.

Insulation

Roof Insulation Board	Unit	Material	Labor	Equip.	Total	Specification
Urethane						
3/4" Urethane						
Demolish	Sq.		54		54	Includes material and labor to install
Install	Sq.	54.50	23.50		78	urethane board insulation 3/4" thick,
Demolish and Install	Sq.	54.50	77.50		132	R-5.30 and C-0.19 values.
Minimum Charge	Job		242		242	
1" Urethane						
Demolish	Sq.		54		54	Includes material and labor to install
Install	Sq.	92.50	25		117.50	urethane board insulation 1" thick,
Demolish and Install	Sq.	92.50	79		171.50	R-6.70 and C-0.15 values.
Minimum Charge	Job		242		242	
1-1/2" Urethane						
Demolish	Sq.		54		54	Includes material and labor to install
Install	Sq.	82	28.50		110.50	urethane board insulation 1-1/2"
Demolish and Install	Sq.	82	82.50		164.50	thick, R-10.1 and C-0.09 values.
Minimum Charge	Job		242		242	
2" Urethane						
Demolish	Sq.		54		54	Includes material and labor to install
Install	Sq.	99	34		133	urethane board insulation 2" thick,
Demolish and Install	Sq.	99	88		187	R-13.4 and C-0.06 values.
Minimum Charge	Job		242		242	

For customer support on your Contractor's Pricing Guide: Residential Repair & Remodeling, call 800.448.8182.

Gypsum Wallboard

Tile Ceiling

Gypsum Wallboard Ceiling

Wall Plaster	Unit	Material	Labor	Equip.	Total	Specification
Interior						
Cement						
Demolish	S.Y.		6.95		6.95	Includes material and labor to install
Install	S.Y.	21	30	1.68	52.68	cement plaster on interior wood
Demolish and Install	S.Y.	21	36.95	1.68	59.63	framing with metal lath.
Clean	S.Y.		3.40		3.40	
Paint	S.Y.	1.93	5.40		7.33	
Minimum Charge	Job		234		234	
Acoustic						
Demolish	S.Y.		6.95		6.95	Includes material and labor to install
Install	S.Y.	13.15	36.50	1.68	51.33	acoustical cement plaster on interior
Demolish and Install	S.Y.	13.15	43.45	1.68	58.28	wood framing with metal lath.
Clean	S.Y.		3.40		3.40	
Paint	S.Y.	1.93	5.40		7.33	
Minimum Charge	Job		234		234	
Gypsum						
Demolish	S.Y.		6.95		6.95	Includes material and labor to install
Install	S.Y.	11	31	1.43	43.43	gypsum wall or ceiling plaster on
Demolish and Install	S.Y.	11	37.95	1.43	50.38	interior wood framing with metal lath.
Clean	S.Y.		3.40		3.40	
Paint	S.Y.	1.93	5.40		7.33	
Minimum Charge	Job		234		234	
Stucco						
Demolish	S.Y.		16.60		16.60	Includes material and labor to install
Install	S.Y.	9.80	43	1.97	54.77	stucco on wire mesh over wood
Demolish and Install	S.Y.	9.80	59.60	1.97	71.37	framing.
Clean	S.Y.		3.40		3.40	
Paint	S.Y.	1.93	5.40		7.33	
Minimum Charge	Job		234		234	
Thin Coat						
Veneer						
Demolish	S.Y.		6.95		6.95	Includes materials and labor to install
Install	S.Y.	1.09	5.55	.31	6.95	thin coat veneer plaster.
Demolish and Install	S.Y.	1.09	12.50	.31	13.90	
Clean	S.Y.		3.40		3.40	
Paint	S.Y.	1.93	5.40		7.33	
Minimum Charge	Job		234		234	

For customer support on your Contractor's Pricing Guide: Residential Repair & Remodeling, call 800.448.8182.

Walls / Ceilings

Wall Plaster

Wall Plaster		Unit	Material	Labor	Equip.	Total	Specification
Lath							
Metal							
	Demolish	S.Y.		1.63		1.63	Includes material and labor to install
	Install	S.Y.	4.63	6.40		11.03	wire lath on steel framed walls.
	Demolish and Install	S.Y.	4.63	8.03		12.66	
	Clean	S.Y.		3.40		3.40	
	Minimum Charge	Job		113		113	
Gypsum							
	Demolish	S.Y.		1.63		1.63	Includes material and labor to install
	Install	S.Y.	3.27	6		9.27	gypsum lath on walls.
	Demolish and Install	S.Y.	3.27	7.63		10.90	
	Clean	S.Y.		3.40		3.40	
	Minimum Charge	Job		234		234	
Repairs							
Fill in Small Cracks							
	Install	L.F.	.09	3.75		3.84	Includes labor and material to fill in
	Minimum Charge	Job		234		234	hairline cracks up to 1/8" with filler,
							sand and prep for paint.
Repair up to 1" Stock							
	Install	L.F.	.34	4.68		5.02	Includes labor and material to fill in
	Minimum Charge	Job		234		234	hairline cracks up to 1" with filler,
							sand and prep for paint.
Patch Small Hole (2")							
	Install	Ea.	.09	4.93		5.02	Includes labor and material to patch
	Minimum Charge	Job		234		234	up to a 2" diameter hole in plaster.
Patch Large Hole							
	Install	Ea.	10.40	156		166.40	Includes labor and material to patch
	Minimum Charge	Job		234		234	up to a 20" diameter hole in plaster.
Patch Plaster Section							
	Install	Ea.	2.92	4.68		7.60	Includes labor and material to repair
	Minimum Charge	Job		234		234	section of plaster wall, area less than
							20 S.F.
Repair Inside Corner							
	Install	L.F.	4.15	11.70		15.85	Includes material and labor to patch
	Minimum Charge	Job		234		234	plaster corners.

Ceiling Plaster

Ceiling Plaster		Unit	Material	Labor	Equip.	Total	Specification
Interior							
Cement							
	Demolish	S.Y.		6.95		6.95	Includes material and labor to install
	Install	S.Y.	21	30	1.68	52.68	cement plaster on interior wood
	Demolish and Install	S.Y.	21	36.95	1.68	59.63	framing with metal lath.
	Clean	S.Y.		3.56		3.56	
	Paint	S.Y.	1.93	5.40		7.33	
	Minimum Charge	Job		234		234	
Acoustic							
	Demolish	S.Y.		6.95		6.95	Includes material and labor to install
	Install	S.Y.	13.15	36.50	1.68	51.33	acoustical cement plaster on interior
	Demolish and Install	S.Y.	13.15	43.45	1.68	58.28	wood framing with metal lath.
	Clean	S.Y.		3.56		3.56	
	Paint	S.Y.	1.93	5.40		7.33	
	Minimum Charge	Job		234		234	

For customer support on your Contractor's Pricing Guide: Residential Repair & Remodeling, call 800.448.8182.

Ceiling Plaster		Unit	Material	Labor	Equip.	Total	Specification
Gypsum							
	Demolish	S.Y.		6.95		6.95	Includes material and labor to install
	Install	S.Y.	11	31	1.43	43.43	gypsum wall or ceiling plaster on
	Demolish and Install	S.Y.	11	37.95	1.43	50.38	interior wood framing with metal lath.
	Clean	S.Y.		3.56		3.56	
	Paint	S.Y.	1.93	5.40		7.33	
	Minimum Charge	Job		234		234	
Stucco							
	Demolish	S.Y.		16.60		16.60	Includes material and labor to install
	Install	S.Y.	9.80	43	1.97	54.77	stucco on wire mesh over wood
	Demolish and Install	S.Y.	9.80	59.60	1.97	71.37	framing.
	Clean	S.Y.		3.56		3.56	
	Paint	S.Y.	1.93	5.40		7.33	
	Minimum Charge	Job		234		234	
Thin Coat							
Veneer							
	Demolish	S.Y.		6.95		6.95	Includes materials and labor to install
	Install	S.Y.	1.09	5.55	.31	6.95	thin coat veneer plaster.
	Demolish and Install	S.Y.	1.09	12.50	.31	13.90	
	Clean	S.Y.		3.56		3.56	
	Paint	S.Y.	1.93	5.40		7.33	
	Minimum Charge	Job		234		234	
Lath							
Metal							
	Demolish	S.Y.		1.63		1.63	Includes material and labor to install
	Install	S.Y.	4.63	8		12.63	wire lath on steel framed ceilings.
	Demolish and Install	S.Y.	4.63	9.63		14.26	
	Minimum Charge	Job		234		234	
Gypsum							
	Demolish	S.Y.		1.63		1.63	Includes material and labor to install
	Install	S.Y.	3.27	5		8.27	gypsum lath on ceilings.
	Demolish and Install	S.Y.	3.27	6.63		9.90	
	Minimum Charge	Job		234		234	
Repairs							
Fill in Small Cracks							
	Install	S.F.	.09	.85		.94	Includes labor and material to fill in
	Minimum Charge	Job		470		470	hairline cracks up to 1/8" with filler, sand and prep for paint.

Gypsum Wallboard		Unit	Material	Labor	Equip.	Total	Specification
Finished							
1/4"							
	Demolish	S.F.		.42		.42	Cost includes material and labor to
	Install	S.F.	.32	1.50		1.82	install 1/4" drywall, including joint
	Demolish and Install	S.F.	.32	1.92		2.24	taping, premixed compound and
	Clean	S.F.		.34		.34	screws.
	Paint	S.F.	.13	.35		.48	
	Minimum Charge	Job		242		242	
3/8"							
	Demolish	S.F.		.42		.42	Cost includes material and labor to
	Install	S.F.	.25	1.52		1.77	install 3/8" drywall including joint
	Demolish and Install	S.F.	.25	1.94		2.19	tape, premixed compound and
	Clean	S.F.		.34		.34	screws.
	Paint	S.F.	.13	.35		.48	
	Minimum Charge	Job		242		242	

177

Gypsum Wallboard	Unit	Material	Labor	Equip.	Total	Specification
1/2″						
Demolish	S.F.		.42		.42	Cost includes material and labor to
Install	S.F.	.25	1.53		1.78	install 1/2″ drywall including joint
Demolish and Install	S.F.	.25	1.95		2.20	tape, premixed compound and
Clean	S.F.		.34		.34	screws.
Paint	S.F.	.13	.35		.48	
Minimum Charge	Job		242		242	
1/2″ Water Resistant						
Demolish	S.F.		.42		.42	Cost includes material and labor to
Install	S.F.	.35	1.53		1.88	install 1/2″ WR drywall including joint
Demolish and Install	S.F.	.35	1.95		2.30	tape, premixed compound and
Clean	S.F.		.34		.34	screws.
Paint	S.F.	.13	.35		.48	
Minimum Charge	Job		242		242	
1/2″ Fire-rated						
Demolish	S.F.		.42		.42	Cost includes material and labor to
Install	S.F.	.42	1.53		1.95	install 1/2″ FR drywall including
Demolish and Install	S.F.	.42	1.95		2.37	screws. Taping and finishing not
Clean	S.F.		.34		.34	included.
Paint	S.F.	.13	.35		.48	
Minimum Charge	Job		242		242	
5/8″						
Demolish	S.F.		.42		.42	Cost includes material and labor to
Install	S.F.	.28	1.48		1.76	install 5/8″ drywall including joint
Demolish and Install	S.F.	.28	1.90		2.18	tape, premixed compound and
Clean	S.F.		.34		.34	screws.
Paint	S.F.	.13	.35		.48	
Minimum Charge	Job		242		242	
5/8″ Fire-rated						
Demolish	S.F.		.42		.42	Cost includes material and labor to
Install	S.F.	.36	1.74		2.10	install 5/8″ FR drywall, including joint
Demolish and Install	S.F.	.36	2.16		2.52	taping, premixed compound and
Clean	S.F.		.34		.34	screws.
Paint	S.F.	.13	.35		.48	
Minimum Charge	Job		242		242	
2-hour Fire-rated						
Demolish	S.F.		1.04		1.04	Cost includes material and labor to
Install	S.F.	.63	3.46		4.09	install 2 layers of 5/8″ type X drywall
Demolish and Install	S.F.	.63	4.50		5.13	including joint taping, premixed
Clean	S.F.		.34		.34	compound and screws.
Paint	S.F.	.13	.35		.48	
Minimum Charge	Job		242		242	
Unfinished						
1/4″						
Demolish	S.F.		.42		.42	Cost includes material and labor to
Install	S.F.	.28	.52		.80	install 1/4″ drywall including screws.
Demolish and Install	S.F.	.28	.94		1.22	Taping and finishing not included.
Clean	S.F.		.34		.34	
Minimum Charge	Job		242		242	
3/8″						
Demolish	S.F.		.42		.42	Cost includes material and labor to
Install	S.F.	.21	.52		.73	install 3/8″ drywall including screws.
Demolish and Install	S.F.	.21	.94		1.15	Tape, finish or texture is not included.
Clean	S.F.		.34		.34	
Minimum Charge	Job		242		242	

For customer support on your Contractor's Pricing Guide: Residential Repair & Remodeling, call 800.448.8182.

Gypsum Wallboard		Unit	Material	Labor	Equip.	Total	Specification
1/2"							
	Demolish	S.F.		.42		.42	Cost includes material and labor to
	Install	S.F.	.21	.53		.74	install 1/2" drywall including screws.
	Demolish and Install	S.F.	.21	.95		1.16	Taping and finishing not included.
	Clean	S.F.		.34		.34	
	Minimum Charge	Job		242		242	
1/2" Water Resistant							
	Demolish	S.F.		.42		.42	Cost includes material and labor to
	Install	S.F.	.31	.53		.84	install 1/2" WR drywall including
	Demolish and Install	S.F.	.31	.95		1.26	screws. taping and finishing not
	Clean	S.F.		.34		.34	included.
	Minimum Charge	Job		242		242	
1/2" Fire-rated							
	Demolish	S.F.		.42		.42	Cost includes labor and material to
	Install	S.F.	.21	.35		.56	install 1/2" FR drywall including
	Demolish and Install	S.F.	.21	.77		.98	screws. Tape, finish or texture is not
	Clean	S.F.		.34		.34	included.
	Minimum Charge	Job		242		242	
5/8"							
	Demolish	S.F.		.42		.42	Cost includes material and labor to
	Install	S.F.	.24	.59		.83	install 5/8" drywall including screws.
	Demolish and Install	S.F.	.24	1.01		1.25	Taping and finishing not included.
	Clean	S.F.		.34		.34	
	Minimum Charge	Job		242		242	
5/8" Fire-rated							
	Demolish	S.F.		.42		.42	Cost includes material and labor to
	Install	S.F.	.31	.59		.90	install 5/8" FR drywall including
	Demolish and Install	S.F.	.31	1.01		1.32	screws. Taping and finishing not
	Clean	S.F.		.34		.34	included.
	Minimum Charge	Job		242		242	
2-hour Fire-rated							
	Demolish	S.F.		1.04		1.04	Cost includes material and labor to
	Install	S.F.	.63	3.46		4.09	install 2 layers of 5/8" type X drywall
	Demolish and Install	S.F.	.63	4.50		5.13	including joint taping, premixed
	Clean	S.F.		.34		.34	compound and screws.
	Minimum Charge	Job		242		242	
Accessories							
Tape & Finish							
	Install	S.F.	.06	.48		.54	Includes labor and material for taping,
	Minimum Charge	Job		242		242	sanding and finish.
Texture							
Sprayed							
	Install	S.F.	.06	.60		.66	Includes labor and material to install
	Minimum Charge	Job		242		242	by spray, texture finish.
Trowel							
	Install	S.F.	.35	1.06		1.41	Includes labor and material to apply
	Minimum Charge	Job		242		242	hand troweled texture.
Patch Hole in Gypsum Wallboard							
	Install	S.F.	.41	66.50		66.91	Includes labor and material to patch a
	Minimum Charge	Job		242		242	one square foot area of damaged
							drywall.
Furring Strips							
	Demolish	S.F.		.10		.10	Cost includes material and labor to
	Install	S.F.	.25	1.08		1.33	install 1" x 2" furring strips, 16" O.C.
	Demolish and Install	S.F.	.25	1.18		1.43	per S.F. of surface area to be covered.
	Minimum Charge	Job		121		121	

Ceramic Wall Tile	Unit	Material	Labor	Equip.	Total	Specification
Thin Set						
2" x 2"						
Demolish	S.F.		1.39		1.39	Cost includes material and labor to
Install	S.F.	4.86	7.80		12.66	install 2" x 2" ceramic tile in organic
Demolish and Install	S.F.	4.86	9.19		14.05	adhesive including cutting and grout.
Clean	S.F.		.71		.71	
Minimum Charge	Job		235		235	
4-1/4" x 4-1/4"						
Demolish	S.F.		1.39		1.39	Cost includes material and labor to
Install	S.F.	1.51	4.37		5.88	install 4-1/4" x 4-1/4" ceramic tile in
Demolish and Install	S.F.	1.51	5.76		7.27	organic adhesive including grout.
Clean	S.F.		.71		.71	
Minimum Charge	Job		235		235	
6" x 6"						
Demolish	S.F.		1.39		1.39	Cost includes material and labor to
Install	S.F.	2.01	4.75		6.76	install 6" x 6" ceramic tile in organic
Demolish and Install	S.F.	2.01	6.14		8.15	adhesive including cutting and grout.
Clean	S.F.		.71		.71	
Minimum Charge	Job		235		235	
8" x 8"						
Demolish	S.F.		1.39		1.39	Cost includes material and labor to
Install	S.F.	5.85	4.89		10.74	install 8" x 8" ceramic tile in organic
Demolish and Install	S.F.	5.85	6.28		12.13	adhesive including cutting and grout.
Clean	S.F.		.71		.71	
Minimum Charge	Job		235		235	
Thick Set						
2" x 2"						
Demolish	S.F.		1.66		1.66	Cost includes material and labor to
Install	S.F.	7.70	9.40		17.10	install 2" x 2" ceramic tile in mortar
Demolish and Install	S.F.	7.70	11.06		18.76	bed including cutting and grout.
Clean	S.F.		.71		.71	
Minimum Charge	Job		235		235	
4-1/4" x 4-1/4"						
Demolish	S.F.		1.66		1.66	Cost includes material and labor to
Install	S.F.	9.90	9.40		19.30	install 4-1/4" x 4-1/4" ceramic tile in
Demolish and Install	S.F.	9.90	11.06		20.96	mortar bed including cutting and
Clean	S.F.		.71		.71	grout.
Minimum Charge	Job		235		235	
6" x 6"						
Demolish	S.F.		1.66		1.66	Cost includes material and labor to
Install	S.F.	9.25	8.55		17.80	install 6" x 6" ceramic tile in mortar
Demolish and Install	S.F.	9.25	10.21		19.46	bed including cutting and grout.
Clean	S.F.		.71		.71	
Minimum Charge	Job		235		235	
8" x 8"						
Demolish	S.F.		1.66		1.66	Cost includes material and labor to
Install	S.F.	5.55	7.80		13.35	install 8" x 8" ceramic tile in mortar
Demolish and Install	S.F.	5.55	9.46		15.01	bed including cutting and grout.
Clean	S.F.		.71		.71	
Minimum Charge	Job		235		235	
Re-grout						
Install	S.F.	.15	4.69		4.84	Includes material and labor to regrout
Minimum Charge	Job		235		235	wall tiles.

For customer support on your Contractor's Pricing Guide: Residential Repair & Remodeling, call 800.448.8182.

Walls / Ceilings

Wall Tile	Unit	Material	Labor	Equip.	Total	Specification
Mirror						
Custom Cut Wall						
Demolish	S.F.		.79		.79	Includes material and labor to install
Install	S.F.	3.14	4.82		7.96	distortion-free float glass mirror, 1/8"
Demolish and Install	S.F.	3.14	5.61		8.75	to 3/16" thick with cut and polished
Clean	S.F.	.01	.22		.23	edges.
Minimum Charge	Job		235		235	
Beveled Wall						
Demolish	S.F.		.79		.79	Includes material and labor to install
Install	S.F.	7.65	8.55		16.20	distortion-free float glass mirror, 1/8"
Demolish and Install	S.F.	7.65	9.34		16.99	to 3/16" thick with beveled edges.
Clean	S.F.	.01	.22		.23	
Minimum Charge	Job		235		235	
Smoked Wall						
Demolish	S.F.		.79		.79	Includes material and labor to install
Install	S.F.	9.10	8.55		17.65	distortion-free smoked float glass
Demolish and Install	S.F.	9.10	9.34		18.44	mirror, 1/8" to 3/16" thick with cut
Clean	S.F.	.01	.22		.23	and polished edges.
Minimum Charge	Job		235		235	
Marble						
Good Grade						
Demolish	S.F.		2.60		2.60	Cost includes material and labor to
Install	S.F.	9.65	20.50		30.15	install 3/8" x 12" x 12" marble tile in
Demolish and Install	S.F.	9.65	23.10		32.75	mortar bed with grout.
Clean	S.F.		.71		.71	
Minimum Charge	Job		235		235	
Better Grade						
Demolish	S.F.		2.60		2.60	Cost includes material and labor to
Install	S.F.	11.35	20.50		31.85	install 3/8" x 12" x 12" marble, in
Demolish and Install	S.F.	11.35	23.10		34.45	mortar bed with grout.
Clean	S.F.		.71		.71	
Minimum Charge	Job		235		235	
Premium Grade						
Demolish	S.F.		2.60		2.60	Cost includes material and labor to
Install	S.F.	20	20.50		40.50	install 3/8" x 12" x 12" marble, in
Demolish and Install	S.F.	20	23.10		43.10	mortar bed with grout.
Clean	S.F.		.71		.71	
Minimum Charge	Job		235		235	
Glass Block						
Demolish	S.F.		2.08		2.08	Cost includes material and labor to
Install	S.F.	29	19.60		48.60	install 3-7/8" thick, 6" x 6" smooth
Demolish and Install	S.F.	29	21.68		50.68	face glass block including mortar and
Clean	S.F.		.71		.71	cleaning.
Minimum Charge	Job		445		445	
Cork						
Demolish	S.F.		.32		.32	Cost includes material and labor to
Install	S.F.	4.44	1.71		6.15	install 3/16" thick cork tile and
Demolish and Install	S.F.	4.44	2.03		6.47	installation.
Clean	S.F.		.71		.71	
Minimum Charge	Job		206		206	
Re-grout						
Install	S.F.	.15	4.69		4.84	Includes material and labor to regrout
Minimum Charge	Job		235		235	wall tiles.

For customer support on your Contractor's Pricing Guide: Residential Repair & Remodeling, call 800.448.8182.

Wall Paneling	Unit	Material	Labor	Equip.	Total	Specification
Plywood						
Good Grade						
Demolish	S.F.		.42		.42	Cost includes material and labor to
Install	S.F.	.97	1.94		2.91	install 4' x 8' x 1/4" paneling,
Demolish and Install	S.F.	.97	2.36		3.33	including nails and adhesive.
Clean	S.F.		.34		.34	
Paint	S.F.	.41	.77		1.18	
Minimum Charge	Job		242		242	
Better Grade						
Demolish	S.F.		.42		.42	Cost includes material and labor to
Install	S.F.	1.68	2.30		3.98	install 4' x 8' x 1/4" plywood wall
Demolish and Install	S.F.	1.68	2.72		4.40	panel, including nails and adhesive.
Clean	S.F.		.34		.34	
Paint	S.F.	.41	.77		1.18	
Minimum Charge	Job		242		242	
Premium Grade						
Demolish	S.F.		.42		.42	Cost includes material and labor to
Install	S.F.	2.20	2.77		4.97	install 4' x 8' x 1/4" plywood wall
Demolish and Install	S.F.	2.20	3.19		5.39	panel, including nails and adhesive.
Clean	S.F.		.34		.34	
Paint	S.F.	.41	.77		1.18	
Minimum Charge	Job		242		242	
Hardboard						
Embossed Paneling						
Demolish	S.F.		.42		.42	Includes material and labor to install
Install	S.F.	.59	1.94		2.53	hardboard embossed paneling
Demolish and Install	S.F.	.59	2.36		2.95	including adhesives.
Clean	S.F.		.34		.34	
Minimum Charge	Job		242		242	
Plastic Molding						
Demolish	L.F.		.42		.42	Includes material and labor to install
Install	L.F.	.46	1.21		1.67	plastic trim for paneling.
Demolish and Install	L.F.	.46	1.63		2.09	
Clean	L.F.	.01	.22		.23	
Minimum Charge	Job		242		242	
Hardboard						
Embossed Brick						
Demolish	S.F.		.42		.42	Cost includes material and labor to
Install	S.F.	1.09	2.15		3.24	install 4' x 8' non-ceramic mineral
Demolish and Install	S.F.	1.09	2.57		3.66	brick-like veneer wall panels including
Clean	S.F.		.34		.34	adhesive.
Paint	S.F.	.22	.61		.83	
Minimum Charge	Job		242		242	
Embossed Ceramic Tile						
Demolish	S.F.		.42		.42	Includes material and labor to install
Install	S.F.	.59	1.94		2.53	hardboard embossed paneling
Demolish and Install	S.F.	.59	2.36		2.95	including adhesives.
Clean	S.F.		.34		.34	
Paint	S.F.	.22	.61		.83	
Minimum Charge	Job		242		242	
Pegboard						
Demolish	S.F.		.42		.42	Cost includes material and labor to
Install	S.F.	.78	1.94		2.72	install 4' x 8' x 1/4" pegboard panel,
Demolish and Install	S.F.	.78	2.36		3.14	including adhesive.
Clean	S.F.		.34		.34	
Paint	S.F.	.22	.61		.83	
Minimum Charge	Job		242		242	

For customer support on your Contractor's Pricing Guide: Residential Repair & Remodeling, call 800.448.8182.

Wall Paneling		Unit	Material	Labor	Equip.	Total	Specification
Furring Strips							
	Demolish	S.F.		.10		.10	Cost includes material and labor to
	Install	S.F.	.25	1.08		1.33	install 1" x 2" furring strips, 16" O.C.
	Demolish and Install	S.F.	.25	1.18		1.43	per S.F. of surface area to be covered.
	Minimum Charge	Job		242		242	
Molding							
Unfinished Base							
	Demolish	L.F.		.69		.69	Cost includes material and labor to
	Install	L.F.	.82	1.46		2.28	install 1/2" x 2-1/2" pine or birch
	Demolish and Install	L.F.	.82	2.15		2.97	base molding.
	Reinstall	L.F.		1.17		1.17	
	Clean	L.F.	.01	.22		.23	
	Paint	L.F.	.08	.44		.52	
	Minimum Charge	Job		242		242	
Unfinished Corner							
	Demolish	L.F.		.69		.69	Includes material and labor to install
	Install	L.F.	2.33	2.02		4.35	unfinished pine corner molding.
	Demolish and Install	L.F.	2.33	2.71		5.04	
	Reinstall	L.F.		1.61		1.61	
	Clean	L.F.	.01	.22		.23	
	Paint	L.F.	.17	.65		.82	
	Minimum Charge	Job		242		242	
Unfinished Cove							
	Demolish	L.F.		.69		.69	Cost includes material and labor to
	Install	L.F.	2.88	1.83		4.71	install 11/16" x 2-3/4" pine cove
	Demolish and Install	L.F.	2.88	2.52		5.40	molding.
	Reinstall	L.F.		1.46		1.46	
	Clean	L.F.	.01	.22		.23	
	Paint	L.F.	.17	.65		.82	
	Minimum Charge	Job		242		242	
Unfinished Crown							
	Demolish	L.F.		.83		.83	Includes material and labor to install
	Install	L.F.	1.69	1.94		3.63	unfinished pine crown molding.
	Demolish and Install	L.F.	1.69	2.77		4.46	
	Reinstall	L.F.		1.55		1.55	
	Clean	L.F.	.02	.29		.31	
	Paint	L.F.	.15	.44		.59	
	Minimum Charge	Job		242		242	
Prefinished Base							
	Demolish	L.F.		.69		.69	Cost includes material and labor to
	Install	L.F.	1.73	1.55		3.28	install 9/16" x 3-5/16" prefinished
	Demolish and Install	L.F.	1.73	2.24		3.97	base molding.
	Reinstall	L.F.		1.24		1.24	
	Clean	L.F.	.01	.22		.23	
	Paint	L.F.	.08	.44		.52	
	Minimum Charge	Job		242		242	
Prefinished Corner							
	Demolish	L.F.		.69		.69	Includes material and labor to install
	Install	L.F.	2.70	2.02		4.72	inside or outside prefinished corner
	Demolish and Install	L.F.	2.70	2.71		5.41	molding.
	Reinstall	L.F.		1.61		1.61	
	Clean	L.F.	.01	.22		.23	
	Paint	L.F.	.17	.65		.82	
	Minimum Charge	Job		242		242	

Walls / Ceilings

Wall Paneling

Wall Paneling	Unit	Material	Labor	Equip.	Total	Specification
Prefinished Crown						
Demolish	L.F.		.83		.83	Cost includes material and labor to
Install	L.F.	2.29	1.94		4.23	install 3/4" x 3-13/16" prefinished
Demolish and Install	L.F.	2.29	2.77		5.06	crown molding.
Reinstall	L.F.		1.55		1.55	
Clean	L.F.	.02	.29		.31	
Paint	L.F.	.15	.44		.59	
Minimum Charge	Job		242		242	

Wallcovering

Wallcovering	Unit	Material	Labor	Equip.	Total	Specification
Surface Prep						
Patch Walls						
Install	S.F.	.14	4.68		4.82	Includes labor and material to repair
Minimum Charge	Job		206		206	section of plaster wall, area less than 20 S.F.
Sizing						
Install	S.F.	.08	.21		.29	Includes labor and materials to apply
Minimum Charge	Job		206		206	wall sizing.
Good Grade						
Walls						
Demolish	S.F.		.76		.76	Includes material and labor to install
Install	S.F.	1.02	.86		1.88	pre-pasted, vinyl-coated wallcovering.
Demolish and Install	S.F.	1.02	1.62		2.64	
Clean	S.F.		.09		.09	
Minimum Charge	Job		206		206	
Ceiling						
Demolish	S.F.		.76		.76	Includes material and labor to install
Install	S.F.	1.02	.86		1.88	pre-pasted, vinyl-coated wallcovering.
Demolish and Install	S.F.	1.02	1.62		2.64	
Clean	S.F.	.01	.22		.23	
Minimum Charge	Job		206		206	
Better Grade						
Walls						
Demolish	S.F.		.76		.76	Includes material and labor to install
Install	S.F.	1.09	.86		1.95	pre-pasted, vinyl-coated wallcovering.
Demolish and Install	S.F.	1.09	1.62		2.71	
Clean	S.F.		.09		.09	
Minimum Charge	Job		206		206	
Ceiling						
Demolish	S.F.		.76		.76	Includes material and labor to install
Install	S.F.	1.09	.86		1.95	pre-pasted, vinyl-coated wallcovering.
Demolish and Install	S.F.	1.09	1.62		2.71	
Clean	S.F.	.01	.22		.23	
Minimum Charge	Job		206		206	
Premium Grade						
Walls						
Demolish	S.F.		.95		.95	Includes material and labor to install
Install	S.F.	1.01	.86		1.87	pre-pasted, vinyl-coated wallcovering.
Demolish and Install	S.F.	1.01	1.81		2.82	
Clean	S.F.		.09		.09	
Minimum Charge	Job		206		206	
Ceiling						
Demolish	S.F.		.95		.95	Includes material and labor to install
Install	S.F.	1.01	.86		1.87	pre-pasted, vinyl-coated wallcovering.
Demolish and Install	S.F.	1.01	1.81		2.82	
Clean	S.F.	.01	.22		.23	
Minimum Charge	Job		206		206	

For customer support on your Contractor's Pricing Guide: Residential Repair & Remodeling, call 800.448.8182.

Walls / Ceilings

Wallcovering		Unit	Material	Labor	Equip.	Total	Specification
Mural							
Pre-printed							
	Demolish	Ea.		130		130	Includes material and labor to install
	Install	Ea.	295	310		605	wallcovering with a pre-printed mural.
	Demolish and Install	Ea.	295	440		735	
	Clean	Ea.		9.35		9.35	
	Minimum Charge	Job		206		206	
Hand-painted							
	Demolish	Ea.		130		130	Includes material and labor to install
	Install	Ea.	565	625		1190	wallcovering with a hand-painted
	Demolish and Install	Ea.	565	755		1320	mural.
	Clean	Ea.		9.35		9.35	
	Minimum Charge	Job		206		206	
Vinyl-coated							
Good Grade							
	Demolish	Ea.		23		23	Includes material and labor to install
	Install	Ea.	19.75	29		48.75	pre-pasted, vinyl-coated wallcovering.
	Demolish and Install	Ea.	19.75	52		71.75	Based upon 30 S.F. rolls.
	Clean	Ea.		9.35		9.35	
	Minimum Charge	Job		206		206	
Better Grade							
	Demolish	Ea.		23		23	Includes material and labor to install
	Install	Ea.	26.50	29		55.50	pre-pasted, vinyl-coated wallcovering.
	Demolish and Install	Ea.	26.50	52		78.50	Based upon 30 S.F. rolls.
	Clean	Ea.		9.35		9.35	
	Minimum Charge	Job		206		206	
Premium Grade							
	Demolish	Ea.		23		23	Includes material and labor to install
	Install	Ea.	38.50	29		67.50	pre-pasted, vinyl-coated wallcovering.
	Demolish and Install	Ea.	38.50	52		90.50	Based upon 30 S.F. rolls.
	Clean	Ea.		9.35		9.35	
	Minimum Charge	Job		206		206	
Blank Underliner							
	Demolish	Ea.		23		23	Includes material and labor to install
	Install	Ea.	10.35	29.50		39.85	blank stock liner paper or bridging
	Demolish and Install	Ea.	10.35	52.50		62.85	paper. Based upon 30 S.F. rolls.
	Clean	Ea.		9.35		9.35	
	Minimum Charge	Job		206		206	
Solid Vinyl							
Good Grade							
	Demolish	Roll		25.50		25.50	Includes material and labor to install
	Install	Ea.	26	29		55	pre-pasted, vinyl wallcovering. Based
	Demolish and Install	Ea.	26	54.50		80.50	upon 30 S.F. rolls.
	Clean	Ea.		9.35		9.35	
	Minimum Charge	Job		206		206	
Better Grade							
	Demolish	Roll		25.50		25.50	Includes material and labor to install
	Install	Ea.	37	29		66	pre-pasted, vinyl wallcovering. Based
	Demolish and Install	Ea.	37	54.50		91.50	upon 30 S.F. rolls.
	Clean	Ea.		9.35		9.35	
	Minimum Charge	Job		206		206	
Premium Grade							
	Demolish	Roll		25.50		25.50	Includes material and labor to install
	Install	Ea.	47.50	29		76.50	pre-pasted, vinyl wallcovering. Based
	Demolish and Install	Ea.	47.50	54.50		102	upon 30 S.F. rolls.
	Clean	Ea.		9.35		9.35	
	Minimum Charge	Job		206		206	

Walls / Ceilings

Wallcovering	Unit	Material	Labor	Equip.	Total	Specification
Blank Underliner						
Demolish	Ea.		23		23	Includes material and labor to install
Install	Ea.	10.35	29.50		39.85	blank stock liner paper or bridging
Demolish and Install	Ea.	10.35	52.50		62.85	paper. Based upon 30 S.F. rolls.
Clean	Ea.		9.35		9.35	
Minimum Charge	Job		206		206	
Grasscloth / String						
Good Grade						
Demolish	Ea.		23		23	Includes material and labor to install
Install	Ea.	50.50	37.50		88	grasscloth or string wallcovering.
Demolish and Install	Ea.	50.50	60.50		111	Based upon 30 S.F. rolls.
Clean	Ea.		12.45		12.45	
Minimum Charge	Job		206		206	
Better Grade						
Demolish	Ea.		23		23	Includes material and labor to install
Install	Ea.	62.50	37.50		100	grasscloth or string wallcovering.
Demolish and Install	Ea.	62.50	60.50		123	Based upon 30 S.F. rolls.
Clean	Ea.		12.45		12.45	
Minimum Charge	Job		206		206	
Premium Grade						
Demolish	Ea.		23		23	Includes material and labor to install
Install	Ea.	66	37.50		103.50	grasscloth or string wallcovering.
Demolish and Install	Ea.	66	60.50		126.50	Based upon 30 S.F. rolls.
Clean	Ea.		12.45		12.45	
Minimum Charge	Job		206		206	
Blank Underliner						
Demolish	Ea.		23		23	Includes material and labor to install
Install	Ea.	10.35	29.50		39.85	blank stock liner paper or bridging
Demolish and Install	Ea.	10.35	52.50		62.85	paper. Based upon 30 S.F. rolls.
Clean	Ea.		9.35		9.35	
Minimum Charge	Job		206		206	
Designer						
Premium Grade						
Demolish	S.F.		.95		.95	Includes material and labor to install
Install	Ea.	135	43.50		178.50	paperback vinyl, untrimmed and
Demolish and Install	Ea.	135	44.45		179.45	hand-painted. Based upon 30 S.F.
Clean	Ea.		9.35		9.35	rolls.
Minimum Charge	Job		206		206	
Blank Underliner						
Demolish	Ea.		23		23	Includes material and labor to install
Install	Ea.	10.35	29.50		39.85	blank stock liner paper or bridging
Demolish and Install	Ea.	10.35	52.50		62.85	paper. Based upon 30 S.F. rolls.
Clean	Ea.		9.35		9.35	
Minimum Charge	Job		206		206	
Foil						
Demolish	Ea.		23		23	Includes material and labor to install
Install	Ea.	93.50	37.50		131	foil coated wallcovering. Based upon
Demolish and Install	Ea.	93.50	60.50		154	30 S.F. rolls.
Clean	Ea.		9.35		9.35	
Minimum Charge	Job		206		206	
Wallcovering Border						
Good Grade						
Demolish	L.F.		1.51		1.51	Includes material and labor install
Install	L.F.	1.95	1.58		3.53	vinyl-coated border 4" to 6" wide.
Demolish and Install	L.F.	1.95	3.09		5.04	
Clean	L.F.		.14		.14	
Minimum Charge	Job		206		206	

For customer support on your Contractor's Pricing Guide: Residential Repair & Remodeling, call 800.448.8182.

Wallcovering		Unit	Material	Labor	Equip.	Total	Specification
Better Grade							
	Demolish	L.F.		1.51		1.51	Includes material and labor install
	Install	L.F.	2.06	1.58		3.64	vinyl-coated border 4" to 6" wide.
	Demolish and Install	L.F.	2.06	3.09		5.15	
	Clean	L.F.		.14		.14	
	Minimum Charge	Job		206		206	
Premium Grade							
	Demolish	L.F.		1.51		1.51	Includes material and labor install
	Install	L.F.	3.69	1.58		5.27	vinyl-coated border 4" to 6" wide.
	Demolish and Install	L.F.	3.69	3.09		6.78	
	Clean	L.F.		.14		.14	
	Minimum Charge	Job		206		206	

Acoustical Ceiling		Unit	Material	Labor	Equip.	Total	Specification
Complete System							
2' x 2' Square Edge							
	Demolish	S.F.		.69		.69	Includes material and labor to install
	Install	S.F.	4.86	1.49		6.35	T-bar grid acoustical ceiling system,
	Demolish and Install	S.F.	4.86	2.18		7.04	random pinhole 5/8" thick square
	Reinstall	S.F.		1.19		1.19	edge tile.
	Clean	S.F.	.02	.28		.30	
	Minimum Charge	Job		242		242	
2' x 2' Reveal Edge							
	Demolish	S.F.		.69		.69	Includes material and labor to install
	Install	S.F.	5.80	1.94		7.74	T-bar grid acoustical ceiling system,
	Demolish and Install	S.F.	5.80	2.63		8.43	random pinhole 5/8" thick reveal
	Reinstall	S.F.		1.55		1.55	edge tile.
	Clean	S.F.	.02	.28		.30	
	Minimum Charge	Job		242		242	
2' x 2' Fire-rated							
	Demolish	S.F.		.69		.69	Includes material and labor to install
	Install	S.F.	5.10	1.49		6.59	T-bar grid FR acoustical ceiling system,
	Demolish and Install	S.F.	5.10	2.18		7.28	random pinhole 5/8" thick square
	Reinstall	S.F.		1.19		1.19	edge tile.
	Clean	S.F.	.02	.28		.30	
	Minimum Charge	Job		242		242	
2' x 4' Square Edge							
	Demolish	S.F.		.69		.69	Cost includes material and labor to
	Install	S.F.	5.85	1.27		7.12	install 2' x 4' T-bar grid acoustical
	Demolish and Install	S.F.	5.85	1.96		7.81	ceiling system, random pinhole 5/8"
	Reinstall	S.F.		1.02		1.02	thick square edge tile.
	Clean	S.F.	.02	.28		.30	
	Minimum Charge	Job		242		242	
2' x 4' Reveal Edge							
	Demolish	S.F.		.69		.69	Includes material and labor to install
	Install	S.F.	5.10	1.76		6.86	T-bar grid acoustical ceiling system,
	Demolish and Install	S.F.	5.10	2.45		7.55	random pinhole 5/8" thick reveal
	Reinstall	S.F.		1.41		1.41	edge tile.
	Clean	S.F.	.02	.28		.30	
	Minimum Charge	Job		242		242	
2' x 4' Fire-rated							
	Demolish	S.F.		.69		.69	Includes material and labor to install
	Install	S.F.	5.10	1.49		6.59	T-bar grid FR acoustical ceiling system,
	Demolish and Install	S.F.	5.10	2.18		7.28	random pinhole 5/8" thick square
	Reinstall	S.F.		1.19		1.19	edge tile.
	Clean	S.F.	.02	.28		.30	
	Minimum Charge	Job		242		242	

Walls / Ceilings

Acoustical Ceiling	Unit	Material	Labor	Equip.	Total	Specification
T-bar Grid						
2' x 2'						
Demolish	S.F.		.81		.81	Includes material and labor to install
Install	S.F.	2.13	.74		2.87	T-bar grid acoustical ceiling
Demolish and Install	S.F.	2.13	1.55		3.68	suspension system.
Reinstall	S.F.		.60		.60	
Clean	S.F.	.02	.28		.30	
Minimum Charge	Job		234		234	
2' x 4'						
Demolish	S.F.		.81		.81	Includes material and labor to install
Install	S.F.	1.66	.61		2.27	T-bar grid acoustical ceiling
Demolish and Install	S.F.	1.66	1.42		3.08	suspension system.
Reinstall	S.F.		.48		.48	
Clean	S.F.	.02	.28		.30	
Minimum Charge	Job		234		234	
Tile Panels						
2' x 2' Square Edge						
Demolish	S.F.		.23		.23	Includes material and labor to install
Install	S.F.	2.48	.77		3.25	square edge panels 5/8" thick with
Demolish and Install	S.F.	2.48	1		3.48	pinhole pattern. T-bar grid not
Reinstall	S.F.		.62		.62	included.
Clean	S.F.	.02	.28		.30	
Minimum Charge	Job		242		242	
2' x 2' Reveal Edge						
Demolish	S.F.		.23		.23	Includes material and labor to install
Install	S.F.	2.81	1.03		3.84	reveal edge panels 5/8" thick with
Demolish and Install	S.F.	2.81	1.26		4.07	pinhole pattern. T-bar grid not
Reinstall	S.F.		.82		.82	included.
Clean	S.F.	.02	.28		.30	
Minimum Charge	Job		242		242	
2' x 2' Fire-rated						
Demolish	S.F.		.23		.23	Includes material and labor to install
Install	S.F.	1.20	.72		1.92	FR square edge panels 5/8" thick with
Demolish and Install	S.F.	1.20	.95		2.15	pinhole pattern. T-bar grid not
Reinstall	S.F.		.57		.57	included.
Clean	S.F.	.02	.28		.30	
Minimum Charge	Job		242		242	
2' x 4' Square Edge						
Demolish	S.F.		.23		.23	Includes material and labor to install
Install	S.F.	2.48	.77		3.25	square edge panels 5/8" thick with
Demolish and Install	S.F.	2.48	1		3.48	pinhole pattern. T-bar grid not
Reinstall	S.F.		.62		.62	included.
Clean	S.F.	.02	.28		.30	
Minimum Charge	Job		242		242	
2' x 4' Reveal Edge						
Demolish	S.F.		.23		.23	Includes material and labor to install
Install	S.F.	2.81	1.03		3.84	reveal edge panels 5/8" thick with
Demolish and Install	S.F.	2.81	1.26		4.07	pinhole pattern. T-bar grid not
Reinstall	S.F.		.82		.82	included.
Clean	S.F.	.02	.28		.30	
Minimum Charge	Job		242		242	
2' x 4' Fire-rated						
Demolish	S.F.		.23		.23	Includes material and labor to install
Install	S.F.	1.20	.72		1.92	FR square edge panels 5/8" thick with
Demolish and Install	S.F.	1.20	.95		2.15	pinhole pattern. T-bar grid not
Reinstall	S.F.		.57		.57	included.
Clean	S.F.	.02	.28		.30	
Minimum Charge	Job		242		242	

188

Acoustical Ceiling

	Unit	Material	Labor	Equip.	Total	Specification
Bleach Out Stains						
Install	S.F.	.21	.37		.58	Includes labor and material to apply
Minimum Charge	Job		242		242	laundry type chlorine bleach to remove light staining of acoustic tile material.

Adhesive Tile

Good Grade

	Unit	Material	Labor	Equip.	Total	Specification
Demolish	S.F.		.92		.92	Cost includes material and labor to
Install	S.F.	2.63	1.61		4.24	install 12″ x 12″ tile with embossed
Demolish and Install	S.F.	2.63	2.53		5.16	texture pattern. Furring strips not
Clean	S.F.	.01	.22		.23	included.
Minimum Charge	Job		242		242	

Better Grade

	Unit	Material	Labor	Equip.	Total	Specification
Demolish	S.F.		.92		.92	Cost includes material and labor to
Install	S.F.	2.73	1.61		4.34	install 12″ x 12″ tile with embossed
Demolish and Install	S.F.	2.73	2.53		5.26	texture pattern. Furring strips not
Clean	S.F.	.01	.22		.23	included.
Minimum Charge	Job		242		242	

Premium Grade

	Unit	Material	Labor	Equip.	Total	Specification
Demolish	S.F.		.92		.92	Cost includes material and labor to
Install	S.F.	7.20	1.61		8.81	install 12″ x 12″ tile with embossed
Demolish and Install	S.F.	7.20	2.53		9.73	texture pattern. Furring strips not
Clean	S.F.	.01	.22		.23	included.
Minimum Charge	Job		242		242	

Bleach Out Stains

	Unit	Material	Labor	Equip.	Total	Specification
Install	S.F.	.21	.37		.58	Includes labor and material to apply
Minimum Charge	Job		242		242	laundry type chlorine bleach to remove light staining of acoustic tile material.

Stapled Tile

Good Grade

	Unit	Material	Labor	Equip.	Total	Specification
Demolish	S.F.		.55		.55	Cost includes material and labor to
Install	S.F.	2.63	1.61		4.24	install 12″ x 12″ tile with embossed
Demolish and Install	S.F.	2.63	2.16		4.79	texture pattern. Furring strips not
Clean	S.F.	.01	.22		.23	included.
Minimum Charge	Job		242		242	

Better Grade

	Unit	Material	Labor	Equip.	Total	Specification
Demolish	S.F.		.55		.55	Cost includes material and labor to
Install	S.F.	2.73	1.61		4.34	install 12″ x 12″ tile with embossed
Demolish and Install	S.F.	2.73	2.16		4.89	texture pattern. Furring strips not
Clean	S.F.	.01	.22		.23	included.
Minimum Charge	Job		242		242	

Premium Grade

	Unit	Material	Labor	Equip.	Total	Specification
Demolish	S.F.		.55		.55	Cost includes material and labor to
Install	S.F.	7.20	1.61		8.81	install 12″ x 12″ tile with embossed
Demolish and Install	S.F.	7.20	2.16		9.36	texture pattern. Furring strips not
Clean	S.F.	.01	.22		.23	included.
Minimum Charge	Job		242		242	

Bleach Out Stains

	Unit	Material	Labor	Equip.	Total	Specification
Install	S.F.	.21	.37		.58	Includes labor and material to apply
Minimum Charge	Job		242		242	laundry type chlorine bleach to remove light staining of acoustic tile material.

Acoustical Ceiling	Unit	Material	Labor	Equip.	Total	Specification
Blown						
Scrape Off Existing						
Install	S.F.		.77		.77	Includes labor and material to remove existing blown acoustical ceiling material.
Minimum Charge	Job		242		242	
Seal Ceiling Gypsum Wallboard						
Install	S.F.	.10	.35		.45	Includes labor and material to seal existing drywall with shellac-based material.
Minimum Charge	Job		202		202	
Reblow Existing						
Install	S.F.	.06	.60		.66	Includes labor and material to install by spray, texture finish.
Minimum Charge	Job		625	76	701	
Bleach Out Stains						
Install	S.F.	.21	.37		.58	Includes labor and material to apply laundry type chlorine bleach to remove light staining of acoustic tile material.
Minimum Charge	Job		242		242	
New Blown						
Install	S.F.	.06	.60		.66	Includes labor and material to install by spray, texture finish.
Minimum Charge	Job		202		202	
Apply Glitter To						
Install	S.F.	.03	.19		.22	Includes labor and material to apply glitter material to existing or new drywall.
Minimum Charge	Job		202		202	
Furring Strips						
Demolish	S.F.		.10		.10	Cost includes material and labor to install 1" x 2" furring strips, 16" O.C. per S.F. of surface area to be covered.
Install	S.F.	.25	1.08		1.33	
Demolish and Install	S.F.	.25	1.18		1.43	
Minimum Charge	Job		121		121	

For customer support on your Contractor's Pricing Guide: Residential Repair & Remodeling, call 800.448.8182.

Finished Stair

Bi-Fold Door

Raised Panel Door

Hollow Core Door	Unit	Material	Labor	Equip.	Total	Specification
Pre-hung Wood						
1′ 6″ x 6′ 8″						
Demolish	Ea.		45		45	Cost includes material and labor to
Install	Ea.	184	27		211	install 1′ 6″ x 6′ 8″ pre-hung 1-3/8″
Demolish and Install	Ea.	184	72		256	lauan door with split pine jamb
Reinstall	Ea.		26.89		26.89	including casing, stops, and hinges.
Clean	Ea.	1.52	15.60		17.12	
Paint	Ea.	7.70	57.50		65.20	
Minimum Charge	Job		242		242	
2′ x 6′ 8″						
Demolish	Ea.		45		45	Cost includes material and labor to
Install	Ea.	193	28.50		221.50	install 2′ x 6′ 8″ pre-hung 1-3/8″
Demolish and Install	Ea.	193	73.50		266.50	lauan door with split pine jamb
Reinstall	Ea.		28.47		28.47	including casing, stops, and hinges.
Clean	Ea.	1.52	15.60		17.12	
Paint	Ea.	7.70	57.50		65.20	
Minimum Charge	Job		242		242	
2′ 4″ x 6′ 8″						
Demolish	Ea.		45		45	Cost includes material and labor to
Install	Ea.	184	28.50		212.50	install 2′ 4″ x 6′ 8″ pre-hung 1-3/8″
Demolish and Install	Ea.	184	73.50		257.50	lauan door with split pine jamb
Reinstall	Ea.		28.47		28.47	including casing, stops, and hinges.
Clean	Ea.	1.52	15.60		17.12	
Paint	Ea.	7.70	57.50		65.20	
Minimum Charge	Job		242		242	
2′ 6″ x 6′ 8″						
Demolish	Ea.		45		45	Cost includes material and labor to
Install	Ea.	202	28.50		230.50	install 2′ 6″ x 6′ 8″ pre-hung 1-3/8″
Demolish and Install	Ea.	202	73.50		275.50	lauan door with split pine jamb
Reinstall	Ea.		28.47		28.47	including casing, stops, and hinges.
Clean	Ea.	1.52	15.60		17.12	
Paint	Ea.	7.70	57.50		65.20	
Minimum Charge	Job		242		242	
2′ 8″ x 6′ 8″						
Demolish	Ea.		45		45	Cost includes material and labor to
Install	Ea.	175	28.50		203.50	install 2′ 8″ x 6′ 8″ pre-hung 1-3/8″
Demolish and Install	Ea.	175	73.50		248.50	lauan door with split pine jamb,
Reinstall	Ea.		28.47		28.47	including casing, stop and hinges.
Clean	Ea.	1.52	15.60		17.12	
Paint	Ea.	7.70	57.50		65.20	
Minimum Charge	Job		242		242	

Finish Carpentry

Hollow Core Door

	Unit	Material	Labor	Equip.	Total	Specification
3' x 6' 8"						
Demolish	Ea.		45		45	Cost includes material and labor to
Install	Ea.	194	30.50		224.50	install 3' x 6' 8" pre-hung 1-3/8"
Demolish and Install	Ea.	194	75.50		269.50	lauan door with split pine jamb
Reinstall	Ea.		30.25		30.25	including casing, stops, and hinges.
Clean	Ea.	1.52	15.60		17.12	
Paint	Ea.	7.70	57.50		65.20	
Minimum Charge	Job		242		242	
Pocket Type						
Demolish	Ea.		67		67	Cost includes material and labor to
Install	Ea.	106	54		160	install 2' 8" x 6' 8" lauan door.
Demolish and Install	Ea.	106	121		227	
Reinstall	Ea.		53.78		53.78	
Clean	Ea.	1.52	15.60		17.12	
Paint	Ea.	7.70	57.50		65.20	
Minimum Charge	Job		242		242	
Stain						
Install	Ea.	9.50	33.50		43	Includes labor and material to stain
Minimum Charge	Job		202		202	single door and trim on both sides.
Pre-hung Masonite						
1' 6" x 6' 8"						
Demolish	Ea.		45		45	Cost includes material and labor to
Install	Ea.	140	27		167	install 1' 6" x 6' 8" pre-hung 1-3/8"
Demolish and Install	Ea.	140	72		212	masonite door with split pine jamb
Reinstall	Ea.		26.89		26.89	including casing, stops, and hinges.
Clean	Ea.	1.52	15.60		17.12	
Paint	Ea.	7.70	57.50		65.20	
Minimum Charge	Job		242		242	
2' x 6' 8"						
Demolish	Ea.		45		45	Cost includes material and labor to
Install	Ea.	251	28.50		279.50	install 2' x 6' 8" pre-hung 1-3/8"
Demolish and Install	Ea.	251	73.50		324.50	masonite door with split pine jamb
Reinstall	Ea.		28.47		28.47	including casing, stops, and hinges.
Clean	Ea.	1.52	15.60		17.12	
Paint	Ea.	7.70	57.50		65.20	
Minimum Charge	Job		242		242	
2' 4" x 6' 8"						
Demolish	Ea.		45		45	Cost includes material and labor to
Install	Ea.	239	28.50		267.50	install 2' 4" x 6' 8" pre-hung 1-3/8"
Demolish and Install	Ea.	239	73.50		312.50	masonite door with split pine jamb
Reinstall	Ea.		28.47		28.47	including casing, stops, and hinges.
Clean	Ea.	1.52	15.60		17.12	
Paint	Ea.	7.70	57.50		65.20	
Minimum Charge	Job		242		242	
2' 8" x 6' 8"						
Demolish	Ea.		45		45	Cost includes material and labor to
Install	Ea.	241	28.50		269.50	install 2' 8" x 6' 8" pre-hung 1-3/8"
Demolish and Install	Ea.	241	73.50		314.50	masonite door with split pine jamb
Reinstall	Ea.		28.47		28.47	including casing, stops, and hinges.
Clean	Ea.	1.52	15.60		17.12	
Paint	Ea.	7.70	57.50		65.20	
Minimum Charge	Job		242		242	
3' x 6' 8"						
Demolish	Ea.		45		45	Cost includes material and labor to
Install	Ea.	285	30.50		315.50	install 3' x 6' 8" pre-hung 1-3/8"
Demolish and Install	Ea.	285	75.50		360.50	masonite door with split pine jamb
Reinstall	Ea.		30.25		30.25	including casing, stops, and hinges.
Clean	Ea.	1.52	15.60		17.12	
Paint	Ea.	7.70	57.50		65.20	
Minimum Charge	Job		242		242	

For customer support on your Contractor's Pricing Guide: Residential Repair & Remodeling, call 800.448.8182.

Finish Carpentry

Hollow Core Door

Hollow Core Door	Unit	Material	Labor	Equip.	Total	Specification
Pocket Type						
Demolish	Ea.		67		67	Cost includes material and labor to
Install	Ea.	106	54		160	install 2' 8" x 6' 8" lauan door.
Demolish and Install	Ea.	106	121		227	
Reinstall	Ea.		53.78		53.78	
Clean	Ea.	1.52	15.60		17.12	
Paint	Ea.	7.70	57.50		65.20	
Minimum Charge	Job		242		242	
Stain						
Install	Ea.	9.50	33.50		43	Includes labor and material to stain
Minimum Charge	Job		202		202	single door and trim on both sides.

Hollow Core Door Only

1' 6" x 6' 8"

	Unit	Material	Labor	Equip.	Total	Specification
Demolish	Ea.		9.60		9.60	Cost includes material and labor to
Install	Ea.	95.50	25.50		121	install 1' 6" x 6' 8" lauan hollow core
Demolish and Install	Ea.	95.50	35.10		130.60	door slab.
Reinstall	Ea.		25.47		25.47	
Clean	Ea.	.64	11.70		12.34	
Paint	Ea.	8.05	50.50		58.55	
Minimum Charge	Job		242		242	

2' x 6' 8"

	Unit	Material	Labor	Equip.	Total	Specification
Demolish	Ea.		9.60		9.60	Cost includes material and labor to
Install	Ea.	83.50	27		110.50	install 2' x 6' 8" lauan hollow core
Demolish and Install	Ea.	83.50	36.60		120.10	door slab.
Reinstall	Ea.		26.89		26.89	
Clean	Ea.	.64	11.70		12.34	
Paint	Ea.	8.05	50.50		58.55	
Minimum Charge	Job		242		242	

2' 4" x 6' 8"

	Unit	Material	Labor	Equip.	Total	Specification
Demolish	Ea.		9.60		9.60	Cost includes material and labor to
Install	Ea.	95	27		122	install 2' 4" x 6' 8" lauan hollow core
Demolish and Install	Ea.	95	36.60		131.60	door slab.
Reinstall	Ea.		26.89		26.89	
Clean	Ea.	.64	11.70		12.34	
Paint	Ea.	8.05	50.50		58.55	
Minimum Charge	Job		242		242	

2' 8" x 6' 8"

	Unit	Material	Labor	Equip.	Total	Specification
Demolish	Ea.		9.60		9.60	Cost includes material and labor to
Install	Ea.	108	27		135	install 2' 8" x 6' 8" lauan hollow core
Demolish and Install	Ea.	108	36.60		144.60	door slab.
Reinstall	Ea.		26.89		26.89	
Clean	Ea.	.64	11.70		12.34	
Paint	Ea.	8.05	50.50		58.55	
Minimum Charge	Job		242		242	

3' x 6' 8"

	Unit	Material	Labor	Equip.	Total	Specification
Demolish	Ea.		9.60		9.60	Cost includes material and labor to
Install	Ea.	82	28.50		110.50	install 3' x 6' 8" lauan hollow core
Demolish and Install	Ea.	82	38.10		120.10	door slab.
Reinstall	Ea.		28.47		28.47	
Clean	Ea.	.64	11.70		12.34	
Paint	Ea.	8.05	50.50		58.55	
Minimum Charge	Job		242		242	

Pocket Type

	Unit	Material	Labor	Equip.	Total	Specification
Demolish	Ea.		9.60		9.60	Cost includes material and labor to
Install	Ea.	106	54		160	install 2' 8" x 6' 8" lauan door.
Demolish and Install	Ea.	106	63.60		169.60	
Reinstall	Ea.		53.78		53.78	
Clean	Ea.	.64	11.70		12.34	
Paint	Ea.	8.05	50.50		58.55	
Minimum Charge	Job		242		242	

For customer support on your Contractor's Pricing Guide: Residential Repair & Remodeling, call 800.448.8182.

Hollow Core Door		Unit	Material	Labor	Equip.	Total	Specification
Stain							
	Install	Ea.	5.25	45		50.25	Includes labor and materials to stain a
	Minimum Charge	Job		202		202	flush style door on both sides by brush.
Casing Trim							
Single Width							
	Demolish	Ea.		8.55		8.55	Cost includes material and labor to
	Install	Opng.	27	40.50		67.50	install 11/16" x 2-1/2" pine ranch
	Demolish and Install	Opng.	27	49.05		76.05	style casing for one side of a standard
	Reinstall	Opng.		32.27		32.27	door opening.
	Clean	Opng.	.22	3.74		3.96	
	Paint	Ea.	7.45	8.05		15.50	
	Minimum Charge	Job		242		242	
Double Width							
	Demolish	Ea.		10.15		10.15	Cost includes material and labor to
	Install	Opng.	31.50	48.50		80	install 11/16" x 2-1/2" pine ranch
	Demolish and Install	Opng.	31.50	58.65		90.15	style door casing for one side of a
	Reinstall	Opng.		38.72		38.72	double door opening.
	Clean	Opng.	.25	7.50		7.75	
	Paint	Ea.	7.45	9.15		16.60	
	Minimum Charge	Job		242		242	
Hardware							
Doorknob w / Lock							
	Demolish	Ea.		19.20		19.20	Includes material and labor to install
	Install	Ea.	55.50	30.50		86	privacy lockset.
	Demolish and Install	Ea.	55.50	49.70		105.20	
	Reinstall	Ea.		24.20		24.20	
	Clean	Ea.	.15	11.70		11.85	
	Minimum Charge	Job		242		242	
Doorknob							
	Demolish	Ea.		19.20		19.20	Includes material and labor to install
	Install	Ea.	54	30.50		84.50	residential passage lockset, keyless
	Demolish and Install	Ea.	54	49.70		103.70	bored type.
	Reinstall	Ea.		24.20		24.20	
	Clean	Ea.	.15	11.70		11.85	
	Minimum Charge	Job		242		242	
Deadbolt							
	Demolish	Ea.		16.80		16.80	Includes material and labor to install a
	Install	Ea.	90	34.50		124.50	deadbolt.
	Demolish and Install	Ea.	90	51.30		141.30	
	Reinstall	Ea.		27.66		27.66	
	Clean	Ea.	.15	11.70		11.85	
	Minimum Charge	Job		242		242	
Lever Handle							
	Demolish	Ea.		19.20		19.20	Includes material and labor to install
	Install	Ea.	241	48.50		289.50	residential passage lockset, keyless
	Demolish and Install	Ea.	241	67.70		308.70	bored type with lever handle.
	Reinstall	Ea.		38.72		38.72	
	Clean	Ea.	.15	11.70		11.85	
	Minimum Charge	Job		242		242	
Closer							
	Demolish	Ea.		10.10		10.10	Includes material and labor to install
	Install	Ea.	171	74.50		245.50	pneumatic light duty closer for interior
	Demolish and Install	Ea.	171	84.60		255.60	type doors.
	Reinstall	Ea.		59.57		59.57	
	Clean	Ea.	.08	15.60		15.68	
	Minimum Charge	Job		242		242	

For customer support on your Contractor's Pricing Guide: Residential Repair & Remodeling, call 800.448.8182.

Finish Carpentry

Hollow Core Door

Hollow Core Door		Unit	Material	Labor	Equip.	Total	Specification
Push Plate							
	Demolish	Ea.		14.95		14.95	Includes material and labor to install
	Install	Ea.	25.50	40.50		66	bronze push-pull plate.
	Demolish and Install	Ea.	25.50	55.45		80.95	
	Reinstall	Ea.		32.27		32.27	
	Clean	Ea.	.15	11.70		11.85	
	Minimum Charge	Job		242		242	
Kickplate							
	Demolish	Ea.		14.95		14.95	Includes material and labor to install
	Install	Ea.	29	32.50		61.50	aluminum kickplate, 10" x 28".
	Demolish and Install	Ea.	29	47.45		76.45	
	Reinstall	Ea.		25.81		25.81	
	Clean	Ea.	.15	11.70		11.85	
	Minimum Charge	Job		242		242	
Exit Sign							
	Demolish	Ea.		28.50		28.50	Includes material and labor to install a
	Install	Ea.	116	71		187	wall mounted interior electric exit sign.
	Demolish and Install	Ea.	116	99.50		215.50	
	Reinstall	Ea.		56.84		56.84	
	Clean	Ea.	.01	9.35		9.36	
	Minimum Charge	Job		242		242	
Jamb							
	Demolish	Ea.		33.50		33.50	Includes material and labor to install
	Install	Ea.	142	13.10		155.10	flat pine jamb with square cut heads
	Demolish and Install	Ea.	142	46.60		188.60	and rabbeted sides for 6' 8" high and
	Clean	Ea.	.04	9.35		9.39	3-9/16" door including trim sets for
	Paint	Ea.	12.25	27		39.25	both sides.
	Minimum Charge	Job		242		242	
Shave & Refit							
	Install	Ea.		40.50		40.50	Includes labor to shave and rework
	Minimum Charge	Job		242		242	door to fit opening at the job site.

Solid Core Door

Solid Core Door		Unit	Material	Labor	Equip.	Total	Specification
Pre-hung Wood							
2' 4" x 6' 8"							
	Demolish	Ea.		45		45	Cost includes material and labor to
	Install	Ea.	267	28.50		295.50	install 2' 4" x 6' 8" pre-hung 1-3/4"
	Demolish and Install	Ea.	267	73.50		340.50	door, 4-1/2" split jamb, including
	Reinstall	Ea.		28.47		28.47	casing and stop, hinges, aluminum sill,
	Clean	Ea.	1.52	15.60		17.12	weatherstripping.
	Paint	Ea.	22	57.50		79.50	
	Minimum Charge	Job		242		242	
2' 6" x 6' 8"							
	Demolish	Ea.		45		45	Cost includes material and labor to
	Install	Ea.	259	28.50		287.50	install 2' 6" x 6' 8" pre-hung 1-3/4"
	Demolish and Install	Ea.	259	73.50		332.50	door, 4-1/2" split jamb, including
	Reinstall	Ea.		28.47		28.47	casing and stop, hinges, aluminum sill,
	Clean	Ea.	1.52	15.60		17.12	weatherstripping.
	Paint	Ea.	22	57.50		79.50	
	Minimum Charge	Job		242		242	

195

Finish Carpentry

Solid Core Door

	Unit	Material	Labor	Equip.	Total	Specification
2′ 8″ x 6′ 8″						
Demolish	Ea.		45		45	Cost includes material and labor to
Install	Ea.	242	28.50		270.50	install 2′ 8″ x 6′ 8″ pre-hung 1-3/4″
Demolish and Install	Ea.	242	73.50		315.50	door, 4-1/2″ split jamb, including
Reinstall	Ea.		28.47		28.47	casing and stop, hinges, aluminum sill,
Clean	Ea.	1.52	15.60		17.12	weatherstripping.
Paint	Ea.	22	57.50		79.50	
Minimum Charge	Job		242		242	
3′ x 6′ 8″						
Demolish	Ea.		45		45	Cost includes material and labor to
Install	Ea.	252	30.50		282.50	install 3′ x 6′ 8″ pre-hung 1-3/4″
Demolish and Install	Ea.	252	75.50		327.50	door, including casing and stop,
Reinstall	Ea.		30.25		30.25	hinges, jamb, aluminum sill,
Clean	Ea.	1.52	15.60		17.12	weatherstripped.
Paint	Ea.	22	57.50		79.50	
Minimum Charge	Job		242		242	
3′ 6″ x 6′ 8″						
Demolish	Ea.		45		45	Cost includes material and labor to
Install	Ea.	440	32.50		472.50	install 3′ 6″ x 6′ 8″ pre-hung 1-3/4″
Demolish and Install	Ea.	440	77.50		517.50	door, 4-1/2″ split jamb, including
Reinstall	Ea.		32.27		32.27	casing and stop, hinges, aluminum sill,
Clean	Ea.	1.52	15.60		17.12	weatherstripping.
Paint	Ea.	22	57.50		79.50	
Minimum Charge	Job		242		242	
Raised Panel						
Demolish	Ea.		45		45	Cost includes material and labor to
Install	Ea.	580	27		607	install 3′ x 6′ 8″ pre-hung 1-3/4″
Demolish and Install	Ea.	580	72		652	door, jamb, including casing and
Reinstall	Ea.		26.89		26.89	stop, hinges, aluminum sill,
Clean	Ea.	1.52	15.60		17.12	weatherstripping.
Paint	Ea.	22	135		157	
Minimum Charge	Job		242		242	
Stain						
Install	Ea.	9.50	33.50		43	Includes labor and material to stain
Minimum Charge	Job		202		202	single door and trim on both sides.

Bifold Door

	Unit	Material	Labor	Equip.	Total	Specification
Wood, Single						
3′						
Demolish	Ea.		33.50		33.50	Cost includes material and labor to
Install	Ea.	209	74.50		283.50	install 3′ wide solid wood bi-fold door
Demolish and Install	Ea.	209	108		317	not including jamb, trim, track, and
Reinstall	Ea.		74.46		74.46	hardware.
Clean	Ea.	1.52	15.60		17.12	
Paint	Ea.	7.70	57.50		65.20	
Minimum Charge	Job		242		242	
5′						
Demolish	Ea.		33.50		33.50	Includes material and labor to install
Install	Ea.	365	88		453	pair of 2′ 6″ wide solid wood bi-fold
Demolish and Install	Ea.	365	121.50		486.50	doors not including jamb, trim, track,
Reinstall	Ea.		88		88	and hardware.
Clean	Ea.	.67	26.50		27.17	
Paint	Ea.	55.50	101		156.50	
Minimum Charge	Job		242		242	

For customer support on your Contractor's Pricing Guide: Residential Repair & Remodeling, call 800.448.8182.

Bifold Door	Unit	Material	Labor	Equip.	Total	Specification
6'						
Demolish	Ea.		33.50		33.50	Includes material and labor to install
Install	Ea.	400	97		497	pair of 3' wide solid wood bi-fold
Demolish and Install	Ea.	400	130.50		530.50	doors not including jamb, trim, track,
Reinstall	Ea.		96.80		96.80	and hardware.
Clean	Ea.	.76	31		31.76	
Paint	Ea.	55.50	101		156.50	
Minimum Charge	Job		242		242	
8'						
Demolish	Ea.		33.50		33.50	Includes material and labor to install
Install	Ea.	234	108		342	pair of 4' wide solid wood bi-fold
Demolish and Install	Ea.	234	141.50		375.50	doors not including jamb, trim, track,
Reinstall	Ea.		107.56		107.56	and hardware.
Clean	Ea.	1.05	37.50		38.55	
Paint	Ea.	29	101		130	
Minimum Charge	Job		242		242	
Mirrored						
6'						
Demolish	Ea.		33.50		33.50	Includes material and labor to install
Install	Ea.	560	54		614	pair of 3' wide mirrored bi-fold doors
Demolish and Install	Ea.	560	87.50		647.50	including jamb, trim, track, and
Reinstall	Ea.		53.78		53.78	hardware.
Clean	Ea.	.76	15.60		16.36	
Minimum Charge	Job		242		242	
8'						
Demolish	Ea.		33.50		33.50	Includes material and labor to install
Install	Ea.	945	80.50		1025.50	pair of 4' wide mirrored bi-fold doors
Demolish and Install	Ea.	945	114		1059	including jamb, trim, track, and
Reinstall	Ea.		80.67		80.67	hardware.
Clean	Ea.	1.05	18.70		19.75	
Minimum Charge	Job		242		242	
Louvered						
3'						
Demolish	Ea.		33.50		33.50	Cost includes material and labor to
Install	Ea.	297	74.50		371.50	install 3' wide louvered bi-fold door
Demolish and Install	Ea.	297	108		405	not including jamb, trim, track, and
Reinstall	Ea.		74.46		74.46	hardware.
Clean	Ea.	1.52	15.60		17.12	
Paint	Ea.	7.70	57.50		65.20	
Minimum Charge	Job		242		242	
5'						
Demolish	Ea.		33.50		33.50	Includes material and labor to install
Install	Ea.	305	88		393	pair of 2' 6" wide louvered wood
Demolish and Install	Ea.	305	121.50		426.50	bi-fold doors not including jamb, trim,
Reinstall	Ea.		88		88	track, and hardware.
Clean	Ea.	.67	26.50		27.17	
Paint	Ea.	55.50	101		156.50	
Minimum Charge	Job		242		242	
6'						
Demolish	Ea.		33.50		33.50	Includes material and labor to install
Install	Ea.	375	97		472	pair of 3' wide louvered wood bi-fold
Demolish and Install	Ea.	375	130.50		505.50	doors not including jamb, trim, track,
Reinstall	Ea.		96.80		96.80	and hardware.
Clean	Ea.	.76	31		31.76	
Paint	Ea.	55.50	101		156.50	
Minimum Charge	Job		242		242	

Finish Carpentry

Bifold Door		Unit	Material	Labor	Equip.	Total	Specification
8'							
	Demolish	Ea.		33.50		33.50	Includes material and labor to install
	Install	Ea.	580	97		677	pair of 4' wide louvered wood bi-fold
	Demolish and Install	Ea.	580	130.50		710.50	doors not including jamb, trim, track,
	Reinstall	Ea.		96.80		96.80	and hardware.
	Clean	Ea.	3.62	26.50		30.12	
	Paint	Ea.	29	101		130	
	Minimum Charge	Job		242		242	
Accordion Type							
	Demolish	S.F.		1.22		1.22	Includes material and labor to install
	Install	Ea.	46	97		143	vinyl folding accordion closet doors,
	Demolish and Install	Ea.	46	98.22		144.22	including track and frame.
	Reinstall	Ea.		96.80		96.80	
	Clean	S.F.		.50		.50	
	Minimum Charge	Job		242		242	
Casing Trim							
Single Width							
	Demolish	Ea.		8.55		8.55	Cost includes material and labor to
	Install	Opng.	27	40.50		67.50	install 11/16" x 2-1/2" pine ranch
	Demolish and Install	Opng.	27	49.05		76.05	style casing for one side of a standard
	Reinstall	Opng.		32.27		32.27	door opening.
	Clean	Opng.	.22	3.74		3.96	
	Paint	Ea.	7.45	8.05		15.50	
	Minimum Charge	Job		242		242	
Double Width							
	Demolish	Ea.		10.15		10.15	Cost includes material and labor to
	Install	Opng.	31.50	48.50		80	install 11/16" x 2-1/2" pine ranch
	Demolish and Install	Opng.	31.50	58.65		90.15	style door casing for one side of a
	Reinstall	Opng.		38.72		38.72	double door opening.
	Clean	Opng.	.25	7.50		7.75	
	Paint	Ea.	7.45	9.15		16.60	
	Minimum Charge	Job		242		242	
Jamb							
	Demolish	Ea.		33.50		33.50	Includes material and labor to install
	Install	Ea.	142	13.10		155.10	flat pine jamb with square cut heads
	Demolish and Install	Ea.	142	46.60		188.60	and rabbeted sides for 6' 8" high and
	Clean	Ea.	.04	9.35		9.39	3-9/16" door including trim sets for
	Paint	Ea.	12.25	27		39.25	both sides.
	Minimum Charge	Job		121		121	
Stain							
	Install	Ea.	9.50	33.50		43	Includes labor and material to stain
	Minimum Charge	Job		202		202	single door and trim on both sides.
Shave & Refit							
	Install	Ea.		40.50		40.50	Includes labor to shave and rework
	Minimum Charge	Job		121		121	door to fit opening at the job site.

For customer support on your Contractor's Pricing Guide: Residential Repair & Remodeling, call 800.448.8182.

Finish Carpentry

Louver Door		Unit	Material	Labor	Equip.	Total	Specification
Full							
2' x 6' 8"							
	Demolish	Ea.		9.60		9.60	Cost includes material and labor to
	Install	Ea.	350	48.50		398.50	install 2' x 6' 8" pine full louver door,
	Demolish and Install	Ea.	350	58.10		408.10	jamb, hinges.
	Reinstall	Ea.		48.40		48.40	
	Clean	Ea.	2.82	47		49.82	
	Paint	Ea.	7.70	57.50		65.20	
	Minimum Charge	Job		242		242	
2' 8" x 6' 8"							
	Demolish	Ea.		9.60		9.60	Cost includes material and labor to
	Install	Ea.	415	48.50		463.50	install 2' 8" x 6' 8" pine full louver
	Demolish and Install	Ea.	415	58.10		473.10	door, jamb, hinges.
	Reinstall	Ea.		48.40		48.40	
	Clean	Ea.	2.82	47		49.82	
	Paint	Ea.	7.70	57.50		65.20	
	Minimum Charge	Job		242		242	
3' x 6' 8"							
	Demolish	Ea.		9.60		9.60	Cost includes material and labor to
	Install	Ea.	380	51		431	install 3' x 6' 8" pine full louver door,
	Demolish and Install	Ea.	380	60.60		440.60	jamb, hinges.
	Reinstall	Ea.		50.95		50.95	
	Clean	Ea.	2.82	47		49.82	
	Paint	Ea.	7.70	57.50		65.20	
	Minimum Charge	Job		242		242	
Half							
2' x 6' 8"							
	Demolish	Ea.		9.60		9.60	Cost includes material and labor to
	Install	Ea.	470	48.50		518.50	install 2' x 6' 8" pine half louver door,
	Demolish and Install	Ea.	470	58.10		528.10	jamb, hinges.
	Reinstall	Ea.		48.40		48.40	
	Clean	Ea.	2.82	47		49.82	
	Paint	Ea.	7.70	57.50		65.20	
	Minimum Charge	Job		242		242	
2' 8" x 6' 8"							
	Demolish	Ea.		9.60		9.60	Cost includes material and labor to
	Install	Ea.	485	48.50		533.50	install 2' 8" x 6' 8" pine half louver
	Demolish and Install	Ea.	485	58.10		543.10	door, jamb, hinges.
	Reinstall	Ea.		48.40		48.40	
	Clean	Ea.	2.82	47		49.82	
	Paint	Ea.	7.70	57.50		65.20	
	Minimum Charge	Job		242		242	
3' x 6' 8"							
	Demolish	Ea.		9.60		9.60	Cost includes material and labor to
	Install	Ea.	515	51		566	install 3' x 6' 8" pine half louver door,
	Demolish and Install	Ea.	515	60.60		575.60	jamb, hinges.
	Reinstall	Ea.		50.95		50.95	
	Clean	Ea.	2.82	47		49.82	
	Paint	Ea.	7.70	57.50		65.20	
	Minimum Charge	Job		242		242	
Jamb							
	Demolish	Ea.		33.50		33.50	Includes material and labor to install
	Install	Ea.	142	13.10		155.10	flat pine jamb with square cut heads
	Demolish and Install	Ea.	142	46.60		188.60	and rabbeted sides for 6' 8" high and
	Clean	Ea.	.04	9.35		9.39	3-9/16" door including trim sets for
	Paint	Ea.	12.25	27		39.25	both sides.
	Minimum Charge	Job		121		121	

For customer support on your Contractor's Pricing Guide: Residential Repair & Remodeling, call 800.448.8182.

Louver Door		Unit	Material	Labor	Equip.	Total	Specification
Stain							
	Install	Ea.	9.50	33.50		43	Includes labor and material to stain
	Minimum Charge	Job		202		202	single door and trim on both sides.
Shave & Refit							
	Install	Ea.		40.50		40.50	Includes labor to shave and rework
	Minimum Charge	Job		202		202	door to fit opening at the job site.

Bypass Sliding Door		Unit	Material	Labor	Equip.	Total	Specification
Wood							
5'							
	Demolish	Ea.		45		45	Includes material and labor to install
	Install	Opng.	284	88		372	pair of 2' 6" wide lauan hollow core
	Demolish and Install	Opng.	284	133		417	wood bi-pass doors with jamb, track
	Reinstall	Opng.		88		88	hardware.
	Clean	Ea.	.67	26.50		27.17	
	Paint	Ea.	55.50	101		156.50	
	Minimum Charge	Job		242		242	
6'							
	Demolish	Ea.		45		45	Includes material and labor to install
	Install	Opng.	210	97		307	pair of 3' wide lauan hollow core
	Demolish and Install	Opng.	210	142		352	wood bi-pass doors with jamb, track
	Reinstall	Opng.		96.80		96.80	and hardware.
	Clean	Ea.	.76	31		31.76	
	Paint	Ea.	55.50	101		156.50	
	Minimum Charge	Job		242		242	
8'							
	Demolish	Ea.		45		45	Includes material and labor to install
	Install	Opng.	415	121		536	pair of 4' wide lauan hollow core
	Demolish and Install	Opng.	415	166		581	wood bi-pass doors with jamb, track
	Reinstall	Opng.		121		121	and hardware.
	Clean	Ea.	1.05	37.50		38.55	
	Paint	Ea.	29	101		130	
	Minimum Charge	Job		242		242	
Mirrored							
6'							
	Demolish	Ea.		45		45	Includes material and labor to install
	Install	Opng.	450	97		547	pair of 3' wide mirrored bi-pass doors
	Demolish and Install	Opng.	450	142		592	with jamb, trim, track hardware.
	Reinstall	Opng.		96.80		96.80	
	Clean	Ea.	.76	15.60		16.36	
	Minimum Charge	Job		242		242	
8'							
	Demolish	Ea.		45		45	Includes material and labor to install
	Install	Opng.	695	108		803	pair of 4' wide mirrored bi-pass doors
	Demolish and Install	Opng.	695	153		848	with jamb, trim, track hardware.
	Reinstall	Opng.		107.56		107.56	
	Clean	Ea.	1.05	18.70		19.75	
	Minimum Charge	Job		242		242	

Finish Carpentry

Bypass Sliding Door

	Unit	Material	Labor	Equip.	Total	Specification
Casing Trim						
Single Width						Cost includes material and labor to
Demolish	Ea.		8.55		8.55	install 11/16" x 2-1/2" pine ranch
Install	Opng.	27	40.50		67.50	style casing for one side of a standard
Demolish and Install	Opng.	27	49.05		76.05	door opening.
Reinstall	Opng.		32.27		32.27	
Clean	Opng.	.22	3.74		3.96	
Paint	Ea.	7.45	8.05		15.50	
Minimum Charge	Job		242		242	
Double Width						Cost includes material and labor to
Demolish	Ea.		10.15		10.15	install 11/16" x 2-1/2" pine ranch
Install	Opng.	31.50	48.50		80	style door casing for one side of a
Demolish and Install	Opng.	31.50	58.65		90.15	double door opening.
Reinstall	Opng.		38.72		38.72	
Clean	Opng.	.25	7.50		7.75	
Paint	Ea.	7.45	9.15		16.60	
Minimum Charge	Job		242		242	
Jamb						Includes material and labor to install
Demolish	Ea.		33.50		33.50	flat pine jamb with square cut heads
Install	Ea.	142	13.10		155.10	and rabbeted sides for 6' 8" high and
Demolish and Install	Ea.	142	46.60		188.60	3-9/16" door including trim sets for
Clean	Ea.	.04	9.35		9.39	both sides.
Paint	Ea.	12.25	27		39.25	
Minimum Charge	Job		121		121	
Stain						Includes labor and material to stain
Install	Ea.	9.50	33.50		43	single door and trim on both sides.
Minimum Charge	Job		202		202	
Shave & Refit						Includes labor to shave and rework
Install	Ea.		40.50		40.50	door to fit opening at the job site.
Minimum Charge	Job		121		121	

Base Molding

	Unit	Material	Labor	Equip.	Total	Specification
One Piece						
1-5/8"						Includes material and labor to install
Demolish	L.F.		.69		.69	pine or birch base molding.
Install	L.F.	.40	1.43		1.83	
Demolish and Install	L.F.	.40	2.12		2.52	
Reinstall	L.F.		1.15		1.15	
Clean	L.F.	.01	.22		.23	
Paint	L.F.	.08	.44		.52	
Minimum Charge	Job		242		242	
2-1/2"						Cost includes material and labor to
Demolish	L.F.		.69		.69	install 1/2" x 2-1/2" pine or birch
Install	L.F.	.82	1.46		2.28	base molding.
Demolish and Install	L.F.	.82	2.15		2.97	
Reinstall	L.F.		1.17		1.17	
Clean	L.F.	.01	.22		.23	
Paint	L.F.	.08	.44		.52	
Minimum Charge	Job		242		242	

For customer support on your Contractor's Pricing Guide: Residential Repair & Remodeling, call 800.448.8182.

Finish Carpentry

Base Molding		Unit	Material	Labor	Equip.	Total	Specification
3-1/2"							
	Demolish	L.F.		.69		.69	Cost includes material and labor to
	Install	L.F.	.92	1.55		2.47	install 9/16" x 3-1/2" pine or birch
	Demolish and Install	L.F.	.92	2.24		3.16	base molding.
	Reinstall	L.F.		1.24		1.24	
	Clean	L.F.	.01	.22		.23	
	Paint	L.F.	.12	.48		.60	
	Minimum Charge	Job		242		242	
6"							
	Demolish	L.F.		.69		.69	Cost includes material and labor to
	Install	L.F.	1.53	1.55		3.08	install 1" x 6" pine or birch base
	Demolish and Install	L.F.	1.53	2.24		3.77	molding.
	Reinstall	L.F.		1.24		1.24	
	Clean	L.F.	.01	.31		.32	
	Paint	L.F.	.17	.61		.78	
	Minimum Charge	Job		242		242	
8"							
	Demolish	L.F.		.83		.83	Cost includes material and labor to
	Install	L.F.	2.30	1.55		3.85	install 1" x 8" pine or birch base
	Demolish and Install	L.F.	2.30	2.38		4.68	molding.
	Reinstall	L.F.		1.24		1.24	
	Clean	L.F.	.02	.37		.39	
	Paint	L.F.	.19	.67		.86	
	Minimum Charge	Job		242		242	
Premium / Custom Grade							
	Demolish	L.F.		.83		.83	Includes material and labor to install
	Install	L.F.	3.70	4.65		8.35	custom 2 or 3 piece oak or other
	Demolish and Install	L.F.	3.70	5.48		9.18	hardwood, stain grade trim.
	Reinstall	L.F.		3.72		3.72	
	Clean	L.F.	.02	.37		.39	
	Paint	L.F.	.20	.67		.87	
	Minimum Charge	Job		242		242	
One Piece Oak							
1-5/8"							
	Demolish	L.F.		.69		.69	Cost includes material and labor to
	Install	L.F.	1.55	1.43		2.98	install 1/2" x 1-5/8" oak base
	Demolish and Install	L.F.	1.55	2.12		3.67	molding.
	Reinstall	L.F.		1.15		1.15	
	Clean	L.F.	.01	.22		.23	
	Paint	L.F.	.15	.60		.75	
	Minimum Charge	Job		242		242	
2-1/2"							
	Demolish	L.F.		.69		.69	Cost includes material and labor to
	Install	L.F.	1.83	1.46		3.29	install 1/2" x 2-1/2" oak base
	Demolish and Install	L.F.	1.83	2.15		3.98	molding.
	Reinstall	L.F.		1.17		1.17	
	Clean	L.F.	.01	.22		.23	
	Paint	L.F.	.15	.61		.76	
	Minimum Charge	Job		242		242	
3-1/2"							
	Demolish	L.F.		.69		.69	Cost includes material and labor to
	Install	L.F.	2.33	1.49		3.82	install 1/2" x 3-1/2" oak base
	Demolish and Install	L.F.	2.33	2.18		4.51	molding.
	Reinstall	L.F.		1.19		1.19	
	Clean	L.F.	.01	.22		.23	
	Paint	L.F.	.17	.62		.79	
	Minimum Charge	Job		242		242	

For customer support on your Contractor's Pricing Guide: Residential Repair & Remodeling, call 800.448.8182.

Finish Carpentry

Base Molding		Unit	Material	Labor	Equip.	Total	Specification
6″							
	Demolish	L.F.		.69		.69	Cost includes material and labor to
	Install	L.F.	2.19	1.55		3.74	install 1″ x 6″ oak base molding.
	Demolish and Install	L.F.	2.19	2.24		4.43	
	Reinstall	L.F.		1.24		1.24	
	Clean	L.F.	.02	.37		.39	
	Paint	L.F.	.17	.65		.82	
	Minimum Charge	Job		242		242	
8″							
	Demolish	L.F.		.83		.83	Cost includes material and labor to
	Install	L.F.	3.20	1.55		4.75	install 1″ x 8″ oak base molding.
	Demolish and Install	L.F.	3.20	2.38		5.58	
	Reinstall	L.F.		1.24		1.24	
	Clean	L.F.	.02	.37		.39	
	Paint	L.F.	.18	.67		.85	
	Minimum Charge	Job		242		242	

Vinyl Base Molding

		Unit	Material	Labor	Equip.	Total	Specification
2-1/2″							
	Demolish	L.F.		.69		.69	Cost includes material and labor to
	Install	L.F.	1.89	1.49		3.38	install 2-1/2″ vinyl or rubber cove
	Demolish and Install	L.F.	1.89	2.18		4.07	base including adhesive.
	Reinstall	L.F.		1.49		1.49	
	Clean	L.F.		.34		.34	
	Minimum Charge	Job		237		237	
4″							
	Demolish	L.F.		.69		.69	Cost includes material and labor to
	Install	Ea.	3.21	1.49		4.70	install 4″ vinyl or rubber cove base
	Demolish and Install	Ea.	3.21	2.18		5.39	including adhesive.
	Reinstall	Ea.		1.49		1.49	
	Clean	L.F.		.34		.34	
	Minimum Charge	Job		237		237	
6″							
	Demolish	L.F.		.69		.69	Cost includes material and labor to
	Install	L.F.	2.97	1.49		4.46	install 6″ vinyl or rubber cove base
	Demolish and Install	L.F.	2.97	2.18		5.15	including adhesive.
	Reinstall	L.F.		1.49		1.49	
	Clean	L.F.		.34		.34	
	Minimum Charge	Job		237		237	

Prefinished

		Unit	Material	Labor	Equip.	Total	Specification
	Demolish	L.F.		.69		.69	Cost includes material and labor to
	Install	L.F.	1.73	1.55		3.28	install 9/16″ x 3-5/16″ prefinished
	Demolish and Install	L.F.	1.73	2.24		3.97	base molding.
	Reinstall	L.F.		1.24		1.24	
	Clean	L.F.	.01	.22		.23	
	Paint	L.F.	.12	.48		.60	
	Minimum Charge	Job		242		242	

Shoe

		Unit	Material	Labor	Equip.	Total	Specification
	Demolish	L.F.		.69		.69	Includes material and labor to install
	Install	L.F.	.40	1.43		1.83	pine or birch base molding.
	Demolish and Install	L.F.	.40	2.12		2.52	
	Reinstall	L.F.		1.15		1.15	
	Clean	L.F.	.01	.22		.23	
	Paint	L.F.	.17	.65		.82	
	Minimum Charge	Job		242		242	

For customer support on your Contractor's Pricing Guide: Residential Repair & Remodeling, call 800.448.8182.

Finish Carpentry

Base Molding

Base Molding	Unit	Material	Labor	Equip.	Total	Specification
Carpeted						
Demolish	L.F.		.28		.28	Includes material and labor to install
Install	L.F.	1.20	2.84		4.04	carpet baseboard with adhesive.
Demolish and Install	L.F.	1.20	3.12		4.32	
Reinstall	L.F.		2.27		2.27	
Clean	L.F.		.14		.14	
Minimum Charge	Job		235		235	
Ceramic Tile						
Demolish	L.F.		1.51		1.51	Includes material and labor to install
Install	L.F.	6.20	6.50		12.70	thin set ceramic tile cove base
Demolish and Install	L.F.	6.20	8.01		14.21	including grout.
Clean	L.F.		.71		.71	
Minimum Charge	Job		235		235	
Quarry Tile						
Demolish	L.F.		1.51		1.51	Includes material and labor to install
Install	L.F.	1.65	7.55		9.20	quarry tile base including grout.
Demolish and Install	L.F.	1.65	9.06		10.71	
Clean	L.F.		.71		.71	
Minimum Charge	Job		235		235	
Caulk and Renail						
Install	L.F.	.02	.76		.78	Includes labor and material to renail
Minimum Charge	Job		242		242	base and caulk where necessary.

Molding

Molding	Unit	Material	Labor	Equip.	Total	Specification
Oak Base						
Demolish	L.F.		.69		.69	Cost includes material and labor to
Install	L.F.	1.83	1.46		3.29	install 1/2" x 2-1/2" oak base
Demolish and Install	L.F.	1.83	2.15		3.98	molding.
Reinstall	L.F.		1.17		1.17	
Clean	L.F.	.01	.22		.23	
Paint	L.F.	.15	.61		.76	
Minimum Charge	Job		242		242	
Shoe						
Demolish	L.F.		.69		.69	Includes material and labor to install
Install	L.F.	.40	1.43		1.83	pine or birch base molding.
Demolish and Install	L.F.	.40	2.12		2.52	
Reinstall	L.F.		1.15		1.15	
Clean	L.F.	.01	.22		.23	
Paint	L.F.	.17	.65		.82	
Minimum Charge	Job		242		242	
Quarter Round						
Demolish	L.F.		.69		.69	Includes material and labor to install
Install	L.F.	.75	1.90		2.65	pine base quarter round molding.
Demolish and Install	L.F.	.75	2.59		3.34	
Reinstall	L.F.		1.52		1.52	
Clean	L.F.	.01	.22		.23	
Paint	L.F.	.19	1.01		1.20	
Minimum Charge	Job		242		242	
Chair Rail						
Demolish	L.F.		.69		.69	Cost includes material and labor to
Install	L.F.	1.68	1.79		3.47	install 5/8" x 2-1/2" oak chair rail
Demolish and Install	L.F.	1.68	2.48		4.16	molding.
Reinstall	L.F.		1.43		1.43	
Clean	L.F.	.01	.22		.23	
Paint	L.F.	.19	1.01		1.20	
Minimum Charge	Job		242		242	

For customer support on your Contractor's Pricing Guide: Residential Repair & Remodeling, call 800.448.8182.

Molding	Unit	Material	Labor	Equip.	Total	Specification
Crown						
Demolish	L.F.		.69		.69	Includes material and labor to install
Install	L.F.	1.69	1.94		3.63	unfinished pine crown molding.
Demolish and Install	L.F.	1.69	2.63		4.32	
Reinstall	L.F.		1.55		1.55	
Clean	L.F.	.02	.29		.31	
Paint	L.F.	.19	1.01		1.20	
Minimum Charge	Job		242		242	
Cove						
Demolish	L.F.		.69		.69	Cost includes material and labor to
Install	L.F.	2.88	1.83		4.71	install 11/16" x 2-3/4" pine cove
Demolish and Install	L.F.	2.88	2.52		5.40	molding.
Reinstall	L.F.		1.46		1.46	
Clean	L.F.	.01	.22		.23	
Paint	L.F.	.19	1.01		1.20	
Minimum Charge	Job		242		242	
Corner						
Demolish	L.F.		.69		.69	Includes material and labor to install
Install	L.F.	1.84	.76		2.60	inside or outside pine corner molding.
Demolish and Install	L.F.	1.84	1.45		3.29	
Reinstall	L.F.		.61		.61	
Clean	L.F.	.01	.22		.23	
Paint	L.F.	.19	1.01		1.20	
Minimum Charge	Job		242		242	
Picture						
Demolish	L.F.		.69		.69	Cost includes material and labor to
Install	L.F.	1.57	2.02		3.59	install 9/16" x 2-1/2" pine casing.
Demolish and Install	L.F.	1.57	2.71		4.28	
Reinstall	L.F.		1.61		1.61	
Clean	L.F.	.01	.22		.23	
Paint	L.F.	.19	1.01		1.20	
Minimum Charge	Job		242		242	
Single Cased Opening						
Demolish	Ea.		33.50		33.50	Includes material and labor to install
Install	Ea.	174	30.50		204.50	flat pine jamb with square cut heads
Demolish and Install	Ea.	174	64		238	and rabbeted sides for 6' 8" high
Clean	Ea.	.11	7.80		7.91	opening including trim sets for both
Paint	Ea.	.72	10.10		10.82	sides.
Minimum Charge	Job		242		242	
Double Cased Opening						
Demolish	Ea.		33.50		33.50	Includes material and labor to install
Install	Ea.	204	34.50		238.50	flat pine jamb with square cut heads
Demolish and Install	Ea.	204	68		272	and rabbeted sides for 6' 8" high
Clean	Ea.	.12	9.10		9.22	opening including trim sets for both
Paint	Ea.	.84	11.85		12.69	sides.
Minimum Charge	Job		242		242	

Window Trim Set	Unit	Material	Labor	Equip.	Total	Specification
Single						
Demolish	Ea.		9.25		9.25	Cost includes material and labor to
Install	Opng.	47	48.50		95.50	install 11/16" x 2-1/2" pine ranch
Demolish and Install	Opng.	47	57.75		104.75	style window casing.
Reinstall	Opng.		38.72		38.72	
Clean	Ea.	.15	5.35		5.50	
Paint	Ea.	.67	9.60		10.27	
Minimum Charge	Job		242		242	

Finish Carpentry

Window Trim Set

	Unit	Material	Labor	Equip.	Total	Specification
Double						
Demolish	Ea.		10.40		10.40	Cost includes material and labor to
Install	Ea.	87.50	60.50		148	install 11/16" x 2-1/2" pine ranch
Demolish and Install	Ea.	87.50	70.90		158.40	style trim for casing one side of a
Reinstall	Ea.		48.40		48.40	double window opening.
Clean	Ea.	.28	6.25		6.53	
Paint	Ea.	1.13	16.15		17.28	
Minimum Charge	Job		242		242	
Triple						
Demolish	Ea.		11.85		11.85	Cost includes material and labor to
Install	Opng.	94.50	80.50		175	install 11/16" x 2-1/2" pine ranch
Demolish and Install	Opng.	94.50	92.35		186.85	style trim for casing one side of a
Reinstall	Opng.		64.53		64.53	triple window opening.
Clean	Ea.	.39	7.50		7.89	
Paint	Ea.	1.58	22.50		24.08	
Minimum Charge	Job		242		242	
Window Casing Per L.F.						
Demolish	L.F.		.69		.69	Cost includes material and labor to
Install	L.F.	1.58	1.94		3.52	install 11/16" x 2-1/2" pine ranch
Demolish and Install	L.F.	1.58	2.63		4.21	style trim for casing.
Reinstall	L.F.		1.55		1.55	
Clean	L.F.	.01	.22		.23	
Paint	L.F.	.19	1.01		1.20	
Minimum Charge	Job		242		242	

Window Sill

	Unit	Material	Labor	Equip.	Total	Specification
Wood						
Demolish	L.F.		.69		.69	Includes material and labor to install
Install	L.F.	3.10	2.42		5.52	flat, wood window stool.
Demolish and Install	L.F.	3.10	3.11		6.21	
Reinstall	L.F.		1.94		1.94	
Clean	L.F.	.01	.22		.23	
Paint	L.F.	.19	1.01		1.20	
Minimum Charge	Job		242		242	
Marble						
Demolish	L.F.		1.04		1.04	Includes material and labor to install
Install	L.F.	13.90	10.45		24.35	marble window sills, 6" x 3/4" thick.
Demolish and Install	L.F.	13.90	11.49		25.39	
Reinstall	L.F.		8.34		8.34	
Clean	L.F.	.01	.22		.23	
Minimum Charge	Job		235		235	

Shelving

	Unit	Material	Labor	Equip.	Total	Specification
Pine						
18"						
Demolish	L.F.		1.19		1.19	Cost includes material and labor to
Install	L.F.	7.80	5.10		12.90	install 1" thick custom shelving board
Demolish and Install	L.F.	7.80	6.29		14.09	18" wide.
Reinstall	L.F.		4.08		4.08	
Clean	L.F.	.02	.62		.64	
Paint	L.F.	2.46	.64		3.10	
Minimum Charge	Job		242		242	

For customer support on your Contractor's Pricing Guide: Residential Repair & Remodeling, call 800.448.8182.

Finish Carpentry

Shelving

Shelving	Unit	Material	Labor	Equip.	Total	Specification
24″						
Demolish	L.F.		1.19		1.19	Cost includes material and labor to
Install	L.F.	10.40	5.70		16.10	install 1″ thick custom shelving board.
Demolish and Install	L.F.	10.40	6.89		17.29	
Reinstall	L.F.		4.56		4.56	
Clean	L.F.	.03	.94		.97	
Paint	L.F.	3.34	.85		4.19	
Minimum Charge	Job		242		242	
Particle Board						
Demolish	L.F.		1.09		1.09	Cost includes material and labor to
Install	L.F.	1.99	2.55		4.54	install 3/4″ thick particle shelving
Demolish and Install	L.F.	1.99	3.64		5.63	board 18″ wide.
Reinstall	L.F.		2.04		2.04	
Clean	L.F.	.02	.62		.64	
Paint	S.F.	.10	.94		1.04	
Minimum Charge	Job		242		242	
Plywood						
Demolish	L.F.		1.09		1.09	Cost includes material and labor to
Install	L.F.	1.13	7.45		8.58	install 3/4″ thick plywood shelving
Demolish and Install	L.F.	1.13	8.54		9.67	board.
Reinstall	L.F.		5.96		5.96	
Clean	L.F.	.02	.62		.64	
Paint	S.F.	.10	.94		1.04	
Minimum Charge	Job		242		242	
Hardwood (oak)						
Demolish	L.F.		1.09		1.09	Includes material and labor to install
Install	L.F.	10.30	3.34		13.64	oak shelving, 1″ x 12″, incl. cleats and
Demolish and Install	L.F.	10.30	4.43		14.73	bracing.
Reinstall	L.F.		2.67		2.67	
Clean	L.F.	.02	.62		.64	
Paint	S.F.	.10	.94		1.04	
Minimum Charge	Job		242		242	
Custom Bookcase						
Demolish	L.F.		13.85		13.85	Includes material and labor to install
Install	L.F.	40	48.50		88.50	custom modular bookcase unit with
Demolish and Install	L.F.	40	62.35		102.35	clear pine faced frames, shelves 12″
Reinstall	L.F.		38.72		38.72	O.C., 7′ high, 8″ deep.
Clean	L.F.	.61	13.35		13.96	
Paint	L.F.	2.27	31		33.27	
Minimum Charge	Job		242		242	
Glass						
Demolish	L.F.		1.09		1.09	Includes material and labor to install
Install	L.F.	51	3.34		54.34	glass shelving, 1″ x 12″, incl. cleats
Demolish and Install	L.F.	51	4.43		55.43	and bracing.
Reinstall	L.F.		2.67		2.67	
Clean	L.F.	.02	.58		.60	
Minimum Charge	Job		235		235	

Closet Shelving

Closet Shelving	Unit	Material	Labor	Equip.	Total	Specification
Closet Shelf and Rod						
Demolish	L.F.		1.09		1.09	Cost includes material and labor to
Install	L.F.	3.36	8.05		11.41	install 1″ thick custom shelving board
Demolish and Install	L.F.	3.36	9.14		12.50	18″ wide and 1″ diameter clothes pole
Reinstall	L.F.		6.45		6.45	with brackets 3′ O.C.
Clean	L.F.	.01	.47		.48	
Paint	L.F.	.39	1.01		1.40	
Minimum Charge	Job		242		242	

207

Finish Carpentry

Closet Shelving

Clothing Rod	Unit	Material	Labor	Equip.	Total	Specification
Demolish	L.F.		.72		.72	Includes material and labor to install
Install	L.F.	2.12	2.42		4.54	fir closet pole, 1-5/8" diameter.
Demolish and Install	L.F.	2.12	3.14		5.26	
Reinstall	L.F.		1.94		1.94	
Clean	L.F.	.01	.63		.64	
Paint	L.F.	.36	1.15		1.51	
Minimum Charge	Job		242		242	

Stair Assemblies

	Unit	Material	Labor	Equip.	Total	Specification
Straight Hardwood						
Demolish	Ea.		480		480	Includes material and labor to install
Install	Flight	28600	325		28925	factory cut and assembled straight
Demolish and Install	Flight	28600	805		29405	closed box stairs with oak treads and
Clean	Flight	1.52	31.50		33.02	prefinished stair rail with balusters.
Paint	Flight	34.50	54		88.50	
Minimum Charge	Job		242		242	
Clear Oak Tread						
Demolish	Ea.		12.20		12.20	Includes material and labor to install
Install	Ea.	62.50	27		89.50	1-1/4" thick clear oak treads per riser.
Demolish and Install	Ea.	62.50	39.20		101.70	
Clean	Ea.	.04	3.15		3.19	
Paint	Ea.	1.14	2.99		4.13	
Minimum Charge	Job		242		242	
Landing						
Demolish	S.F.		4.33		4.33	Includes material and labor to install
Install	S.F.	11.70	27		38.70	clear oak landing.
Demolish and Install	S.F.	11.70	31.33		43.03	
Clean	S.F.		.47		.47	
Paint	S.F.	.63	1		1.63	
Minimum Charge	Job		242		242	
Refinish / Stain						
Install	S.F.	.26	2.88		3.14	Includes labor and material to refinish
Minimum Charge	Job		202		202	wood floor.
Spiral Hardwood						
Demolish	Ea.		480		480	Cost includes material and labor to
Install	Flight	2550	645		3195	install 4' - 6' diameter spiral stairs with
Demolish and Install	Flight	2550	1125		3675	oak treads, factory cut and assembled
Clean	Flight	1.52	31.50		33.02	with double handrails.
Paint	Flight	34.50	54		88.50	
Minimum Charge	Job		242		242	

Stairs

Disappearing	Unit	Material	Labor	Equip.	Total	Specification
Demolish	Ea.		26		26	Includes material and labor to install a
Install	Ea.	210	138		348	folding staircase.
Demolish and Install	Ea.	210	164		374	
Reinstall	Ea.		110.63		110.63	
Clean	Ea.	.23	23.50		23.73	
Minimum Charge	Job		242		242	

For customer support on your Contractor's Pricing Guide: Residential Repair & Remodeling, call 800.448.8182.

Finish Carpentry

Stair Components

	Unit	Material	Labor	Equip.	Total	Specification
Handrail w / Balusters						
Demolish	L.F.		3.46		3.46	Includes material and labor to install
Install	L.F.	67	10.10		77.10	prefinished assembled stair rail with
Demolish and Install	L.F.	67	13.56		80.56	brackets, turned balusters and newel.
Clean	L.F.	.31	2.58		2.89	
Paint	L.F.	3.23	4.75		7.98	
Minimum Charge	Job		242		242	
Bannister						
Demolish	L.F.		3.46		3.46	Includes material and labor to install
Install	L.F.	18.60	8.05		26.65	built-up oak railings.
Demolish and Install	L.F.	18.60	11.51		30.11	
Clean	L.F.	.31	2.34		2.65	
Paint	L.F.	3.23	3.67		6.90	
Minimum Charge	Job		242		242	

For customer support on your Contractor's Pricing Guide: Residential Repair & Remodeling, call 800.448.8182.

Cabinets and Countertops

Wall and Base Cabinets

Countertop

Wall Cabinets	Unit	Material	Labor	Equip.	Total	Specification
Good Grade Laminated						
12″ Wide, 1 Door						
Demolish	Ea.		10.40		10.40	Includes material and labor to install
Install	Ea.	89	44		133	good modular unit with melamine
Demolish and Install	Ea.	89	54.40		143.40	laminated to particle board, in
Reinstall	Ea.		35.20		35.20	textured colors or wood grain print
Clean	Ea.	.23	7.20		7.43	finish, hinges and pulls.
Paint	Ea.	.97	14.40		15.37	
Minimum Charge	Job		242		242	
15″ Wide, 1 Door						
Demolish	Ea.		13		13	Includes material and labor to install
Install	Ea.	202	45		247	good modular unit with melamine
Demolish and Install	Ea.	202	58		260	laminated to particle board, in
Reinstall	Ea.		36.19		36.19	textured colors or wood grain print
Clean	Ea.	.29	8.90		9.19	finish, hinges and pulls.
Paint	Ea.	1.21	18.35		19.56	
Minimum Charge	Job		242		242	
18″ Wide, 1 Door						
Demolish	Ea.		15.70		15.70	Includes material and labor to install
Install	Ea.	223	46.50		269.50	good modular unit with melamine
Demolish and Install	Ea.	223	62.20		285.20	laminated to particle board, in
Reinstall	Ea.		37.05		37.05	textured colors or wood grain print
Clean	Ea.	.34	10.70		11.04	finish, hinges and pulls.
Paint	Ea.	1.45	21		22.45	
Minimum Charge	Job		242		242	
21″ Wide, 1 Door						
Demolish	Ea.		18.05		18.05	Includes material and labor to install
Install	Ea.	290	47		337	good modular unit with melamine
Demolish and Install	Ea.	290	65.05		355.05	laminated to particle board, in
Reinstall	Ea.		37.59		37.59	textured colors or wood grain print
Clean	Ea.	.40	12.45		12.85	finish, hinges and pulls.
Paint	Ea.	1.71	25		26.71	
Minimum Charge	Job		242		242	
24″ w / Blind Corner						
Demolish	Ea.		21		21	Includes material and labor to install
Install	Ea.	260	47.50		307.50	good modular unit with melamine
Demolish and Install	Ea.	260	68.50		328.50	laminated to particle board, in
Reinstall	Ea.		38.15		38.15	textured colors or wood grain print
Clean	Ea.	.46	14.40		14.86	finish, hinges and pulls.
Paint	Ea.	1.94	29		30.94	
Minimum Charge	Job		242		242	

211

Cabinets and Countertops

Wall Cabinets	Unit	Material	Labor	Equip.	Total	Specification
27″ Wide, 2 Door						
Demolish	Ea.		23		23	Includes material and labor to install
Install	Ea.	83.50	49		132.50	good modular unit with melamine
Demolish and Install	Ea.	83.50	72		155.50	laminated to particle board, in
Reinstall	Ea.		39.11		39.11	textured colors or wood grain print
Clean	Ea.	.52	16.25		16.77	finish, hinges and pulls.
Paint	Ea.	2.20	33.50		35.70	
Minimum Charge	Job		242		242	
30″ Wide, 2 Door						
Demolish	Ea.		26		26	Includes material and labor to install
Install	Ea.	335	50		385	good modular unit with melamine
Demolish and Install	Ea.	335	76		411	laminated to particle board, in
Reinstall	Ea.		40.12		40.12	textured colors or wood grain print
Clean	Ea.	.57	17.80		18.37	finish, hinges and pulls.
Paint	Ea.	2.43	36.50		38.93	
Minimum Charge	Job		242		242	
36″ Wide, 2 Door						
Demolish	Ea.		31		31	Includes material and labor to install
Install	Ea.	320	51.50		371.50	good modular unit with melamine
Demolish and Install	Ea.	320	82.50		402.50	laminated to particle board, in
Reinstall	Ea.		41.19		41.19	textured colors or wood grain print
Clean	Ea.	.68	22		22.68	finish, hinges and pulls.
Paint	Ea.	2.94	45		47.94	
Minimum Charge	Job		242		242	
48″ Wide, 2 Door						
Demolish	Ea.		41.50		41.50	Includes material and labor to install
Install	Ea.	645	52.50		697.50	good modular unit with melamine
Demolish and Install	Ea.	645	94		739	laminated to particle board, in
Reinstall	Ea.		42.09		42.09	textured colors or wood grain print
Clean	Ea.	.91	29		29.91	finish, hinges and pulls.
Paint	Ea.	3.89	57.50		61.39	
Minimum Charge	Job		242		242	
Above-Appliance						
Demolish	Ea.		26		26	Includes material and labor to install
Install	Ea.	183	39		222	good modular unit with melamine
Demolish and Install	Ea.	183	65		248	laminated to particle board, in
Reinstall	Ea.		31.23		31.23	textured colors or wood grain print
Clean	Ea.	.57	17.80		18.37	finish, hinges and pulls.
Paint	Ea.	2.43	36.50		38.93	
Minimum Charge	Job		242		242	
Better Grade Wood						
12″ Wide, 1 Door						
Demolish	Ea.		10.40		10.40	Includes material and labor to install
Install	Ea.	119	44		163	custom modular unit with solid
Demolish and Install	Ea.	119	54.40		173.40	hardwood faced frames, hardwood
Reinstall	Ea.		35.20		35.20	door frames and drawer fronts,
Clean	Ea.	.23	7.20		7.43	hardwood veneer on raised door
Paint	Ea.	.97	15.50		16.47	panels, hinges and pulls.
Minimum Charge	Job		242		242	
15″ Wide, 1 Door						
Demolish	Ea.		13		13	Includes material and labor to install
Install	Ea.	271	45		316	custom modular unit with solid
Demolish and Install	Ea.	271	58		329	hardwood faced frames, hardwood
Reinstall	Ea.		36.19		36.19	door frames and drawer fronts,
Clean	Ea.	.29	8.90		9.19	hardwood veneer on raised door
Paint	Ea.	1.21	19.20		20.41	panels, hinges and pulls.
Minimum Charge	Job		242		242	

Cabinets and Countertops

Wall Cabinets	Unit	Material	Labor	Equip.	Total	Specification
18" Wide, 1 Door						
Demolish	Ea.		15.70		15.70	Includes material and labor to install
Install	Ea.	297	46.50		343.50	custom modular unit with solid
Demolish and Install	Ea.	297	62.20		359.20	hardwood faced frames, hardwood
Reinstall	Ea.		37.05		37.05	door frames and drawer fronts,
Clean	Ea.	.34	10.70		11.04	hardwood veneer on raised door
Paint	Ea.	1.45	23.50		24.95	panels, hinges and pulls.
Minimum Charge	Job		242		242	
21" Wide, 1 Door						
Demolish	Ea.		18.05		18.05	Includes material and labor to install
Install	Ea.	385	47		432	custom modular unit with solid
Demolish and Install	Ea.	385	65.05		450.05	hardwood faced frames, hardwood
Reinstall	Ea.		37.59		37.59	door frames and drawer fronts,
Clean	Ea.	.40	12.45		12.85	hardwood veneer on raised door
Paint	Ea.	1.71	27		28.71	panels, hinges and pulls.
Minimum Charge	Job		242		242	
24" w / Blind Corner						
Demolish	Ea.		21		21	Includes material and labor to install
Install	Ea.	345	47.50		392.50	custom modular unit with solid
Demolish and Install	Ea.	345	68.50		413.50	hardwood faced frames, hardwood
Reinstall	Ea.		38.15		38.15	door frames and drawer fronts,
Clean	Ea.	.46	14.40		14.86	hardwood veneer on raised door
Paint	Ea.	1.94	31		32.94	panels, hinges and pulls.
Minimum Charge	Job		242		242	
27" Wide, 2 Door						
Demolish	Ea.		23		23	Includes material and labor to install
Install	Ea.	111	49		160	custom modular unit with solid
Demolish and Install	Ea.	111	72		183	hardwood faced frames, hardwood
Reinstall	Ea.		39.11		39.11	door frames and drawer fronts,
Clean	Ea.	.52	16.25		16.77	hardwood veneer on raised door
Paint	Ea.	2.20	33.50		35.70	panels, hinges and pulls.
Minimum Charge	Job		242		242	
30" Wide, 2 Door						
Demolish	Ea.		26		26	Includes material and labor to install
Install	Ea.	445	50		495	custom modular unit with solid
Demolish and Install	Ea.	445	76		521	hardwood faced frames, hardwood
Reinstall	Ea.		40.12		40.12	door frames and drawer fronts,
Clean	Ea.	.57	17.80		18.37	hardwood veneer on raised door
Paint	Ea.	2.43	40.50		42.93	panels, hinges and pulls.
Minimum Charge	Job		242		242	
36" Wide, 2 Door						
Demolish	Ea.		31		31	Includes material and labor to install
Install	Ea.	425	51.50		476.50	custom modular unit with solid
Demolish and Install	Ea.	425	82.50		507.50	hardwood faced frames, hardwood
Reinstall	Ea.		41.19		41.19	door frames and drawer fronts,
Clean	Ea.	.68	22		22.68	hardwood veneer on raised door
Paint	Ea.	2.94	45		47.94	panels, hinges and pulls.
Minimum Charge	Job		242		242	
48" Wide, 2 Door						
Demolish	Ea.		41.50		41.50	Includes material and labor to install
Install	Ea.	860	52.50		912.50	custom modular unit with solid
Demolish and Install	Ea.	860	94		954	hardwood faced frames, hardwood
Reinstall	Ea.		42.09		42.09	door frames and drawer fronts,
Clean	Ea.	.91	29		29.91	hardwood veneer on raised door
Paint	Ea.	3.89	57.50		61.39	panels, hinges and pulls.
Minimum Charge	Job		242		242	

For customer support on your Contractor's Pricing Guide: Residential Repair & Remodeling, call 800.448.8182.

Cabinets and Countertops

Wall Cabinets		Unit	Material	Labor	Equip.	Total	Specification
Above-appliance							
	Demolish	Ea.		26		26	Includes material and labor to install
	Install	Ea.	244	39		283	custom modular unit with solid
	Demolish and Install	Ea.	244	65		309	hardwood faced frames, hardwood
	Reinstall	Ea.		31.23		31.23	door frames and drawer fronts,
	Clean	Ea.	.57	17.80		18.37	hardwood veneer on raised door
	Paint	Ea.	2.43	40.50		42.93	panels, hinges and pulls.
	Minimum Charge	Job		242		242	
Premium Grade Wood							
12″ Wide, 1 Door							
	Demolish	Ea.		10.40		10.40	Includes material and labor to install
	Install	Ea.	143	44		187	premium modular unit with solid
	Demolish and Install	Ea.	143	54.40		197.40	hardwood faced frames, hardwood
	Reinstall	Ea.		35.20		35.20	door frames and drawer fronts,
	Clean	Ea.	.23	7.20		7.43	hardwood veneer on raised door
	Paint	Ea.	.97	16.80		17.77	panels, hinges and pulls.
	Minimum Charge	Job		242		242	
15″ Wide, 1 Door							
	Demolish	Ea.		13		13	Includes material and labor to install
	Install	Ea.	360	45		405	premium modular unit with solid
	Demolish and Install	Ea.	360	58		418	hardwood faced frames, hardwood
	Reinstall	Ea.		36.19		36.19	door frames and drawer fronts,
	Clean	Ea.	.29	8.90		9.19	hardwood veneer on raised door
	Paint	Ea.	1.21	21		22.21	panels, hinges and pulls.
	Minimum Charge	Job		242		242	
18″ Wide, 1 Door							
	Demolish	Ea.		15.70		15.70	Includes material and labor to install
	Install	Ea.	360	46.50		406.50	premium modular unit with solid
	Demolish and Install	Ea.	360	62.20		422.20	hardwood faced frames, hardwood
	Reinstall	Ea.		37.05		37.05	door frames and drawer fronts,
	Clean	Ea.	.34	10.70		11.04	hardwood veneer on raised door
	Paint	Ea.	1.45	25		26.45	panels, hinges and pulls.
	Minimum Charge	Job		242		242	
21″ Wide, 1 Door							
	Demolish	Ea.		18.05		18.05	Includes material and labor to install
	Install	Ea.	460	47		507	premium modular unit with solid
	Demolish and Install	Ea.	460	65.05		525.05	hardwood faced frames, hardwood
	Reinstall	Ea.		37.59		37.59	door frames and drawer fronts,
	Clean	Ea.	.40	12.45		12.85	hardwood veneer on raised door
	Paint	Ea.	1.71	29		30.71	panels, hinges and pulls.
	Minimum Charge	Job		242		242	
24″ w / Blind Corner							
	Demolish	Ea.		21		21	Includes material and labor to install
	Install	Ea.	415	47.50		462.50	premium modular unit with solid
	Demolish and Install	Ea.	415	68.50		483.50	hardwood faced frames, hardwood
	Reinstall	Ea.		38.15		38.15	door frames and drawer fronts,
	Clean	Ea.	.46	14.40		14.86	hardwood veneer on raised door
	Paint	Ea.	1.94	33.50		35.44	panels, hinges and pulls.
	Minimum Charge	Job		242		242	
27″ Wide, 2 Door							
	Demolish	Ea.		23		23	Includes material and labor to install
	Install	Ea.	199	49		248	premium modular unit with solid
	Demolish and Install	Ea.	199	72		271	hardwood faced frames, hardwood
	Reinstall	Ea.		39.11		39.11	door frames and drawer fronts,
	Clean	Ea.	.52	16.25		16.77	hardwood veneer on raised door
	Paint	Ea.	2.20	36.50		38.70	panels, hinges and pulls.
	Minimum Charge	Job		242		242	

For customer support on your Contractor's Pricing Guide: Residential Repair & Remodeling, call 800.448.8182.

Cabinets and Countertops

Wall Cabinets		Unit	Material	Labor	Equip.	Total	Specification
30" Wide, 2 Door							
	Demolish	Ea.		26		26	Includes material and labor to install
	Install	Ea.	535	50		585	premium modular unit with solid
	Demolish and Install	Ea.	535	76		611	hardwood faced frames, hardwood
	Reinstall	Ea.		40.12		40.12	door frames and drawer fronts,
	Clean	Ea.	.57	17.80		18.37	hardwood veneer on raised door
	Paint	Ea.	2.43	40.50		42.93	panels, hinges and pulls.
	Minimum Charge	Job		242		242	
36" Wide, 2 Door							
	Demolish	Ea.		31		31	Includes material and labor to install
	Install	Ea.	510	51.50		561.50	premium modular unit with solid
	Demolish and Install	Ea.	510	82.50		592.50	hardwood faced frames, hardwood
	Reinstall	Ea.		41.19		41.19	door frames and drawer fronts,
	Clean	Ea.	.68	22		22.68	hardwood veneer on raised door
	Paint	Ea.	2.94	50.50		53.44	panels, hinges and pulls.
	Minimum Charge	Job		242		242	
48" Wide, 2 Door							
	Demolish	Ea.		41.50		41.50	Includes material and labor to install
	Install	Ea.	1025	52.50		1077.50	premium modular unit with solid
	Demolish and Install	Ea.	1025	94		1119	hardwood faced frames, hardwood
	Reinstall	Ea.		42.09		42.09	door frames and drawer fronts,
	Clean	Ea.	.91	29		29.91	hardwood veneer on raised door
	Paint	Ea.	3.89	67.50		71.39	panels, hinges and pulls.
	Minimum Charge	Job		242		242	
Above-appliance							
	Demolish	Ea.		26		26	Includes material and labor to install
	Install	Ea.	293	39		332	premium modular unit with solid
	Demolish and Install	Ea.	293	65		358	hardwood faced frames, hardwood
	Reinstall	Ea.		31.23		31.23	door frames and drawer fronts,
	Clean	Ea.	.57	17.80		18.37	hardwood veneer on raised door
	Paint	Ea.	2.43	40.50		42.93	panels, hinges, pulls.
	Minimum Charge	Job		242		242	
Premium Hardwood							
12" Wide, 1 Door							
	Demolish	Ea.		10.40		10.40	Includes material and labor to install
	Install	Ea.	156	44		200	premium hardwood modular unit with
	Demolish and Install	Ea.	156	54.40		210.40	solid hardwood faced frames, drawer
	Reinstall	Ea.		35.20		35.20	fronts, and door panels, steel drawer
	Clean	Ea.	.23	7.20		7.43	guides, hinges, pulls, laminated
	Paint	Ea.	.97	18.35		19.32	interior.
	Minimum Charge	Job		242		242	
15" Wide, 1 Door							
	Demolish	Ea.		13		13	Includes material and labor to install
	Install	Ea.	475	45		520	premium hardwood modular unit with
	Demolish and Install	Ea.	475	58		533	solid hardwood faced frames, drawer
	Reinstall	Ea.		36.19		36.19	fronts, and door panels, steel drawer
	Clean	Ea.	.29	8.90		9.19	guides, hinges, pulls, laminated
	Paint	Ea.	1.21	22.50		23.71	interior.
	Minimum Charge	Job		242		242	
18" Wide, 1 Door							
	Demolish	Ea.		15.70		15.70	Includes material and labor to install
	Install	Ea.	390	46.50		436.50	premium hardwood modular unit with
	Demolish and Install	Ea.	390	62.20		452.20	solid hardwood faced frames, drawer
	Reinstall	Ea.		37.05		37.05	fronts, and door panels, steel drawer
	Clean	Ea.	.34	10.70		11.04	guides, hinges, pulls, laminated
	Paint	Ea.	1.45	27		28.45	interior.
	Minimum Charge	Job		242		242	

For customer support on your Contractor's Pricing Guide: Residential Repair & Remodeling, call 800.448.8182.

Cabinets and Countertops

Wall Cabinets	Unit	Material	Labor	Equip.	Total	Specification
21″ Wide, 1 Door						
Demolish	Ea.		18.05		18.05	Includes material and labor to install
Install	Ea.	505	47		552	premium hardwood modular unit with
Demolish and Install	Ea.	505	65.05		570.05	solid hardwood faced frames, drawer
Reinstall	Ea.		37.59		37.59	fronts, and door panels, steel drawer
Clean	Ea.	.40	12.45		12.85	guides, hinges, pulls, laminated
Paint	Ea.	1.71	31		32.71	interior.
Minimum Charge	Job		242		242	
24″ w / Blind Corner						
Demolish	Ea.		21		21	Includes material and labor to install
Install	Ea.	450	47.50		497.50	premium hardwood modular unit with
Demolish and Install	Ea.	450	68.50		518.50	solid hardwood faced frames, drawer
Reinstall	Ea.		38.15		38.15	fronts, and door panels, steel drawer
Clean	Ea.	.46	14.40		14.86	guides, hinges, pulls, laminated
Paint	Ea.	1.94	36.50		38.44	interior.
Minimum Charge	Job		242		242	
27″ Wide, 2 Door						
Demolish	Ea.		23		23	Includes material and labor to install
Install	Ea.	146	49		195	premium hardwood modular unit with
Demolish and Install	Ea.	146	72		218	solid hardwood faced frames, drawer
Reinstall	Ea.		39.11		39.11	fronts, and door panels, steel drawer
Clean	Ea.	.52	16.25		16.77	guides, hinges, pulls, laminated
Paint	Ea.	2.20	40.50		42.70	interior.
Minimum Charge	Job		242		242	
30″ Wide, 2 Door						
Demolish	Ea.		26		26	Includes material and labor to install
Install	Ea.	585	50		635	premium hardwood modular unit with
Demolish and Install	Ea.	585	76		661	solid hardwood faced frames, drawer
Reinstall	Ea.		40.12		40.12	fronts, and door panels, steel drawer
Clean	Ea.	.57	17.80		18.37	guides, hinges, pulls, laminated
Paint	Ea.	2.43	45		47.43	interior.
Minimum Charge	Job		242		242	
36″ Wide, 2 Door						
Demolish	Ea.		31		31	Includes material and labor to install
Install	Ea.	555	51.50		606.50	premium hardwood modular unit with
Demolish and Install	Ea.	555	82.50		637.50	solid hardwood faced frames, drawer
Reinstall	Ea.		41.19		41.19	fronts, and door panels, steel drawer
Clean	Ea.	.68	22		22.68	guides, hinges, pulls, laminated
Paint	Ea.	2.94	57.50		60.44	interior.
Minimum Charge	Job		242		242	
48″ Wide, 2 Door						
Demolish	Ea.		41.50		41.50	Includes material and labor to install
Install	Ea.	1125	52.50		1177.50	premium hardwood modular unit with
Demolish and Install	Ea.	1125	94		1219	solid hardwood faced frames, drawer
Reinstall	Ea.		42.09		42.09	fronts, and door panels, steel drawer
Clean	Ea.	.91	29		29.91	guides, hinges, pulls, laminated
Paint	Ea.	3.89	67.50		71.39	interior.
Minimum Charge	Job		242		242	
Above-appliance						
Demolish	Ea.		26		26	Includes material and labor to install
Install	Ea.	320	39		359	premium hardwood modular unit with
Demolish and Install	Ea.	320	65		385	solid hardwood faced frames, drawer
Reinstall	Ea.		31.23		31.23	fronts, and door panels, steel drawer
Clean	Ea.	.57	17.80		18.37	guides, hinges, pulls, laminated
Paint	Ea.	2.43	45		47.43	interior.
Minimum Charge	Job		242		242	

For customer support on your Contractor's Pricing Guide: Residential Repair & Remodeling, call 800.448.8182.

Cabinets and Countertops

Wall Cabinets	Unit	Material	Labor	Equip.	Total	Specification
Medicine Cabinet						
Good Grade						
Demolish	Ea.		17.30		17.30	Cost includes material and labor to
Install	Ea.	325	34.50		359.50	install 14" x 18" recessed medicine
Demolish and Install	Ea.	325	51.80		376.80	cabinet with stainless steel frame,
Reinstall	Ea.		27.66		27.66	mirror. Light not included.
Clean	Ea.	.01	18.65		18.66	
Paint	Ea.	3.34	25		28.34	
Minimum Charge	Job		242		242	
Good Grade Lighted						
Demolish	Ea.		17.30		17.30	Cost includes material and labor to
Install	Ea.	635	40.50		675.50	install 14" x 18" recessed medicine
Demolish and Install	Ea.	635	57.80		692.80	cabinet with stainless steel frame,
Reinstall	Ea.		32.27		32.27	mirror and light.
Clean	Ea.	.01	18.65		18.66	
Paint	Ea.	3.34	25		28.34	
Minimum Charge	Job		242		242	
Better Grade						
Demolish	Ea.		17.30		17.30	Includes material and labor to install
Install	Ea.	390	34.50		424.50	recessed medicine cabinet with
Demolish and Install	Ea.	390	51.80		441.80	polished chrome frame and mirror.
Reinstall	Ea.		27.66		27.66	Light not included.
Clean	Ea.	.01	18.65		18.66	
Paint	Ea.	3.34	25		28.34	
Minimum Charge	Job		242		242	
Better Grade Lighted						
Demolish	Ea.		17.30		17.30	Includes material and labor to install
Install	Ea.	700	40.50		740.50	recessed medicine cabinet with
Demolish and Install	Ea.	700	57.80		757.80	polished chrome frame, mirror and
Reinstall	Ea.		32.27		32.27	light.
Clean	Ea.	.01	18.65		18.66	
Paint	Ea.	3.34	25		28.34	
Minimum Charge	Job		242		242	
Premium Grade						
Demolish	Ea.		17.30		17.30	Cost includes material and labor to
Install	Ea.	360	34.50		394.50	install 14" x 24" recessed medicine
Demolish and Install	Ea.	360	51.80		411.80	cabinet with beveled mirror, frameless
Reinstall	Ea.		27.66		27.66	swing door. Light not included.
Clean	Ea.	.01	18.65		18.66	
Paint	Ea.	3.34	25		28.34	
Minimum Charge	Job		242		242	
Premium Grade Lighted						
Demolish	Ea.		17.30		17.30	Cost includes material and labor to
Install	Ea.	665	40.50		705.50	install 36" x 30" surface mounted
Demolish and Install	Ea.	665	57.80		722.80	medicine cabinet with beveled tri-view
Reinstall	Ea.		32.27		32.27	mirror, frameless swing door and light.
Clean	Ea.	.01	18.65		18.66	
Paint	Ea.	3.34	25		28.34	
Minimum Charge	Job		242		242	
Bath Mirror						
Demolish	Ea.		13		13	Cost includes material and labor to
Install	Ea.	47.50	24		71.50	install 18" x 24" surface mount mirror
Demolish and Install	Ea.	47.50	37		84.50	with stainless steel frame.
Reinstall	Ea.		19.36		19.36	
Clean	Ea.	.01	1.24		1.25	
Minimum Charge	Job		242		242	

For customer support on your Contractor's Pricing Guide: Residential Repair & Remodeling, call 800.448.8182.

Cabinets and Countertops

Wall Cabinets	Unit	Material	Labor	Equip.	Total	Specification
Garage						
Demolish	L.F.		10.40		10.40	Includes material and labor to install
Install	L.F.	83	24		107	good modular unit with melamine
Demolish and Install	L.F.	83	34.40		117.40	laminated to particle board, in
Reinstall	L.F.		19.36		19.36	textured colors or wood grain print
Clean	L.F.	.23	7.20		7.43	finish, hinges and pulls.
Paint	L.F.	.97	14.40		15.37	
Minimum Charge	Job		242		242	
Strip and Refinish						
Install	L.F.	5.60	50.50		56.10	Includes labor and material to strip,
Minimum Charge	Job		202		202	prep and refinish exterior of cabinets.

Tall Cabinets	Unit	Material	Labor	Equip.	Total	Specification
Good Grade Laminated						
Demolish	L.F.		18.45		18.45	Includes material and labor to install
Install	L.F.	251	54		305	custom modular unit with solid
Demolish and Install	L.F.	251	72.45		323.45	hardwood faced frames, hardwood
Reinstall	L.F.		43.02		43.02	door frames and drawer fronts,
Clean	L.F.	.61	13.35		13.96	hardwood veneer on raised door
Paint	L.F.	2.27	29		31.27	panels, hinges and pulls.
Minimum Charge	Job		242		242	
Better Grade Wood						
Demolish	L.F.		18.45		18.45	Includes material and labor to install
Install	L.F.	365	54		419	custom modular unit with solid
Demolish and Install	L.F.	365	72.45		437.45	hardwood faced frames, hardwood
Reinstall	L.F.		43.02		43.02	door frames and drawer fronts,
Clean	L.F.	.61	13.35		13.96	hardwood veneer on raised door
Paint	L.F.	2.27	31		33.27	panels, hinges and pulls.
Minimum Charge	Job		242		242	
Premium Grade Wood						
Demolish	L.F.		18.45		18.45	Includes material and labor to install
Install	L.F.	510	54		564	premium modular unit with solid
Demolish and Install	L.F.	510	72.45		582.45	hardwood faced frames, hardwood
Reinstall	L.F.		43.02		43.02	door frames and drawer fronts,
Clean	L.F.	.61	13.35		13.96	hardwood veneer on raised door
Paint	L.F.	2.27	33.50		35.77	panels, hinges and pulls.
Minimum Charge	Job		242		242	
Premium Hardwood						
Demolish	L.F.		18.45		18.45	Includes material and labor to install
Install	L.F.	720	54		774	premium hardwood modular unit with
Demolish and Install	L.F.	720	72.45		792.45	solid hardwood faced frames, drawer
Reinstall	L.F.		43.02		43.02	fronts, and door panels, steel drawer
Clean	L.F.	.61	13.35		13.96	guides, hinges and pulls, laminated
Paint	L.F.	2.27	36.50		38.77	interior.
Minimum Charge	Job		242		242	

For customer support on your Contractor's Pricing Guide: Residential Repair & Remodeling, call 800.448.8182.

Tall Cabinets

Tall Cabinets	Unit	Material	Labor	Equip.	Total	Specification
Single Oven, 27″ Wide						
Good Grade Laminated						
Demolish	Ea.		41.50		41.50	Includes material and labor to install
Install	Ea.	870	121		991	good modular unit with melamine
Demolish and Install	Ea.	870	162.50		1032.50	laminated to particle board, in
Reinstall	Ea.		96.80		96.80	textured colors or wood grain print
Clean	Ea.	1.90	23.50		25.40	finish, hinges and pulls.
Paint	Ea.	3.81	64.50		68.31	
Minimum Charge	Job		242		242	
Better Grade Veneer						
Demolish	Ea.		41.50		41.50	Includes material and labor to install
Install	Ea.	1550	121		1671	custom modular unit with solid
Demolish and Install	Ea.	1550	162.50		1712.50	hardwood faced frames, hardwood
Reinstall	Ea.		96.80		96.80	door frames and drawer fronts,
Clean	Ea.	1.90	23.50		25.40	hardwood veneer on raised door
Paint	Ea.	3.81	69.50		73.31	panels, hinges and pulls.
Minimum Charge	Job		242		242	
Premium Grade Veneer						
Demolish	Ea.		41.50		41.50	Includes material and labor to install
Install	Ea.	1400	121		1521	premium modular unit with solid
Demolish and Install	Ea.	1400	162.50		1562.50	hardwood faced frames, hardwood
Reinstall	Ea.		96.80		96.80	door frames and drawer fronts,
Clean	Ea.	1.90	23.50		25.40	hardwood veneer on raised door
Paint	Ea.	3.81	76		79.81	panels, hinges and pulls.
Minimum Charge	Job		242		242	
Premium Hardwood						
Demolish	Ea.		41.50		41.50	Includes material and labor to install
Install	Ea.	1525	121		1646	premium hardwood modular unit with
Demolish and Install	Ea.	1525	162.50		1687.50	solid hardwood faced frames, drawer
Reinstall	Ea.		96.80		96.80	fronts, and door panels, steel drawer
Clean	Ea.	1.90	23.50		25.40	guides, hinges and pulls, laminated
Paint	Ea.	3.81	82.50		86.31	interior.
Minimum Charge	Job		242		242	
Single Oven, 30″ Wide						
Good Grade Laminated						
Demolish	Ea.		41.50		41.50	Includes material and labor to install
Install	Ea.	1075	121		1196	good modular unit with melamine
Demolish and Install	Ea.	1075	162.50		1237.50	laminated to particle board, in
Reinstall	Ea.		96.80		96.80	textured colors or wood grain print
Clean	Ea.	1.90	23.50		25.40	finish, hinges and pulls.
Paint	Ea.	3.81	64.50		68.31	
Minimum Charge	Job		242		242	
Better Grade Veneer						
Demolish	Ea.		41.50		41.50	Includes material and labor to install
Install	Ea.	1150	121		1271	custom modular unit with solid
Demolish and Install	Ea.	1150	162.50		1312.50	hardwood faced frames, hardwood
Reinstall	Ea.		96.80		96.80	door frames and drawer fronts,
Clean	Ea.	1.90	23.50		25.40	hardwood veneer on raised door
Paint	Ea.	3.81	64.50		68.31	panels, hinges and pulls.
Minimum Charge	Job		242		242	

Cabinets and Countertops

Tall Cabinets	Unit	Material	Labor	Equip.	Total	Specification
Premium Grade Veneer						
Demolish	Ea.		41.50		41.50	Includes material and labor to install
Install	Ea.	1275	121		1396	premium modular unit with solid
Demolish and Install	Ea.	1275	162.50		1437.50	hardwood faced frames, hardwood
Reinstall	Ea.		96.80		96.80	door frames and drawer fronts,
Clean	Ea.	1.90	23.50		25.40	hardwood veneer on raised door
Paint	Ea.	3.81	64.50		68.31	panels, hinges and pulls.
Minimum Charge	Job		242		242	
Premium Hardwood						
Demolish	Ea.		41.50		41.50	Includes material and labor to install
Install	Ea.	1625	121		1746	premium hardwood modular unit with
Demolish and Install	Ea.	1625	162.50		1787.50	solid hardwood faced frames, drawer
Reinstall	Ea.		96.80		96.80	fronts, and door panels, steel drawer
Clean	Ea.	1.90	23.50		25.40	guides, hinges and pulls, laminated
Paint	Ea.	3.81	64.50		68.31	interior.
Minimum Charge	Job		242		242	

Utility 18″ W x 12″ D

	Unit	Material	Labor	Equip.	Total	Specification
Good Grade Laminated						
Demolish	Ea.		41.50		41.50	Includes material and labor to install
Install	Ea.	860	97		957	good modular unit with melamine
Demolish and Install	Ea.	860	138.50		998.50	laminated to particle board, in
Reinstall	Ea.		77.44		77.44	textured colors or wood grain print
Clean	Ea.	.45	11.35		11.80	finish, hinges and pulls.
Paint	Ea.	1.45	25		26.45	
Minimum Charge	Job		242		242	
Better Grade Veneer						
Demolish	Ea.		41.50		41.50	Includes material and labor to install
Install	Ea.	1150	97		1247	custom modular unit with solid
Demolish and Install	Ea.	1150	138.50		1288.50	hardwood faced frames, hardwood
Reinstall	Ea.		77.44		77.44	door frames and drawer fronts,
Clean	Ea.	.45	11.35		11.80	hardwood veneer on raised door
Paint	Ea.	1.45	27		28.45	panels, hinges and pulls.
Minimum Charge	Job		242		242	
Premium Grade Veneer						
Demolish	Ea.		41.50		41.50	Includes material and labor to install
Install	Ea.	1375	97		1472	premium modular unit with solid
Demolish and Install	Ea.	1375	138.50		1513.50	hardwood faced frames, hardwood
Reinstall	Ea.		77.44		77.44	door frames and drawer fronts,
Clean	Ea.	.45	11.35		11.80	hardwood veneer on raised door
Paint	Ea.	1.45	31		32.45	panels, hinges and pulls.
Minimum Charge	Job		242		242	
Premium Hardwood						
Demolish	Ea.		41.50		41.50	Includes material and labor to install
Install	Ea.	1525	97		1622	premium hardwood modular unit with
Demolish and Install	Ea.	1525	138.50		1663.50	solid hardwood faced frames, drawer
Reinstall	Ea.		77.44		77.44	fronts, and door panels, steel drawer
Clean	Ea.	.45	11.35		11.80	guides, hinges and pulls, laminated
Paint	Ea.	1.45	33.50		34.95	interior.
Minimum Charge	Job		242		242	

Tall Cabinets	Unit	Material	Labor	Equip.	Total	Specification
Utility 24″ W x 24″ D						
Good Grade Laminated						
Demolish	Ea.		41.50		41.50	Includes material and labor to install
Install	Ea.	895	121		1016	good modular unit with melamine
Demolish and Install	Ea.	895	162.50		1057.50	laminated to particle board, in
Reinstall	Ea.		96.80		96.80	textured colors or wood grain print
Clean	Ea.	.58	14.95		15.53	finish, hinges and pulls.
Paint	Ea.	1.94	33.50		35.44	
Minimum Charge	Job		242		242	
Better Grade Veneer						
Demolish	Ea.		41.50		41.50	Includes material and labor to install
Install	Ea.	1175	121		1296	custom modular unit with solid
Demolish and Install	Ea.	1175	162.50		1337.50	hardwood faced frames, hardwood
Reinstall	Ea.		96.80		96.80	door frames and drawer fronts,
Clean	Ea.	.58	14.95		15.53	hardwood veneer on raised door
Paint	Ea.	1.94	36.50		38.44	panels, hinges and pulls.
Minimum Charge	Job		242		242	
Premium Grade Veneer						
Demolish	Ea.		41.50		41.50	Includes material and labor to install
Install	Ea.	1425	121		1546	premium modular unit with solid
Demolish and Install	Ea.	1425	162.50		1587.50	hardwood faced frames, hardwood
Reinstall	Ea.		96.80		96.80	door frames and drawer fronts,
Clean	Ea.	.58	14.95		15.53	hardwood veneer on raised door
Paint	Ea.	1.94	40.50		42.44	panels, hinges and pulls.
Minimum Charge	Job		242		242	
Premium Hardwood						
Demolish	Ea.		41.50		41.50	Includes material and labor to install
Install	Ea.	1575	121		1696	premium hardwood modular unit with
Demolish and Install	Ea.	1575	162.50		1737.50	solid hardwood faced frames, drawer
Reinstall	Ea.		96.80		96.80	fronts, and door panels, steel drawer
Clean	Ea.	.58	14.95		15.53	guides, hinges and pulls, laminated
Paint	Ea.	1.94	45		46.94	interior.
Minimum Charge	Job		242		242	
Plain Shelves						
Install	Ea.	61	27		88	Includes labor and material to install
Rotating Shelves						tall cabinet shelving, per set.
Install	Ea.	99.50	24		123.50	Includes labor and material to install
Strip and Refinish						lazy susan shelving for wall cabinet.
Install	L.F.	11.20	80.50		91.70	Includes labor and material to strip,
Minimum Charge	Job		202		202	prep and refinish exterior of cabinets.
Stain						
Install	L.F.	.37	3.67		4.04	Includes labor and material to replace
Minimum Charge	Job		202		202	normal prep and stain on doors and
						exterior cabinets.

221

Base Cabinets	Unit	Material	Labor	Equip.	Total	Specification
Good Grade Laminated						
12″ w, 1 Door, 1 Drawer						
Demolish	Ea.		10.40		10.40	Includes material and labor to install
Install	Ea.	320	39		359	good modular unit with melamine
Demolish and Install	Ea.	320	49.40		369.40	laminate, textured colors or wood
Reinstall	Ea.		31.23		31.23	grain print finish including hinges and
Clean	Ea.	.31	7.50		7.81	pulls.
Paint	Ea.	.97	16.80		17.77	
Minimum Charge	Job		242		242	
15″ w, 1 Door, 1 Drawer						
Demolish	Ea.		13		13	Includes material and labor to install
Install	Ea.	330	40.50		370.50	good modular unit with melamine
Demolish and Install	Ea.	330	53.50		383.50	laminate, textured colors or wood
Reinstall	Ea.		32.27		32.27	grain print finish including hinges and
Clean	Ea.	.39	9.35		9.74	pulls.
Paint	Ea.	1.21	21		22.21	
Minimum Charge	Job		242		242	
18″ w, 1 Door, 1 Drawer						
Demolish	Ea.		15.40		15.40	Includes material and labor to install
Install	Ea.	340	41.50		381.50	good modular unit with melamine
Demolish and Install	Ea.	340	56.90		396.90	laminate, textured colors or wood
Reinstall	Ea.		33.24		33.24	grain print finish including hinges and
Clean	Ea.	.45	11.35		11.80	pulls.
Paint	Ea.	1.45	25		26.45	
Minimum Charge	Job		242		242	
21″ w, 1 Door, 1 Drawer						
Demolish	Ea.		18.05		18.05	Includes material and labor to install
Install	Ea.	350	42.50		392.50	good modular unit with melamine
Demolish and Install	Ea.	350	60.55		410.55	laminate, textured colors or wood
Reinstall	Ea.		34.11		34.11	grain print finish including hinges and
Clean	Ea.	.54	13.35		13.89	pulls.
Paint	Ea.	1.71	29		30.71	
Minimum Charge	Job		242		242	
24″ w, 1 Door, 1 Drawer						
Demolish	Ea.		21		21	Includes material and labor to install
Install	Ea.	385	43.50		428.50	good modular unit with melamine
Demolish and Install	Ea.	385	64.50		449.50	laminate, textured colors or wood
Reinstall	Ea.		34.73		34.73	grain print finish including hinges and
Clean	Ea.	.58	14.95		15.53	pulls.
Paint	Ea.	1.94	33.50		35.44	
Minimum Charge	Job		242		242	
30″ Wide Sink Base						
Demolish	Ea.		26		26	Includes material and labor to install
Install	Ea.	305	45		350	good modular unit with melamine
Demolish and Install	Ea.	305	71		376	laminate, textured colors or wood
Reinstall	Ea.		36.19		36.19	grain print finish including hinges and
Clean	Ea.	.76	18.70		19.46	pulls.
Paint	Ea.	2.43	40.50		42.93	
Minimum Charge	Job		242		242	
36″ Wide Sink Base						
Demolish	Ea.		32		32	Includes material and labor to install
Install	Ea.	335	47.50		382.50	good modular unit with melamine
Demolish and Install	Ea.	335	79.50		414.50	laminate, textured colors or wood
Reinstall	Ea.		38.15		38.15	grain print finish including hinges and
Clean	Ea.	.85	22		22.85	pulls.
Paint	Ea.	2.94	50.50		53.44	
Minimum Charge	Job		242		242	

Cabinets and Countertops

Base Cabinets		Unit	Material	Labor	Equip.	Total	Specification
36" Blind Corner							
	Demolish	Ea.		32		32	Includes material and labor to install
	Install	Ea.	315	54		369	good modular unit with melamine
	Demolish and Install	Ea.	315	86		401	laminate, textured colors or wood
	Reinstall	Ea.		43.02		43.02	grain print finish including hinges and
	Clean	Ea.	.85	22		22.85	pulls.
	Paint	Ea.	2.94	50.50		53.44	
	Minimum Charge	Job		242		242	
42" w, 2 Door, 2 Drawer							
	Demolish	Ea.		38		38	Includes material and labor to install
	Install	Ea.	545	49		594	good modular unit with melamine
	Demolish and Install	Ea.	545	87		632	laminate, textured colors or wood
	Reinstall	Ea.		39.11		39.11	grain print finish including hinges and
	Clean	Ea.	1.09	26.50		27.59	pulls.
	Paint	Ea.	3.37	57.50		60.87	
	Minimum Charge	Job		242		242	
Clean Interior							
	Clean	L.F.	.31	5.65		5.96	Includes labor and materials to clean
	Minimum Charge	Job		187		187	the interior of a 27" cabinet.
Better Grade Wood							
12" w, 1 Door, 1 Drawer							
	Demolish	Ea.		10.40		10.40	Includes material and labor to install
	Install	Ea.	430	39		469	custom modular unit including solid
	Demolish and Install	Ea.	430	49.40		479.40	hardwood faced frames, hardwood
	Reinstall	Ea.		31.23		31.23	door frames and drawer fronts,
	Clean	Ea.	.31	7.50		7.81	hardwood veneer on raised door
	Paint	Ea.	.97	18.35		19.32	panels, hinges and pulls.
	Minimum Charge	Job		242		242	
15" w, 1 Door, 1 Drawer							
	Demolish	Ea.		13		13	Includes material and labor to install
	Install	Ea.	440	40.50		480.50	custom modular unit including solid
	Demolish and Install	Ea.	440	53.50		493.50	hardwood faced frames, hardwood
	Reinstall	Ea.		32.27		32.27	door frames and drawer fronts,
	Clean	Ea.	.39	9.35		9.74	hardwood veneer on raised door
	Paint	Ea.	1.21	23.50		24.71	panels, hinges and pulls.
	Minimum Charge	Job		242		242	
18" w, 1 Door, 1 Drawer							
	Demolish	Ea.		15.40		15.40	Includes material and labor to install
	Install	Ea.	455	41.50		496.50	custom modular unit including solid
	Demolish and Install	Ea.	455	56.90		511.90	hardwood faced frames, hardwood
	Reinstall	Ea.		33.24		33.24	door frames and drawer fronts,
	Clean	Ea.	.45	11.35		11.80	hardwood veneer on raised door
	Paint	Ea.	1.45	27		28.45	panels, hinges and pulls.
	Minimum Charge	Job		242		242	
21" w, 1 Door, 1 Drawer							
	Demolish	Ea.		18.05		18.05	Includes material and labor to install
	Install	Ea.	470	42.50		512.50	custom modular unit including solid
	Demolish and Install	Ea.	470	60.55		530.55	hardwood faced frames, hardwood
	Reinstall	Ea.		34.11		34.11	door frames and drawer fronts,
	Clean	Ea.	.54	13.35		13.89	hardwood veneer on raised door
	Paint	Ea.	1.71	31		32.71	panels, hinges and pulls.
	Minimum Charge	Job		242		242	

For customer support on your Contractor's Pricing Guide: Residential Repair & Remodeling, call 800.448.8182.

Cabinets and Countertops

Base Cabinets	Unit	Material	Labor	Equip.	Total	Specification
24″ w, 1 Door, 1 Drawer						
Demolish	Ea.		21		21	Includes material and labor to install
Install	Ea.	515	43.50		558.50	custom modular unit including solid
Demolish and Install	Ea.	515	64.50		579.50	hardwood faced frames, hardwood
Reinstall	Ea.		34.73		34.73	door frames and drawer fronts,
Clean	Ea.	.58	14.95		15.53	hardwood veneer on raised door
Paint	Ea.	1.94	36.50		38.44	panels, hinges and pulls.
Minimum Charge	Job		242		242	
30″ Wide Sink Base						
Demolish	Ea.		26		26	Includes material and labor to install
Install	Ea.	400	45		445	custom modular unit including solid
Demolish and Install	Ea.	400	71		471	hardwood faced frames, hardwood
Reinstall	Ea.		36.19		36.19	door frames and drawer fronts,
Clean	Ea.	.76	18.70		19.46	hardwood veneer on raised door
Paint	Ea.	2.43	45		47.43	panels, hinges and pulls.
Minimum Charge	Job		242		242	
36″ Wide Sink Base						
Demolish	Ea.		32		32	Includes material and labor to install
Install	Ea.	445	47.50		492.50	custom modular unit including solid
Demolish and Install	Ea.	445	79.50		524.50	hardwood faced frames, hardwood
Reinstall	Ea.		38.15		38.15	door frames and drawer fronts,
Clean	Ea.	.85	22		22.85	hardwood veneer on raised door
Paint	Ea.	2.94	57.50		60.44	panels, hinges and pulls.
Minimum Charge	Job		242		242	
36″ Blind Corner						
Demolish	Ea.		32		32	Includes material and labor to install
Install	Ea.	425	54		479	custom modular unit including solid
Demolish and Install	Ea.	425	86		511	hardwood faced frames, hardwood
Reinstall	Ea.		43.02		43.02	door frames and drawer fronts,
Clean	Ea.	.85	22		22.85	hardwood veneer on raised door
Paint	Ea.	2.94	57.50		60.44	panels, hinges and pulls.
Minimum Charge	Job		242		242	
42″ w, 2 Door, 2 Drawer						
Demolish	Ea.		38		38	Includes material and labor to install
Install	Ea.	725	49		774	modular unit including solid hardwood
Demolish and Install	Ea.	725	87		812	faced frames, hardwood door frames
Reinstall	Ea.		39.11		39.11	and drawer fronts, hardwood veneer
Clean	Ea.	1.09	26.50		27.59	on raised door panels, hinges and
Paint	Ea.	3.37	67.50		70.87	pulls.
Minimum Charge	Job		242		242	
Clean Interior						
Clean	L.F.	.31	5.65		5.96	Includes labor and materials to clean
Minimum Charge	Job		187		187	the interior of a 27″ cabinet.
Premium Grade Wood						
12″ w, 1 Door, 1 Drawer						
Demolish	Ea.		10.40		10.40	Includes material and labor to install
Install	Ea.	510	19.50		529.50	premium hardwood modular unit with
Demolish and Install	Ea.	510	29.90		539.90	solid hardwood faced frames, drawer
Reinstall	Ea.		15.61		15.61	fronts, and door panels, steel drawer
Clean	Ea.	.31	7.50		7.81	guides, hinges, pulls and laminated
Paint	Ea.	.97	20		20.97	interior.
Minimum Charge	Job		242		242	

Cabinets and Countertops

Base Cabinets	Unit	Material	Labor	Equip.	Total	Specification
15″ w, 1 Door, 1 Drawer						
Demolish	Ea.		13		13	Includes material and labor to install
Install	Ea.	530	40.50		570.50	premium modular unit including solid
Demolish and Install	Ea.	530	53.50		583.50	hardwood faced frames, hardwood
Reinstall	Ea.		32.27		32.27	door frames and drawer fronts,
Clean	Ea.	.39	9.35		9.74	hardwood veneer on raised door
Paint	Ea.	1.21	25		26.21	panels, hinges and pulls.
Minimum Charge	Job		242		242	
18″ w, 1 Door, 1 Drawer						
Demolish	Ea.		15.40		15.40	Includes material and labor to install
Install	Ea.	545	41.50		586.50	premium modular unit including solid
Demolish and Install	Ea.	545	56.90		601.90	hardwood faced frames, hardwood
Reinstall	Ea.		33.24		33.24	door frames and drawer fronts,
Clean	Ea.	.45	11.35		11.80	hardwood veneer on raised door
Paint	Ea.	1.45	31		32.45	panels, hinges and pulls.
Minimum Charge	Job		242		242	
21″ w, 1 Door, 1 Drawer						
Demolish	Ea.		18.05		18.05	Includes material and labor to install
Install	Ea.	565	42.50		607.50	premium modular unit including solid
Demolish and Install	Ea.	565	60.55		625.55	hardwood faced frames, hardwood
Reinstall	Ea.		34.11		34.11	door frames and drawer fronts,
Clean	Ea.	.54	13.35		13.89	hardwood veneer on raised door
Paint	Ea.	1.71	33.50		35.21	panels, hinges and pulls.
Minimum Charge	Job		242		242	
24″ w, 1 Door, 1 Drawer						
Demolish	Ea.		21		21	Includes material and labor to install
Install	Ea.	615	43.50		658.50	premium modular unit including solid
Demolish and Install	Ea.	615	64.50		679.50	hardwood faced frames, hardwood
Reinstall	Ea.		34.73		34.73	door frames and drawer fronts,
Clean	Ea.	.58	14.95		15.53	hardwood veneer on raised door
Paint	Ea.	1.94	40.50		42.44	panels, hinges and pulls.
Minimum Charge	Job		242		242	
30″ Wide Sink Base						
Demolish	Ea.		26		26	Includes material and labor to install
Install	Ea.	485	45		530	premium modular unit including solid
Demolish and Install	Ea.	485	71		556	hardwood faced frames, hardwood
Reinstall	Ea.		36.19		36.19	door frames and drawer fronts,
Clean	Ea.	.76	18.70		19.46	hardwood veneer on raised door
Paint	Ea.	2.43	50.50		52.93	panels, hinges and pulls.
Minimum Charge	Job		242		242	
36″ Wide Sink Base						
Demolish	Ea.		32		32	Includes material and labor to install
Install	Ea.	535	47.50		582.50	premium modular unit including solid
Demolish and Install	Ea.	535	79.50		614.50	hardwood faced frames, hardwood
Reinstall	Ea.		38.15		38.15	door frames and drawer fronts,
Clean	Ea.	.85	22		22.85	hardwood veneer on raised door
Paint	Ea.	2.94	57.50		60.44	panels, hinges and pulls.
Minimum Charge	Job		242		242	
36″ Blind Corner						
Demolish	Ea.		32		32	Includes material and labor to install
Install	Ea.	505	54		559	premium modular unit including solid
Demolish and Install	Ea.	505	86		591	hardwood faced frames, hardwood
Reinstall	Ea.		43.02		43.02	door frames and drawer fronts,
Clean	Ea.	.85	22		22.85	hardwood veneer on raised door
Paint	Ea.	2.94	57.50		60.44	panels, hinges and pulls.
Minimum Charge	Job		242		242	

Cabinets and Countertops

Base Cabinets

Base Cabinets	Unit	Material	Labor	Equip.	Total	Specification
42" w, 2 Door, 2 Drawer						
Demolish	Ea.		38		38	Includes material and labor to install
Install	Ea.	965	49		1014	premium modular unit including solid
Demolish and Install	Ea.	965	87		1052	hardwood faced frames, hardwood
Reinstall	Ea.		39.11		39.11	door frames and drawer fronts,
Clean	Ea.	1.09	26.50		27.59	hardwood veneer on raised door
Paint	Ea.	3.37	67.50		70.87	panels, hinges and pulls.
Minimum Charge	Job		242		242	
Clean Interior						
Clean	L.F.	.31	5.65		5.96	Includes labor and materials to clean
Minimum Charge	Job		187		187	the interior of a 27" cabinet.

Premium Hardwood

Premium Hardwood	Unit	Material	Labor	Equip.	Total	Specification
12" w, 1 Door, 1 Drawer						
Demolish	Ea.		10.40		10.40	Includes material and labor to install
Install	Ea.	560	39		599	premium hardwood modular unit with
Demolish and Install	Ea.	560	49.40		609.40	solid hardwood faced frames, drawer
Reinstall	Ea.		31.23		31.23	fronts, and door panels, steel drawer
Clean	Ea.	.31	7.50		7.81	guides, hinges, pulls and laminated
Paint	Ea.	.97	22.50		23.47	interior.
Minimum Charge	Job		242		242	
15" w, 1 Door, 1 Drawer						
Demolish	Ea.		13		13	Includes material and labor to install
Install	Ea.	570	40.50		610.50	premium hardwood modular unit with
Demolish and Install	Ea.	570	53.50		623.50	solid hardwood faced frames, drawer
Reinstall	Ea.		32.27		32.27	fronts, and door panels, steel drawer
Clean	Ea.	.39	9.35		9.74	guides, hinges, pulls and laminated
Paint	Ea.	1.21	29		30.21	interior.
Minimum Charge	Job		242		242	
18" w, 1 Door, 1 Drawer						
Demolish	Ea.		15.40		15.40	Includes material and labor to install
Install	Ea.	595	41.50		636.50	premium hardwood modular unit with
Demolish and Install	Ea.	595	56.90		651.90	solid hardwood faced frames, drawer
Reinstall	Ea.		33.24		33.24	fronts, and door panels, steel drawer
Clean	Ea.	.45	11.35		11.80	guides, hinges, pulls and laminated
Paint	Ea.	1.45	33.50		34.95	interior.
Minimum Charge	Job		242		242	
21" w, 1 Door, 1 Drawer						
Demolish	Ea.		18.05		18.05	Includes material and labor to install
Install	Ea.	615	42.50		657.50	premium hardwood modular unit with
Demolish and Install	Ea.	615	60.55		675.55	solid hardwood faced frames, drawer
Reinstall	Ea.		34.11		34.11	fronts, and door panels, steel drawer
Clean	Ea.	.54	13.35		13.89	guides, hinges, pulls and laminated
Paint	Ea.	1.71	36.50		38.21	interior.
Minimum Charge	Job		242		242	
24" w, 1 Door, 1 Drawer						
Demolish	Ea.		21		21	Includes material and labor to install
Install	Ea.	675	43.50		718.50	premium hardwood modular unit with
Demolish and Install	Ea.	675	64.50		739.50	solid hardwood faced frames, drawer
Reinstall	Ea.		34.73		34.73	fronts, and door panels, steel drawer
Clean	Ea.	.58	14.95		15.53	guides, hinges and pulls, laminated
Paint	Ea.	1.94	45		46.94	interior.
Minimum Charge	Job		242		242	

For customer support on your Contractor's Pricing Guide: Residential Repair & Remodeling, call 800.448.8182.

Base Cabinets		Unit	Material	Labor	Equip.	Total	Specification
30" Wide Sink Base							
	Demolish	Ea.		26		26	Includes material and labor to install
	Install	Ea.	530	45		575	premium hardwood modular unit with
	Demolish and Install	Ea.	530	71		601	solid hardwood faced frames, drawer
	Reinstall	Ea.		36.19		36.19	fronts, and door panels, steel drawer
	Clean	Ea.	.76	18.70		19.46	guides, hinges, pulls and laminated
	Paint	Ea.	2.43	50.50		52.93	interior.
	Minimum Charge	Job		242		242	
36" Wide Sink Base							
	Demolish	Ea.		32		32	Includes material and labor to install
	Install	Ea.	585	47.50		632.50	premium hardwood modular unit with
	Demolish and Install	Ea.	585	79.50		664.50	solid hardwood faced frames, drawer
	Reinstall	Ea.		38.15		38.15	fronts, and door panels, steel drawer
	Clean	Ea.	.85	22		22.85	guides, hinges, pulls and laminated
	Paint	Ea.	2.94	67.50		70.44	interior.
	Minimum Charge	Job		242		242	
36" Blind Corner							
	Demolish	Ea.		32		32	Includes material and labor to install
	Install	Ea.	555	54		609	premium hardwood modular unit with
	Demolish and Install	Ea.	555	86		641	solid hardwood faced frames, drawer
	Reinstall	Ea.		43.02		43.02	fronts, and door panels, steel drawer
	Clean	Ea.	.85	22		22.85	guides, hinges, pulls and laminated
	Paint	Ea.	2.94	67.50		70.44	interior.
	Minimum Charge	Job		242		242	
42" w, 2 Door, 2 Drawer							
	Demolish	Ea.		38		38	Includes material and labor to install
	Install	Ea.	1275	49		1324	premium hardwood modular unit with
	Demolish and Install	Ea.	1275	87		1362	solid hardwood faced frames, drawer
	Reinstall	Ea.		39.11		39.11	fronts, and door panels, steel drawer
	Clean	Ea.	1.09	26.50		27.59	guides, hinges, pulls and laminated
	Paint	Ea.	3.37	80.50		83.87	interior.
	Minimum Charge	Job		242		242	
Clean Interior							
	Clean	L.F.	.31	5.65		5.96	Includes labor and materials to clean
	Minimum Charge	Job		187		187	the interior of a 27" cabinet.
Island							
24" Wide							
	Demolish	Ea.		21		21	Includes material and labor to install
	Install	Ea.	680	40.50		720.50	custom modular unit with solid
	Demolish and Install	Ea.	680	61.50		741.50	hardwood faced frames, hardwood
	Reinstall	Ea.		32.27		32.27	door frames and drawer fronts,
	Clean	Ea.	.85	22		22.85	hardwood veneer on raised door
	Paint	Ea.	2.90	57.50		60.40	panels, hinges and pulls.
	Minimum Charge	Job		242		242	
30" Wide							
	Demolish	Ea.		26		26	Includes material and labor to install
	Install	Ea.	715	50.50		765.50	custom modular unit with solid
	Demolish and Install	Ea.	715	76.50		791.50	hardwood faced frames, hardwood
	Reinstall	Ea.		40.33		40.33	door frames and drawer fronts,
	Clean	Ea.	1.09	26.50		27.59	hardwood veneer on raised door
	Paint	Ea.	3.64	67.50		71.14	panels, hinges and pulls.
	Minimum Charge	Job		242		242	

Cabinets and Countertops

Base Cabinets		Unit	Material	Labor	Equip.	Total	Specification
36″ Wide							
	Demolish	Ea.		32		32	Includes material and labor to install
	Install	Ea.	805	60.50		865.50	custom modular unit with solid
	Demolish and Install	Ea.	805	92.50		897.50	hardwood faced frames, hardwood
	Reinstall	Ea.		48.40		48.40	door frames and drawer fronts,
	Clean	Ea.	1.27	34		35.27	hardwood veneer on raised door
	Paint	Ea.	4.41	80.50		84.91	panels, hinges and pulls.
	Minimum Charge	Job		242		242	
48″ Wide							
	Demolish	Ea.		41.50		41.50	Includes material and labor to install
	Install	Ea.	1375	80.50		1455.50	custom modular unit with solid
	Demolish and Install	Ea.	1375	122		1497	hardwood faced frames, hardwood
	Reinstall	Ea.		64.53		64.53	door frames and drawer fronts,
	Clean	Ea.	1.52	41.50		43.02	hardwood veneer on raised door
	Paint	Ea.	5.05	101		106.05	panels, hinges and pulls.
	Minimum Charge	Job		242		242	
60″ Wide							
	Demolish	Ea.		46		46	Includes material and labor to install
	Install	Ea.	1425	101		1526	custom modular unit with solid
	Demolish and Install	Ea.	1425	147		1572	hardwood faced frames, hardwood
	Reinstall	Ea.		80.67		80.67	door frames and drawer fronts,
	Clean	Ea.	1.90	53.50		55.40	hardwood veneer on raised door
	Paint	Ea.	5.05	135		140.05	panels, hinges and pulls.
	Minimum Charge	Job		242		242	
Clean Interior							
	Clean	L.F.	.31	5.65		5.96	Includes labor and materials to clean
	Minimum Charge	Job		187		187	the interior of a 27″ cabinet.
Liquor Bar							
	Demolish	L.F.		10.40		10.40	Includes material and labor to install
	Install	L.F.	275	48.50		323.50	custom modular unit with solid
	Demolish and Install	L.F.	275	58.90		333.90	hardwood faced frames, hardwood
	Reinstall	L.F.		38.72		38.72	door frames and drawer fronts,
	Clean	L.F.	.31	7.80		8.11	hardwood veneer on raised door
	Paint	L.F.	7.75	18.35		26.10	panels, hinges and pulls.
	Minimum Charge	Job		242		242	
Clean Interior							
	Clean	L.F.	.31	5.65		5.96	Includes labor and materials to clean
	Minimum Charge	Job		187		187	the interior of a 27″ cabinet.
Seal Interior							
	Install	L.F.	.83	6.60		7.43	Includes labor and materials to seal a
	Minimum Charge	Job		202		202	cabinet interior.
Back Bar							
	Demolish	L.F.		10.40		10.40	Includes material and labor to install
	Install	L.F.	275	48.50		323.50	custom modular unit with solid
	Demolish and Install	L.F.	275	58.90		333.90	hardwood faced frames, hardwood
	Reinstall	L.F.		38.72		38.72	door frames and drawer fronts,
	Clean	L.F.	.31	7.80		8.11	hardwood veneer on raised door
	Paint	L.F.	7.75	18.35		26.10	panels, hinges and pulls.
	Minimum Charge	Job		242		242	
Clean Interior							
	Clean	L.F.	.31	5.65		5.96	Includes labor and materials to clean
	Minimum Charge	Job		187		187	the interior of a 27″ cabinet.

Cabinets and Countertops

Base Cabinets		Unit	Material	Labor	Equip.	Total	Specification
Seal Interior							
	Install	L.F.	.83	6.60		7.43	Includes labor and materials to seal a
	Minimum Charge	Job		202		202	cabinet interior.
Storage (garage)							
	Demolish	L.F.		10.40		10.40	Includes material and labor to install
	Install	L.F.	206	48.50		254.50	good modular unit with melamine
	Demolish and Install	L.F.	206	58.90		264.90	laminate, textured colors or wood
	Reinstall	L.F.		38.72		38.72	grain print finish including hinges and
	Clean	L.F.	.31	7.80		8.11	pulls.
	Paint	L.F.	7.75	18.35		26.10	
	Minimum Charge	Job		242		242	
Clean Interior							
	Clean	L.F.	.31	5.65		5.96	Includes labor and materials to clean
	Minimum Charge	Job		187		187	the interior of a 27" cabinet.
Seal Interior							
	Install	L.F.	.83	6.60		7.43	Includes labor and materials to seal a
	Minimum Charge	Job		202		202	cabinet interior.
Built-in Desk							
	Demolish	L.F.		10.40		10.40	Includes material and labor to install
	Install	L.F.	275	48.50		323.50	custom modular unit with solid
	Demolish and Install	L.F.	275	58.90		333.90	hardwood faced frames, hardwood
	Reinstall	L.F.		38.72		38.72	door frames and drawer fronts,
	Clean	L.F.	.31	7.80		8.11	hardwood veneer on raised door
	Paint	L.F.	7.75	18.35		26.10	panels, hinges and pulls.
	Minimum Charge	Job		242		242	
Built-in Bookcase							
	Demolish	L.F.		13.85		13.85	Includes material and labor to install
	Install	L.F.	40	48.50		88.50	custom modular bookcase unit with
	Demolish and Install	L.F.	40	62.35		102.35	clear pine faced frames, shelves 12"
	Reinstall	L.F.		38.72		38.72	O.C., 7' high, 8" deep.
	Clean	L.F.	.61	13.35		13.96	
	Paint	L.F.	2.27	31		33.27	
	Minimum Charge	Job		242		242	
Strip and Refinish							
	Install	L.F.	5.60	50.50		56.10	Includes labor and material to strip,
	Minimum Charge	Job		202		202	prep and refinish exterior of cabinets.
Stain							
	Install	L.F.	.84	1.88		2.72	Includes labor and material to stain
	Minimum Charge	Job		242		242	door and exterior of cabinets.

Countertop		Unit	Material	Labor	Equip.	Total	Specification
Laminated							
With Splash							
	Demolish	L.F.		6.95		6.95	Includes material and labor to install
	Install	L.F.	52.50	17.30		69.80	one piece laminated top with 4"
	Demolish and Install	L.F.	52.50	24.25		76.75	backsplash.
	Reinstall	L.F.		13.83		13.83	
	Clean	L.F.	.03	.94		.97	
	Minimum Charge	Job		242		242	

For customer support on your Contractor's Pricing Guide: Residential Repair & Remodeling, call 800.448.8182.

Cabinets and Countertops

Countertop		Unit	Material	Labor	Equip.	Total	Specification
Roll Top							
	Demolish	L.F.		6.95		6.95	Includes material and labor to install
	Install	L.F.	14.65	16.15		30.80	one piece laminated top with rolled
	Demolish and Install	L.F.	14.65	23.10		37.75	drip edge (post formed) and
	Reinstall	L.F.		12.91		12.91	backsplash.
	Clean	L.F.	.06	1.42		1.48	
	Minimum Charge	Job		242		242	
Ceramic Tile							
	Demolish	S.F.		3.46		3.46	Cost includes material and labor to
	Install	S.F.	26	9.70		35.70	install 4-1/4" x 4-1/4" to 6" x 6"
	Demolish and Install	S.F.	26	13.16		39.16	glazed tile set in mortar bed and grout
	Clean	S.F.	.02	.71		.73	on particle board substrate.
	Minimum Charge	Job		235		235	
Cultured Marble							
	Demolish	L.F.		6.95		6.95	Includes material and labor to install
	Install	L.F.	109	48		157	marble countertop, 24" wide, no
	Demolish and Install	L.F.	109	54.95		163.95	backsplash.
	Reinstall	L.F.		38.30		38.30	
	Clean	L.F.	.06	1.42		1.48	
	Minimum Charge	Job		235		235	
With Splash							
	Demolish	L.F.		6.95		6.95	Includes material and labor to install
	Install	L.F.	119	48		167	marble countertop, 24" wide, with
	Demolish and Install	L.F.	119	54.95		173.95	backsplash.
	Reinstall	L.F.		38.30		38.30	
	Clean	L.F.	.06	1.42		1.48	
	Minimum Charge	Job		235		235	
Quarry Tile							
	Demolish	S.F.		3.46		3.46	Includes material and labor to install
	Install	S.F.	8.85	9.70		18.55	quarry tile countertop with no
	Demolish and Install	S.F.	8.85	13.16		22.01	backsplash.
	Clean	S.F.	.02	.71		.73	
	Minimum Charge	Job		235		235	
With Splash							
	Demolish	S.F.		3.46		3.46	Includes material and labor to install
	Install	S.F.	9.35	10.75		20.10	quarry tile countertop with backsplash.
	Demolish and Install	S.F.	9.35	14.21		23.56	
	Clean	S.F.	.02	.71		.73	
	Minimum Charge	Job		235		235	
Butcher Block							
	Demolish	S.F.		3.46		3.46	Includes material and labor to install
	Install	S.F.	51.50	8.65		60.15	solid laminated maple countertop with
	Demolish and Install	S.F.	51.50	12.11		63.61	no backsplash.
	Reinstall	S.F.		6.91		6.91	
	Clean	S.F.	.02	.29		.31	
	Minimum Charge	Job		242		242	
Solid Surface							
	Demolish	L.F.		6.95		6.95	Includes material and labor to install
	Install	L.F.	64.50	34.50		99	solid surface counter top 22" deep.
	Demolish and Install	L.F.	64.50	41.45		105.95	
	Reinstall	L.F.		27.66		27.66	
	Clean	L.F.	.06	1.42		1.48	
	Minimum Charge	Job		242		242	

For customer support on your Contractor's Pricing Guide: Residential Repair & Remodeling, call 800.448.8182.

Cabinets and Countertops

Vanity Cabinets		Unit	Material	Labor	Equip.	Total	Specification
Good Grade Laminated							
24" Wide, 2 Door							
	Demolish	Ea.		21		21	Includes material and labor to install
	Install	Ea.	375	48.50		423.50	modular unit with melamine laminated
	Demolish and Install	Ea.	375	69.50		444.50	to particle board, textured colors or
	Reinstall	Ea.		38.72		38.72	wood grain print finish, hinges and
	Clean	Ea.	.58	14.95		15.53	pulls.
	Paint	Ea.	1.94	33.50		35.44	
	Minimum Charge	Job		242		242	
30" Wide, 2 Door							
	Demolish	Ea.		26		26	Includes material and labor to install
	Install	Ea.	395	60.50		455.50	modular unit with melamine laminated
	Demolish and Install	Ea.	395	86.50		481.50	to particle board, textured colors or
	Reinstall	Ea.		48.40		48.40	wood grain print finish, hinges and
	Clean	Ea.	.76	18.70		19.46	pulls.
	Paint	Ea.	2.43	40.50		42.93	
	Minimum Charge	Job		242		242	
36" Wide, 2 Door							
	Demolish	Ea.		32		32	Includes material and labor to install
	Install	Ea.	335	72.50		407.50	modular unit with melamine laminated
	Demolish and Install	Ea.	335	104.50		439.50	to particle board, textured colors or
	Reinstall	Ea.		58.09		58.09	wood grain print finish, hinges and
	Clean	Ea.	.85	22		22.85	pulls.
	Paint	Ea.	2.94	50.50		53.44	
	Minimum Charge	Job		242		242	
42" w, 3 Doors / Drawer							
	Demolish	Ea.		38		38	Includes material and labor to install
	Install	Ea.	670	78		748	modular unit with melamine laminated
	Demolish and Install	Ea.	670	116		786	to particle board, textured colors or
	Reinstall	Ea.		62.55		62.55	wood grain print finish, hinges and
	Clean	Ea.	1.09	26.50		27.59	pulls.
	Paint	Ea.	3.37	57.50		60.87	
	Minimum Charge	Job		242		242	
48" w, 3 Doors / Drawer							
	Demolish	Ea.		41.50		41.50	Includes material and labor to install
	Install	Ea.	435	84.50		519.50	modular unit with melamine laminated
	Demolish and Install	Ea.	435	126		561	to particle board, textured colors or
	Reinstall	Ea.		67.75		67.75	wood grain print finish, hinges and
	Clean	Ea.	1.52	41.50		43.02	pulls.
	Paint	Ea.	5.05	101		106.05	
	Minimum Charge	Job		242		242	
Clean Interior							
	Install	L.F.	.31	7.80		8.11	Includes labor and materials to clean
	Minimum Charge	Job		187		187	base cabinetry.
Better Grade Wood							
24" Wide, 2 Door							
	Demolish	Ea.		21		21	Includes material and labor to install
	Install	Ea.	500	48.50		548.50	custom modular unit with solid
	Demolish and Install	Ea.	500	69.50		569.50	hardwood faced frames, hardwood
	Reinstall	Ea.		38.72		38.72	door frames and drawer fronts,
	Clean	Ea.	.58	14.95		15.53	hardwood veneer on raised door
	Paint	Ea.	1.94	36.50		38.44	panels, hinges and pulls.
	Minimum Charge	Job		242		242	

231

Cabinets and Countertops

Vanity Cabinets	Unit	Material	Labor	Equip.	Total	Specification
30" Wide, 2 Door						
Demolish	Ea.		26		26	Includes material and labor to install
Install	Ea.	530	60.50		590.50	custom modular unit with solid
Demolish and Install	Ea.	530	86.50		616.50	hardwood faced frames, hardwood
Reinstall	Ea.		48.40		48.40	door frames and drawer fronts,
Clean	Ea.	.76	18.70		19.46	hardwood veneer on raised door
Paint	Ea.	2.43	45		47.43	panels, hinges and pulls.
Minimum Charge	Job		242		242	
36" Wide, 2 Door						
Demolish	Ea.		32		32	Includes material and labor to install
Install	Ea.	445	72.50		517.50	custom modular unit with solid
Demolish and Install	Ea.	445	104.50		549.50	hardwood faced frames, hardwood
Reinstall	Ea.		58.09		58.09	door frames and drawer fronts,
Clean	Ea.	.85	22		22.85	hardwood veneer on raised door
Paint	Ea.	2.94	57.50		60.44	panels, hinges and pulls.
Minimum Charge	Job		242		242	
42" w, 3 Doors / Drawer						
Demolish	Ea.		38		38	Includes material and labor to install
Install	Ea.	670	78		748	custom modular unit with solid
Demolish and Install	Ea.	670	116		786	hardwood faced frames, hardwood
Reinstall	Ea.		62.55		62.55	door frames and drawer fronts,
Clean	Ea.	1.09	26.50		27.59	hardwood veneer on raised door
Paint	Ea.	3.37	67.50		70.87	panels, hinges and pulls.
Minimum Charge	Job		242		242	
48" w, 3 Doors / Drawer						
Demolish	Ea.		41.50		41.50	Includes material and labor to install
Install	Ea.	585	84.50		669.50	custom modular unit with solid
Demolish and Install	Ea.	585	126		711	hardwood faced frames, hardwood
Reinstall	Ea.		67.75		67.75	door frames and drawer fronts,
Clean	Ea.	1.52	41.50		43.02	hardwood veneer on raised door
Paint	Ea.	5.05	101		106.05	panels, hinges and pulls.
Minimum Charge	Job		242		242	
Clean Interior						
Install	L.F.	.31	7.80		8.11	Includes labor and materials to clean
Minimum Charge	Job		187		187	base cabinetry.
Premium Grade Wood						
24" Wide, 2 Door						
Demolish	Ea.		21		21	Includes material and labor to install
Install	Ea.	495	48.50		543.50	premium modular unit with solid
Demolish and Install	Ea.	495	69.50		564.50	hardwood faced frames, hardwood
Reinstall	Ea.		38.72		38.72	door frames and drawer fronts,
Clean	Ea.	.58	14.95		15.53	hardwood veneer on raised door
Paint	Ea.	1.94	40.50		42.44	panels, hinges and pulls.
Minimum Charge	Job		242		242	
30" Wide, 2 Door						
Demolish	Ea.		26		26	Includes material and labor to install
Install	Ea.	660	60.50		720.50	premium modular unit with solid
Demolish and Install	Ea.	660	86.50		746.50	hardwood faced frames, hardwood
Reinstall	Ea.		48.40		48.40	door frames and drawer fronts,
Clean	Ea.	.76	18.70		19.46	hardwood veneer on raised door
Paint	Ea.	2.43	50.50		52.93	panels, hinges and pulls.
Minimum Charge	Job		242		242	

For customer support on your Contractor's Pricing Guide: Residential Repair & Remodeling, call 800.448.8182.

Vanity Cabinets		Unit	Material	Labor	Equip.	Total	Specification
36″ Wide, 2 Door							
	Demolish	Ea.		32		32	Includes material and labor to install
	Install	Ea.	530	72.50		602.50	premium modular unit with solid
	Demolish and Install	Ea.	530	104.50		634.50	hardwood faced frames, hardwood
	Reinstall	Ea.		58.09		58.09	door frames and drawer fronts,
	Clean	Ea.	.85	22		22.85	hardwood veneer on raised door
	Paint	Ea.	2.94	57.50		60.44	panels, hinges and pulls.
	Minimum Charge	Job		242		242	
42″ w, 3 Doors / Drawer							
	Demolish	Ea.		38		38	Includes material and labor to install
	Install	Ea.	595	78		673	premium modular unit with solid
	Demolish and Install	Ea.	595	116		711	hardwood faced frames, hardwood
	Reinstall	Ea.		62.55		62.55	door frames and drawer fronts,
	Clean	Ea.	1.09	26.50		27.59	hardwood veneer on raised door
	Paint	Ea.	3.37	67.50		70.87	panels, hinges and pulls.
	Minimum Charge	Job		242		242	
48″ w, 3 Doors / Drawer							
	Demolish	Ea.		41.50		41.50	Includes material and labor to install
	Install	Ea.	780	84.50		864.50	premium modular unit with solid
	Demolish and Install	Ea.	780	126		906	hardwood faced frames, hardwood
	Reinstall	Ea.		67.75		67.75	door frames and drawer fronts,
	Clean	Ea.	1.52	41.50		43.02	hardwood veneer on raised door
	Paint	Ea.	5.05	101		106.05	panels, hinges and pulls.
	Minimum Charge	Job		242		242	
Clean Interior							
	Install	L.F.	.31	7.80		8.11	Includes labor and materials to clean
	Minimum Charge	Job		187		187	base cabinetry.
Premium Hardwood							
24″ Wide, 2 Door							
	Demolish	Ea.		21		21	Includes material and labor to install
	Install	Ea.	590	48.50		638.50	premium modular unit with solid
	Demolish and Install	Ea.	590	69.50		659.50	hardwood faced frames, hardwood
	Reinstall	Ea.		38.72		38.72	door frames and drawer fronts,
	Clean	Ea.	.58	14.95		15.53	hardwood veneer on raised door
	Paint	Ea.	1.94	45		46.94	panels, hinges and pulls.
	Minimum Charge	Job		242		242	
30″ Wide, 2 Door							
	Demolish	Ea.		26		26	Includes material and labor to install
	Install	Ea.	810	60.50		870.50	premium modular unit with solid
	Demolish and Install	Ea.	810	86.50		896.50	hardwood faced frames, hardwood
	Reinstall	Ea.		48.40		48.40	door frames and drawer fronts,
	Clean	Ea.	.76	18.70		19.46	hardwood veneer on raised door
	Paint	Ea.	2.43	50.50		52.93	panels, hinges and pulls.
	Minimum Charge	Job		242		242	
36″ Wide, 2 Door							
	Demolish	Ea.		32		32	Includes material and labor to install
	Install	Ea.	950	72.50		1022.50	premium modular unit with solid
	Demolish and Install	Ea.	950	104.50		1054.50	hardwood faced frames, hardwood
	Reinstall	Ea.		58.09		58.09	door frames and drawer fronts,
	Clean	Ea.	.85	22		22.85	hardwood veneer on raised door
	Paint	Ea.	2.94	67.50		70.44	panels, hinges and pulls.
	Minimum Charge	Job		242		242	

Cabinets and Countertops

Vanity Cabinets

Vanity Cabinets	Unit	Material	Labor	Equip.	Total	Specification
42" w, 3 Doors / Drawer						Includes material and labor to install
Demolish	Ea.		38		38	premium modular unit with solid
Install	Ea.	695	78		773	hardwood faced frames, hardwood
Demolish and Install	Ea.	695	116		811	door frames and drawer fronts,
Reinstall	Ea.		62.55		62.55	hardwood veneer on raised door
Clean	Ea.	1.09	26.50		27.59	panels, hinges and pulls.
Paint	Ea.	3.37	80.50		83.87	
Minimum Charge	Job		242		242	
48" w, 3 Doors / Drawer						Includes material and labor to install
Demolish	Ea.		41.50		41.50	premium modular unit with solid
Install	Ea.	775	84.50		859.50	hardwood faced frames, hardwood
Demolish and Install	Ea.	775	126		901	door frames and drawer fronts,
Reinstall	Ea.		67.75		67.75	hardwood veneer on raised door
Clean	Ea.	1.52	41.50		43.02	panels, hinges and pulls.
Paint	Ea.	5.05	101		106.05	
Minimum Charge	Job		242		242	
Clean Interior						
Install	L.F.	.31	7.80		8.11	Includes labor and materials to clean
Minimum Charge	Job		187		187	base cabinetry.
Strip and Refinish						
Install	L.F.	5.60	50.50		56.10	Includes labor and material to strip,
Minimum Charge	Job		202		202	prep and refinish exterior of cabinets.

Vanity Cabinet Tops

Vanity Cabinet Tops	Unit	Material	Labor	Equip.	Total	Specification
Cultured Marble Top						
Demolish	L.F.		6.95		6.95	Includes material and labor to install
Install	L.F.	61.50	28		89.50	cultured marble vanity top with
Demolish and Install	L.F.	61.50	34.95		96.45	integral sink, 22" deep.
Reinstall	L.F.		22.53		22.53	
Clean	L.F.	.06	1.42		1.48	
Minimum Charge	Job		235		235	
Laminated Top						
Demolish	L.F.		4.16		4.16	Includes material and labor to install
Install	L.F.	52.50	17.30		69.80	one piece laminated top with 4"
Demolish and Install	L.F.	52.50	21.46		73.96	backsplash.
Reinstall	L.F.		13.83		13.83	
Clean	L.F.	.03	.94		.97	
Minimum Charge	Job		242		242	
Solid Surface						
Demolish	L.F.		6.95		6.95	Includes material and labor to install
Install	L.F.	95.50	36		131.50	solid surface vanity top with integral
Demolish and Install	L.F.	95.50	42.95		138.45	sink installed, 22" deep.
Reinstall	L.F.		28.68		28.68	
Clean	L.F.	.06	1.42		1.48	
Minimum Charge	Job		242		242	

For customer support on your Contractor's Pricing Guide: Residential Repair & Remodeling, call 800.448.8182.

Roller Handle

Paint Brush

Paint Preparation, Walls	Unit	Material	Labor	Equip.	Total	Specification
Cover / Protect Floors						
Install	S.F.	.04	.16		.20	Includes material and labor to install
Minimum Charge	Job		206		206	plastic masking sheet on large uninterrupted areas.
Cover / Protect Walls						
Install	S.F.	.04	.16		.20	Includes material and labor to install
Minimum Charge	Job		206		206	plastic masking sheet on walls.
Clean Walls						
Light						
Clean	S.F.		.13		.13	Includes labor to wash gypsum
Minimum Charge	Job		206		206	drywall or plaster wall surfaces.
Heavy						
Clean	S.F.		.16		.16	Includes labor and material for heavy
Minimum Charge	Job		206		206	cleaning (multiple applications) with detergent and solvent.
Prep Walls						
Paint	S.F.	.06	.20		.26	Includes labor and material to prepare
Minimum Charge	Job		206		206	for painting including scraping, patching and puttying.
Sand Walls						
Paint	S.F.		.13		.13	Includes labor to sand gypsum drywall
Minimum Charge	Job		206		206	wall surfaces.
Seal Walls						
Large Area						
Paint	S.F.	.08	.30		.38	Includes labor and material to paint
Minimum Charge	Job		206		206	with primer / sealer by roller.
Spot						
Paint	S.F.	.10	.35		.45	Includes labor and material to seal
Minimum Charge	Job		206		206	existing drywall with shellac-based material.

For customer support on your Contractor's Pricing Guide: Residential Repair & Remodeling, call 800.448.8182.

Painting

Paint Preparation, Ceilings		Unit	Material	Labor	Equip.	Total	Specification
Cover / Protect Floors							
	Install	S.F.	.04	.16		.20	Includes material and labor to install
	Minimum Charge	Job		206		206	plastic masking sheet on large uninterrupted areas.
Cover / Protect Walls							
	Install	S.F.	.04	.16		.20	Includes material and labor to install
	Minimum Charge	Job		206		206	plastic masking sheet on walls.
Prep Ceiling							
	Paint	S.F.	.09	.85		.94	Includes labor and material to fill in
	Minimum Charge	Job		206		206	hairline cracks up to 1/8" with filler, sand and prep for paint.
Sand Ceiling							
	Paint	S.F.		.19		.19	Includes labor to sand gypsum drywall
	Minimum Charge	Job		206		206	ceiling surfaces.
Seal Ceiling							
Large Area							
	Paint	S.F.	.20	.31		.51	Includes labor and material to paint
	Minimum Charge	Job		202		202	ceiling, one coat flat latex.
Spot							
	Paint	S.F.	.10	.35		.45	Includes labor and material to seal
	Minimum Charge	Job		202		202	existing drywall with shellac-based material.

Paint / Texture, Walls		Unit	Material	Labor	Equip.	Total	Specification
Texture Walls							
Spray							
	Install	S.F.	.06	.60		.66	Includes labor and material to install
	Minimum Charge	Job		202		202	by spray, texture finish.
Trowel							
	Install	S.F.	.35	1.06		1.41	Includes labor and material to apply
	Minimum Charge	Job		202		202	hand troweled texture.
Paint Walls 1 Coat							
	Paint	S.F.	.20	.31		.51	Includes labor and material to paint,
	Minimum Charge	Job		202		202	one coat flat latex.
Paint Walls 2 Coats							
	Paint	S.F.	.40	.50		.90	Includes labor and material to paint,
	Minimum Charge	Job		202		202	two coats flat latex.
Paint Walls 3 Coats							
	Paint	S.F.	.59	.62		1.21	Includes labor and material to paint,
	Minimum Charge	Job		202		202	three coats flat latex.
Clean and Seal							
	Paint	S.F.	.34	1.08		1.42	Includes labor and material to do light
	Minimum Charge	Job		202		202	cleaning, sealing with 2 coats of latex paint.

For customer support on your Contractor's Pricing Guide: Residential Repair & Remodeling, call 800.448.8182.

Paint / Texture, Ceilings		Unit	Material	Labor	Equip.	Total	Specification
Texture Ceiling							
Spray							
	Install	S.F.	.06	.60		.66	Includes labor and material to install
	Minimum Charge	Job		202		202	by spray, texture finish.
Trowel							
	Install	S.F.	.35	1.06		1.41	Includes labor and material to apply
	Minimum Charge	Job		202		202	hand troweled texture.
Paint Ceiling, 1 Coat							
	Paint	S.F.	.20	.31		.51	Includes labor and material to paint
	Minimum Charge	Job		202		202	ceiling, one coat flat latex.
Paint Ceiling, 2 Coats							
	Paint	S.F.	.21	.50		.71	Includes labor and material to paint
	Minimum Charge	Job		202		202	two coats flat latex paint.
Paint Ceiling, 3 Coats							
	Paint	S.F.	.31	.59		.90	Includes labor and material to paint
	Minimum Charge	Job		202		202	ceiling, three coats flat latex.
Clean and Seal							
	Paint	S.F.	.31	.85		1.16	Includes labor and material for light
	Minimum Charge	Job		202		202	cleaning, sealing and paint 3 coats of
							flat latex paint.

For customer support on your Contractor's Pricing Guide: Residential Repair & Remodeling, call 800.448.8182.

Flooring

Plank Flooring   **Wood Strip Flooring**   **Tile Flooring**

Wood Plank Flooring	Unit	Material	Labor	Equip.	Total	Specification
Maple						
2-1/4″						
Demolish	S.F.		1.66		1.66	Includes material and labor to install
Install	S.F.	4.83	2.41		7.24	unfinished T&G maple flooring,
Demolish and Install	S.F.	4.83	4.07		8.90	25/32″ thick, 2-1/4″ wide, random
Reinstall	S.F.		1.93		1.93	3′ to 16′ lengths including felt
Clean	S.F.		.47		.47	underlayment, nailed in place over
Paint	S.F.	.26	2.88		3.14	prepared subfloor.
Minimum Charge	Job		242		242	
3-1/4″						
Demolish	S.F.		1.66		1.66	Includes material and labor to install
Install	S.F.	7.25	2.41		9.66	unfinished T&G maple flooring,
Demolish and Install	S.F.	7.25	4.07		11.32	33/32″ thick, 3-1/4″ wide, random
Reinstall	S.F.		1.93		1.93	3′ to 16′ lengths including felt
Clean	S.F.		.47		.47	underlayment, nailed in place over
Paint	S.F.	.26	2.88		3.14	prepared subfloor.
Minimum Charge	Job		242		242	
Sand and Finish						
Install	S.F.	.26	.94		1.20	Includes labor and material to refinish
Minimum Charge	Job		242		242	wood floor.
Oak						
2-1/4″						
Demolish	S.F.		1.66		1.66	Includes material and labor to install
Install	S.F.	5.90	2.41		8.31	unfinished T&G oak flooring, 25/32″
Demolish and Install	S.F.	5.90	4.07		9.97	thick, 2-1/4″ wide, random 3′ to 16′
Reinstall	S.F.		1.93		1.93	lengths including felt underlayment,
Clean	S.F.		.47		.47	nailed in place over prepared
Paint	S.F.	.26	2.88		3.14	subfloor.
Minimum Charge	Job		242		242	
3-1/4″						
Demolish	S.F.		1.66		1.66	Includes material and labor to install
Install	S.F.	6.05	1.71		7.76	unfinished T&G oak flooring, 25/32″
Demolish and Install	S.F.	6.05	3.37		9.42	thick, 3-1/4″ wide, random 3′ to 16′
Reinstall	S.F.		1.37		1.37	lengths including felt underlayment,
Clean	S.F.		.47		.47	nailed in place over prepared
Paint	S.F.	.26	2.88		3.14	subfloor.
Minimum Charge	Job		242		242	
Sand and Finish						
Install	S.F.	.26	2.88		3.14	Includes labor and material to refinish
Minimum Charge	Job		242		242	wood floor.

Wood Plank Flooring	Unit	Material	Labor	Equip.	Total	Specification
Walnut						
3" to 7"						
Demolish	S.F.		1.66		1.66	Includes material and labor to install
Install	S.F.	18.05	1.71		19.76	unfinished T&G walnut flooring,
Demolish and Install	S.F.	18.05	3.37		21.42	25/32" thick, random 3' to 16'
Reinstall	S.F.		1.37		1.37	lengths including felt underlayment,
Clean	S.F.		.47		.47	nailed in place over prepared
Paint	S.F.	.26	2.88		3.14	subfloor.
Minimum Charge	Job		242		242	
Sand and Finish						
Install	S.F.	.26	2.88		3.14	Includes labor and material to refinish
Minimum Charge	Job		242		242	wood floor.
Pine						
2-1/4"						
Demolish	S.F.		1.66		1.66	Includes material and labor to install
Install	S.F.	1.37	1.82		3.19	unfinished T&G pine flooring, 25/32"
Demolish and Install	S.F.	1.37	3.48		4.85	thick, 2-1/4" wide, random 3' to 16'
Reinstall	S.F.		1.46		1.46	lengths including felt underlayment,
Clean	S.F.		.47		.47	nailed in place over prepared
Paint	S.F.	.26	.94		1.20	subfloor.
Minimum Charge	Job		242		242	
Sand and Finish						
Install	S.F.	.26	2.88		3.14	Includes labor and material to refinish
Minimum Charge	Job		242		242	wood floor.
Prefinished Oak						
2-1/4"						
Demolish	S.F.		1.66		1.66	Includes material and labor to install
Install	S.F.	4.05	2.41		6.46	prefinished T&G oak flooring, 2-1/4"
Demolish and Install	S.F.	4.05	4.07		8.12	wide, random 3' to 16' lengths
Reinstall	S.F.		1.93		1.93	including felt underlayment, nailed in
Clean	S.F.		.47		.47	place over prepared subfloor.
Minimum Charge	Job		242		242	
3-1/4"						
Demolish	S.F.		1.66		1.66	Includes material and labor to install
Install	S.F.	3.55	2.22		5.77	prefinished T&G oak flooring, 25/32"
Demolish and Install	S.F.	3.55	3.88		7.43	thick, 3" to 7" wide, random 3' to 16'
Reinstall	S.F.		1.77		1.77	lengths including felt underlayment,
Clean	S.F.		.47		.47	nailed in place over prepared
Minimum Charge	Job		242		242	subfloor.
Sand and Finish New						
Install	S.F.	.26	1.27		1.53	Includes labor and material to refinish
Minimum Charge	Job		242		242	wood floor.
Refinish Existing						
Install	S.F.	.26	.94		1.20	Includes labor and material to refinish
Minimum Charge	Job		242		242	wood floor.

Flooring

Wood Plank Flooring

Wood Plank Flooring		Unit	Material	Labor	Equip.	Total	Specification
Urethane Coat							
	Install	S.F.	.17	.21		.38	Includes labor and material to install
	Minimum Charge	Job		242		242	two coats of urethane, on existing floor.
Clean and Wax							
	Clean	S.F.	.02	.29		.31	Includes labor and material to wash
	Minimum Charge	Job		242		242	and wax hardwood.
Underlayment Per S.F.							
	Demolish	S.F.		.71		.71	Cost includes material and labor to
	Install	S.F.	2.04	.65		2.69	install 1/4" lauan subfloor, standard
	Demolish and Install	S.F.	2.04	1.36		3.40	interior grade, nailed every 6".
	Clean	S.F.		.36		.36	
	Minimum Charge	Job		242		242	

Wood Parquet Tile Floor

Wood Parquet Tile Floor		Unit	Material	Labor	Equip.	Total	Specification
Oak							
9" x 9"							
	Demolish	S.F.		1.19		1.19	Includes material and labor to install
	Install	S.F.	18.65	3.87		22.52	prefinished oak 9" x 9" parquet block
	Demolish and Install	S.F.	18.65	5.06		23.71	flooring, 5/16" thick, installed in
	Reinstall	S.F.		3.10		3.10	mastic.
	Clean	S.F.		.47		.47	
	Paint	S.F.	.26	2.88		3.14	
	Minimum Charge	Job		242		242	
13" x 13"							
	Demolish	S.F.		1.19		1.19	Includes material and labor to install
	Install	S.F.	14.30	3.03		17.33	prefinished oak 13" x 13" parquet
	Demolish and Install	S.F.	14.30	4.22		18.52	block flooring, 5/16" thick, installed in
	Reinstall	S.F.		2.42		2.42	mastic.
	Clean	S.F.		.47		.47	
	Paint	S.F.	.26	2.88		3.14	
	Minimum Charge	Job		242		242	
Cherry							
9" x 9"							
	Demolish	S.F.		1.19		1.19	Includes material and labor to install
	Install	S.F.	23.50	3.87		27.37	prefinished cherry 9" x 9" parquet
	Demolish and Install	S.F.	23.50	5.06		28.56	block flooring, 5/16" thick, installed in
	Reinstall	S.F.		3.87		3.87	mastic.
	Clean	S.F.		.47		.47	
	Paint	S.F.	.26	2.88		3.14	
	Minimum Charge	Job		242		242	
13" x 13"							
	Demolish	S.F.		1.19		1.19	Includes material and labor to install
	Install	S.F.	24	3.03		27.03	prefinished cherry 13" x 13" parquet
	Demolish and Install	S.F.	24	4.22		28.22	block flooring, 5/16" thick, installed in
	Reinstall	S.F.		3.03		3.03	mastic.
	Clean	S.F.		.47		.47	
	Paint	S.F.	.26	2.88		3.14	
	Minimum Charge	Job		242		242	

For customer support on your Contractor's Pricing Guide: Residential Repair & Remodeling, call 800.448.8182.

Wood Parquet Tile Floor		Unit	Material	Labor	Equip.	Total	Specification
Walnut							
9″ x 9″							
	Demolish	S.F.		1.19		1.19	Includes material and labor to install
	Install	S.F.	19.75	3.87		23.62	prefinished walnut 9″ x 9″ parquet
	Demolish and Install	S.F.	19.75	5.06		24.81	flooring in mastic.
	Reinstall	S.F.		3.10		3.10	
	Clean	S.F.		.47		.47	
	Paint	S.F.	.26	2.88		3.14	
	Minimum Charge	Job		242		242	
13″ x 13″							
	Demolish	S.F.		1.19		1.19	Includes material and labor to install
	Install	S.F.	16.55	3.03		19.58	prefinished walnut 13″ x 13″ parquet
	Demolish and Install	S.F.	16.55	4.22		20.77	block flooring, 5/16″ thick, installed in
	Reinstall	S.F.		3.03		3.03	mastic.
	Clean	S.F.		.47		.47	
	Paint	S.F.	.26	2.88		3.14	
	Minimum Charge	Job		242		242	
Acrylic Impregnated							
Oak 12″ x 12″							
	Demolish	S.F.		1.19		1.19	Includes material and labor to install
	Install	S.F.	15.75	3.46		19.21	red oak 12″ x 12″ acrylic
	Demolish and Install	S.F.	15.75	4.65		20.40	impregnated parquet block flooring,
	Reinstall	S.F.		3.46		3.46	5/16″ thick, installed in mastic.
	Clean	S.F.		.47		.47	
	Minimum Charge	Job		242		242	
Cherry 12″ x 12″							
	Demolish	S.F.		1.19		1.19	Includes material and labor to install
	Install	S.F.	29	3.46		32.46	prefinished cherry 12″ x 12″ acrylic
	Demolish and Install	S.F.	29	4.65		33.65	impregnated parquet block flooring,
	Reinstall	S.F.		3.46		3.46	5/16″ thick, installed in mastic.
	Clean	S.F.		.47		.47	
	Minimum Charge	Job		242		242	
Ash 12″ x 12″							
	Demolish	S.F.		1.19		1.19	Includes material and labor to install
	Install	S.F.	30.50	3.46		33.96	ash 12″ x 12″ acrylic impregnated
	Demolish and Install	S.F.	30.50	4.65		35.15	parquet block flooring, 5/16″ thick in
	Reinstall	S.F.		3.46		3.46	mastic.
	Clean	S.F.		.47		.47	
	Minimum Charge	Job		242		242	
Teak							
	Demolish	S.F.		1.19		1.19	Includes material and labor to install
	Install	S.F.	15.60	3.03		18.63	prefinished teak parquet block
	Demolish and Install	S.F.	15.60	4.22		19.82	flooring, 5/16″ thick, and installed in
	Reinstall	S.F.		3.03		3.03	mastic.
	Clean	S.F.		.47		.47	
	Paint	S.F.	.26	2.88		3.14	
	Minimum Charge	Job		242		242	
Clean and Wax							
	Clean	S.F.	.02	.29		.31	Includes labor and material to wash
	Minimum Charge	Job		242		242	and wax hardwood.

Flooring

Ceramic Tile Flooring

Ceramic Tile Flooring		Unit	Material	Labor	Equip.	Total	Specification
Economy Grade							
	Demolish	S.F.		1.23		1.23	Cost includes material and labor to
	Install	S.F.	3.33	9.20		12.53	install 4-1/4" x 4-1/4" ceramic tile in
	Demolish and Install	S.F.	3.33	10.43		13.76	mortar bed including grout.
	Clean	S.F.		.71		.71	
	Minimum Charge	Job		235		235	
Average Grade							
	Demolish	S.F.		1.23		1.23	Cost includes material and labor to
	Install	S.F.	5.15	9.40		14.55	install 4-1/4" x 4-1/4" ceramic tile in
	Demolish and Install	S.F.	5.15	10.63		15.78	mortar bed including grout.
	Clean	S.F.		.71		.71	
	Minimum Charge	Job		235		235	
Premium Grade							
	Demolish	S.F.		1.23		1.23	Cost includes material and labor to
	Install	S.F.	9.40	9.40		18.80	install 4-1/4" x 4-1/4" ceramic tile in
	Demolish and Install	S.F.	9.40	10.63		20.03	mortar bed including grout.
	Clean	S.F.		.71		.71	
	Minimum Charge	Job		235		235	
Re-grout							
	Install	S.F.	.06	3.75		3.81	Includes material and labor to regrout
	Minimum Charge	Job		235		235	tile floors.

Hard Tile Flooring

Hard Tile Flooring		Unit	Material	Labor	Equip.	Total	Specification
Thick Set Paver							
Brick							
	Demolish	S.F.		2.60		2.60	Cost includes material and labor to
	Install	S.F.	1.43	21		22.43	install 6" x 12" adobe brick paver with
	Demolish and Install	S.F.	1.43	23.60		25.03	1/2" mortar joints.
	Clean	S.F.		.71		.71	
	Minimum Charge	Job		235		235	
Mexican Red							
	Demolish	S.F.		2.60		2.60	Cost includes material and labor to
	Install	S.F.	1.98	9.80		11.78	install 12" x 12" red Mexican paver
	Demolish and Install	S.F.	1.98	12.40		14.38	tile in mortar bed with grout.
	Clean	S.F.		.71		.71	
	Minimum Charge	Job		235		235	
Saltillo							
	Demolish	S.F.		1.23		1.23	Cost includes material and labor to
	Install	S.F.	1.52	9.80		11.32	install 12" x 12" saltillo tile in mortar
	Demolish and Install	S.F.	1.52	11.03		12.55	bed with grout.
	Clean	S.F.		.71		.71	
	Minimum Charge	Job		235		235	
Marble							
Premium Grade							
	Demolish	S.F.		1.23		1.23	Cost includes material and labor to
	Install	S.F.	20	20.50		40.50	install 3/8" x 12" x 12" marble, in
	Demolish and Install	S.F.	20	21.73		41.73	mortar bed with grout.
	Clean	S.F.		.71		.71	
	Minimum Charge	Job		235		235	

For customer support on your Contractor's Pricing Guide: Residential Repair & Remodeling, call 800.448.8182.

Hard Tile Flooring		Unit	Material	Labor	Equip.	Total	Specification
Terrazzo							
Gray Cement							
	Demolish	S.F.		1.65		1.65	Cost includes material and labor to
	Install	S.F.	4.50	11.40	3.51	19.41	install 1-3/4" terrazzo , #1 and #2
	Demolish and Install	S.F.	4.50	13.05	3.51	21.06	chips in gray Portland cement.
	Clean	S.F.		.71		.71	
	Minimum Charge	Job		425	132	557	
White Cement							
	Demolish	S.F.		1.65		1.65	Cost includes material and labor to
	Install	S.F.	5.30	11.40	3.51	20.21	install 1-3/4" terrazzo, #1 and #2
	Demolish and Install	S.F.	5.30	13.05	3.51	21.86	chips in white Portland cement.
	Clean	S.F.		.71		.71	
	Minimum Charge	Job		425	132	557	
Non-skid Gray							
	Demolish	S.F.		1.65		1.65	Cost includes material and labor to
	Install	S.F.	8.35	14.25	4.39	26.99	install 1-3/4" terrazzo, #1 and #2
	Demolish and Install	S.F.	8.35	15.90	4.39	28.64	chips in gray Portland cement with
	Clean	S.F.		.71		.71	light non-skid abrasive.
	Minimum Charge	Job		425	132	557	
Non-skid White							
	Demolish	S.F.		1.65		1.65	Cost includes material and labor to
	Install	S.F.	9.25	14.25	4.39	27.89	install 1-3/4" terrazzo, #1 and #2
	Demolish and Install	S.F.	9.25	15.90	4.39	29.54	chips in white Portland cement with
	Clean	S.F.		.71		.71	light non-skid abrasive.
	Minimum Charge	Job		425	132	557	
Gray w / Brass Divider							
	Demolish	S.F.		1.66		1.66	Cost includes material and labor to
	Install	S.F.	13.05	15.50	4.79	33.34	install 1-3/4" terrazzo, #1 and #2
	Demolish and Install	S.F.	13.05	17.16	4.79	35	chips in gray Portland cement with
	Clean	S.F.		.71		.71	brass strips 2' O.C. each way.
	Minimum Charge	Job		425	132	557	
Clean							
	Clean	S.F.	.07	.47		.54	Includes labor and material to clean
	Minimum Charge	Job		425	132	557	terrazzo flooring.
Slate							
	Demolish	S.F.		1.48		1.48	Includes material and labor to install
	Install	S.F.	8.60	4.62		13.22	slate tile flooring in thin set with grout.
	Demolish and Install	S.F.	8.60	6.10		14.70	
	Clean	S.F.		.71		.71	
	Minimum Charge	Job		235		235	
Granite							
	Demolish	S.F.		1.22		1.22	Includes material and labor to install
	Install	S.F.	28	4.69		32.69	granite tile flooring in thin set with
	Demolish and Install	S.F.	28	5.91		33.91	grout.
	Reinstall	S.F.		4.69		4.69	
	Clean	S.F.		.71		.71	
	Minimum Charge	Job		235		235	
Re-grout							
	Install	S.F.	.06	3.75		3.81	Includes material and labor to regrout
	Minimum Charge	Job		235		235	tile floors.

For customer support on your Contractor's Pricing Guide: Residential Repair & Remodeling, call 800.448.8182.

Flooring

Sheet Vinyl Per S.F.

Sheet Vinyl Per S.F.	Unit	Material	Labor	Equip.	Total	Specification
Economy Grade						
Demolish	S.F.		.59		.59	Includes material and labor to install
Install	S.F.	2.40	1.46		3.86	resilient sheet vinyl flooring.
Demolish and Install	S.F.	2.40	2.05		4.45	
Clean	S.F.		.47		.47	
Minimum Charge	Job		235		235	
Average Grade						
Demolish	S.F.		.59		.59	Includes material and labor to install
Install	S.F.	3.41	1.46		4.87	resilient sheet vinyl flooring.
Demolish and Install	S.F.	3.41	2.05		5.46	
Clean	S.F.		.47		.47	
Minimum Charge	Job		235		235	
Premium Grade						
Demolish	S.F.		.59		.59	Includes material and labor to install
Install	S.F.	3.33	1.46		4.79	resilient sheet vinyl flooring.
Demolish and Install	S.F.	3.33	2.05		5.38	
Clean	S.F.		.47		.47	
Minimum Charge	Job		235		235	
Clean and Wax						
Clean	S.F.	.02	.52		.54	Includes labor and material to wash
Minimum Charge	Job		235		235	and wax a vinyl tile floor.

Sheet Vinyl

Sheet Vinyl	Unit	Material	Labor	Equip.	Total	Specification
Economy Grade						
Demolish	S.Y.		5.35		5.35	Includes material and labor to install
Install	S.Y.	21.50	13.20		34.70	no wax sheet vinyl flooring.
Demolish and Install	S.Y.	21.50	18.55		40.05	
Clean	S.Y.		4.20		4.20	
Minimum Charge	Job		235		235	
Average Grade						
Demolish	S.Y.		5.35		5.35	Includes material and labor to install
Install	S.Y.	31	13.20		44.20	no wax sheet vinyl flooring.
Demolish and Install	S.Y.	31	18.55		49.55	
Clean	S.Y.		4.20		4.20	
Minimum Charge	Job		235		235	
Premium Grade						
Demolish	S.Y.		5.35		5.35	Includes material and labor to install
Install	S.Y.	30	13.20		43.20	no wax sheet vinyl flooring.
Demolish and Install	S.Y.	30	18.55		48.55	
Clean	S.Y.		4.20		4.20	
Minimum Charge	Job		235		235	
Clean and Wax						
Clean	S.F.	.02	.52		.54	Includes labor and material to wash
Minimum Charge	Job		235		235	and wax a vinyl tile floor.

Vinyl Tile

Vinyl Tile	Unit	Material	Labor	Equip.	Total	Specification
Economy Grade						
Demolish	S.F.		.83		.83	Includes material and labor to install
Install	S.F.	1.50	.94		2.44	no wax 12" x 12" vinyl tile.
Demolish and Install	S.F.	1.50	1.77		3.27	
Clean	S.F.		.47		.47	
Minimum Charge	Job		235		235	

245

Flooring

Vinyl Tile

Vinyl Tile	Unit	Material	Labor	Equip.	Total	Specification
Average Grade						
Demolish	S.F.		.83		.83	Includes material and labor to install
Install	S.F.	1.21	.94		2.15	no wax 12" x 12" vinyl tile.
Demolish and Install	S.F.	1.21	1.77		2.98	
Clean	S.F.		.47		.47	
Minimum Charge	Job		235		235	
Premium Grade						
Demolish	S.F.		.83		.83	Includes material and labor to install
Install	S.F.	3.58	.94		4.52	no wax 12" x 12" vinyl tile.
Demolish and Install	S.F.	3.58	1.77		5.35	
Clean	S.F.		.47		.47	
Minimum Charge	Job		235		235	
Clean and Wax						
Clean	S.F.	.02	.52		.54	Includes labor and material to wash
Minimum Charge	Job		235		235	and wax a vinyl tile floor.

Carpeting Per S.F.

Carpeting Per S.F.	Unit	Material	Labor	Equip.	Total	Specification
Economy Grade						
Demolish	S.F.		.09		.09	Includes material and labor to install
Install	S.F.	1.24	.70		1.94	carpet including tack strips and hot
Demolish and Install	S.F.	1.24	.79		2.03	melt tape on seams.
Clean	S.F.		.65		.65	
Minimum Charge	Job		235		235	
Average Grade						
Demolish	S.F.		.09		.09	Includes material and labor to install
Install	S.F.	1.33	.70		2.03	carpet including tack strips and hot
Demolish and Install	S.F.	1.33	.79		2.12	melt tape on seams.
Clean	S.F.		.65		.65	
Minimum Charge	Job		235		235	
Premium Grade						
Demolish	S.F.		.09		.09	Includes material and labor to install
Install	S.F.	2.10	.70		2.80	carpet including tack strips and hot
Demolish and Install	S.F.	2.10	.79		2.89	melt tape on seams.
Clean	S.F.		.65		.65	
Minimum Charge	Job		235		235	
Indoor/Outdoor						
Demolish	S.F.		.09		.09	Includes material and labor to install
Install	S.F.	1.09	.70		1.79	indoor-outdoor carpet.
Demolish and Install	S.F.	1.09	.79		1.88	
Clean	S.F.		.65		.65	
Minimum Charge	Job		235		235	
100% Wool						
Average Grade						
Demolish	S.F.		.09		.09	Includes material and labor to install
Install	S.F.	8.25	.93		9.18	carpet including tack strips and hot
Demolish and Install	S.F.	8.25	1.02		9.27	melt tape on seams.
Clean	S.F.		.65		.65	
Minimum Charge	Job		235		235	
Premium Grade						
Demolish	S.F.		.09		.09	Includes material and labor to install
Install	S.F.	11.95	.93		12.88	carpet including tack strips and hot
Demolish and Install	S.F.	11.95	1.02		12.97	melt tape on seams.
Clean	S.F.		.65		.65	
Minimum Charge	Job		235		235	

For customer support on your Contractor's Pricing Guide: Residential Repair & Remodeling, call 800.448.8182.

Flooring

Carpeting Per S.F.		Unit	Material	Labor	Equip.	Total	Specification
Luxury Grade							
	Demolish	S.F.		.09		.09	Includes material and labor to install
	Install	S.F.	12.30	.93		13.23	carpet including tack strips and hot
	Demolish and Install	S.F.	12.30	1.02		13.32	melt tape on seams.
	Clean	S.F.		.65		.65	
	Minimum Charge	Job		235		235	
Berber							
Average Grade							
	Demolish	S.F.		.09		.09	Includes material and labor to install
	Install	S.F.	7.30	1.11		8.41	carpet including tack strips and hot
	Demolish and Install	S.F.	7.30	1.20		8.50	melt tape on seams.
	Clean	S.F.		.65		.65	
	Minimum Charge	Job		235		235	
Premium Grade							
	Demolish	S.F.		.09		.09	Includes material and labor to install
	Install	S.F.	8.25	1.11		9.36	carpet including tack strips and hot
	Demolish and Install	S.F.	8.25	1.20		9.45	melt tape on seams.
	Clean	S.F.		.65		.65	
	Minimum Charge	Job		235		235	

Carpeting		Unit	Material	Labor	Equip.	Total	Specification
Indoor/Outdoor							
	Demolish	S.Y.		1.39		1.39	Includes material and labor to install
	Install	S.Y.	9.80	6.25		16.05	indoor-outdoor carpet.
	Demolish and Install	S.Y.	9.80	7.64		17.44	
	Reinstall	S.Y.		5		5	
	Clean	S.Y.		5.85		5.85	
	Minimum Charge	Job		235		235	
Economy Grade							
	Demolish	S.Y.		1.39		1.39	Includes material and labor to install
	Install	S.Y.	11.15	6.25		17.40	carpet including tack strips and hot
	Demolish and Install	S.Y.	11.15	7.64		18.79	melt tape on seams.
	Reinstall	S.Y.		5		5	
	Clean	S.Y.		5.85		5.85	
	Minimum Charge	Job		235		235	
Average Grade							
	Demolish	S.Y.		1.39		1.39	Includes material and labor to install
	Install	S.Y.	11.95	6.25		18.20	carpet including tack strips and hot
	Demolish and Install	S.Y.	11.95	7.64		19.59	melt tape on seams.
	Reinstall	S.Y.		5		5	
	Clean	S.Y.		5.85		5.85	
	Minimum Charge	Job		235		235	
Premium Grade							
	Demolish	S.Y.		1.39		1.39	Includes material and labor to install
	Install	S.Y.	18.85	6.25		25.10	carpet including tack strips and hot
	Demolish and Install	S.Y.	18.85	7.64		26.49	melt tape on seams.
	Reinstall	S.Y.		5		5	
	Clean	S.Y.		5.85		5.85	
	Minimum Charge	Job		235		235	

For customer support on your Contractor's Pricing Guide: Residential Repair & Remodeling, call 800.448.8182.

Flooring

Carpeting	Unit	Material	Labor	Equip.	Total	Specification
100% Wool						
Average Grade						
Demolish	S.Y.		1.39		1.39	Includes material and labor to install
Install	S.Y.	74.50	8.40		82.90	wool carpet including tack strips and
Demolish and Install	S.Y.	74.50	9.79		84.29	hot melt tape on seams.
Reinstall	S.Y.		6.70		6.70	
Clean	S.Y.		5.85		5.85	
Minimum Charge	Job		235		235	
Premium Grade						
Demolish	S.Y.		1.39		1.39	Includes material and labor to install
Install	S.Y.	107	8.40		115.40	carpet including tack strips and hot
Demolish and Install	S.Y.	107	9.79		116.79	melt tape on seams.
Reinstall	S.Y.		6.70		6.70	
Clean	S.Y.		5.85		5.85	
Minimum Charge	Job		235		235	
Luxury Grade						
Demolish	S.Y.		1.39		1.39	Includes material and labor to install
Install	S.Y.	111	8.40		119.40	carpet including tack strips and hot
Demolish and Install	S.Y.	111	9.79		120.79	melt tape on seams.
Reinstall	S.Y.		6.70		6.70	
Clean	S.Y.		5.85		5.85	
Minimum Charge	Job		235		235	
Berber						
Average Grade						
Demolish	S.Y.		1.39		1.39	Includes material and labor to install
Install	S.Y.	66	10		76	berber carpet including tack strips and
Demolish and Install	S.Y.	66	11.39		77.39	hot melt tape on seams.
Reinstall	S.Y.		7.99		7.99	
Clean	S.Y.		5.85		5.85	
Minimum Charge	Job		235		235	
Premium Grade						
Demolish	S.Y.		1.39		1.39	Includes material and labor to install
Install	S.Y.	74.50	10		84.50	carpet including tack strips and hot
Demolish and Install	S.Y.	74.50	11.39		85.89	melt tape on seams.
Reinstall	S.Y.		7.99		7.99	
Clean	S.Y.		5.85		5.85	
Minimum Charge	Job		235		235	

Carpet Pad Per S.F.	Unit	Material	Labor	Equip.	Total	Specification
Urethane						
Demolish	S.F.		.03		.03	Includes material and labor to install
Install	S.F.	.48	.35		.83	urethane carpet pad.
Demolish and Install	S.F.	.48	.38		.86	
Minimum Charge	Job		235		235	
Foam Rubber Slab						
Demolish	S.F.		.03		.03	Includes material and labor to install
Install	S.F.	.72	.35		1.07	carpet pad, 3/8" thick foam rubber.
Demolish and Install	S.F.	.72	.38		1.10	
Minimum Charge	Job		235		235	
Waffle						
Demolish	S.F.		.03		.03	Includes material and labor to install
Install	S.F.	.88	.04		.92	rubber waffle carpet pad.
Demolish and Install	S.F.	.88	.07		.95	
Minimum Charge	Job		235		235	

For customer support on your Contractor's Pricing Guide: Residential Repair & Remodeling, call 800.448.8182.

Flooring

Carpet Pad Per S.F.

Carpet Pad Per S.F.		Unit	Material	Labor	Equip.	Total	Specification
Jute							
	Demolish	S.F.		.03		.03	Includes material and labor to install
	Install	S.F.	.84	.31		1.15	jute hair carpet pad.
	Demolish and Install	S.F.	.84	.34		1.18	
	Minimum Charge	Job		235		235	
Rebound							
	Demolish	S.F.		.03		.03	Includes material and labor to install
	Install	S.F.	.48	.26		.74	rebound carpet pad.
	Demolish and Install	S.F.	.48	.29		.77	
	Minimum Charge	Job		235		235	

Carpet Pad

Carpet Pad		Unit	Material	Labor	Equip.	Total	Specification
Urethane							
	Demolish	S.Y.		.24		.24	Includes material and labor to install
	Install	S.Y.	4.39	3.13		7.52	urethane carpet pad.
	Demolish and Install	S.Y.	4.39	3.37		7.76	
	Reinstall	S.Y.		2.50		2.50	
	Minimum Charge	Job		235		235	
Foam Rubber Slab							
	Demolish	S.Y.		.24		.24	Includes material and labor to install
	Install	S.Y.	6.45	3.13		9.58	rubber slab carpet pad.
	Demolish and Install	S.Y.	6.45	3.37		9.82	
	Reinstall	S.Y.		2.50		2.50	
	Minimum Charge	Job		235		235	
Waffle							
	Demolish	S.Y.		.24		.24	Includes material and labor to install
	Install	S.Y.	7.85	.35		8.20	rubber waffle carpet pad.
	Demolish and Install	S.Y.	7.85	.59		8.44	
	Reinstall	S.Y.		.28		.28	
	Minimum Charge	Job		235		235	
Jute							
	Demolish	S.Y.		.24		.24	Includes material and labor to install
	Install	S.Y.	7.55	2.76		10.31	jute hair carpet pad.
	Demolish and Install	S.Y.	7.55	3		10.55	
	Reinstall	S.Y.		2.21		2.21	
	Minimum Charge	Job		235		235	
Rebound							
	Demolish	S.Y.		.24		.24	Includes material and labor to install
	Install	S.Y.	4.30	2.34		6.64	rebound carpet pad.
	Demolish and Install	S.Y.	4.30	2.58		6.88	
	Reinstall	S.Y.		1.87		1.87	
	Minimum Charge	Job		235		235	

Stair Components

Stair Components		Unit	Material	Labor	Equip.	Total	Specification
Carpeting							
	Demolish	Ea.		2.52		2.52	Includes material and labor to install
	Install	Riser	59	12.35		71.35	carpet and pad on stairs.
	Demolish and Install	Riser	59	14.87		73.87	
	Reinstall	Riser		9.88		9.88	
	Clean	Ea.		5.85		5.85	
	Minimum Charge	Job		235		235	

For customer support on your Contractor's Pricing Guide: Residential Repair & Remodeling, call 800.448.8182.

Finish Mechanical

Tub/Shower

Stainless Steel Double Sink

Water Closet

Commode	Unit	Material	Labor	Equip.	Total	Specification
Floor Mounted w / Tank						
Demolish	Ea.		54.50		54.50	Includes material and labor to install a
Install	Ea.	219	186		405	toilet with valve, seat and cover.
Demolish and Install	Ea.	219	240.50		459.50	
Reinstall	Ea.		148.53		148.53	
Clean	Ea.	.15	18.70		18.85	
Minimum Charge	Job		273		273	
Wall w / Flush Valve						
Demolish	Ea.		54.50		54.50	Includes material and labor to install a
Install	Ea.	1075	170		1245	wall hung toilet with flush valve, seat
Demolish and Install	Ea.	1075	224.50		1299.50	and cover.
Reinstall	Ea.		135.72		135.72	
Clean	Ea.	.15	18.70		18.85	
Minimum Charge	Job		273		273	
Designer Floor Mount						
Demolish	Ea.		54.50		54.50	Includes material and labor to install a
Install	Ea.	1025	186		1211	toilet with valve, seat and cover.
Demolish and Install	Ea.	1025	240.50		1265.50	
Reinstall	Ea.		148.53		148.53	
Clean	Ea.	.15	18.70		18.85	
Minimum Charge	Job		273		273	
Urinal						
Demolish	Ea.		54.50		54.50	Includes material and labor to install a
Install	Ea.	375	330		705	wall hung urinal with flush valve.
Demolish and Install	Ea.	375	384.50		759.50	
Reinstall	Ea.		262.40		262.40	
Clean	Ea.	.15	18.70		18.85	
Minimum Charge	Job		273		273	
Bidet						
Vitreous China						
Demolish	Ea.		54.50		54.50	Includes material and labor to install a
Install	Ea.	735	197		932	vitreous china bidet complete with
Demolish and Install	Ea.	735	251.50		986.50	trim.
Reinstall	Ea.		157.44		157.44	
Clean	Ea.	.15	18.70		18.85	
Minimum Charge	Job		273		273	

For customer support on your Contractor's Pricing Guide: Residential Repair & Remodeling, call 800.448.8182.

Finish Mechanical

Commode

	Unit	Material	Labor	Equip.	Total	Specification
Chrome Fittings						
Install	Ea.	263	68.50		331.50	Includes labor and materials to install
Brass Fittings						chrome fittings for a bidet.
Install	Ea.	289	68.50		357.50	Includes labor and materials to install brass fittings for a bidet.
Seat						
Demolish	Ea.		23		23	Includes material and labor to install a
Install	Ea.	33	23		56	toilet seat and cover.
Demolish and Install	Ea.	33	46		79	
Reinstall	Ea.		18.23		18.23	
Clean	Ea.	.08	7.80		7.88	
Minimum Charge	Job		273		273	
Rough-in						
Install	Ea.	460	345		805	Includes labor and materials to install water supply pipe with valves and drain, waste and vent pipe with all couplings, hangers and fasteners necessary.

Sink (assembly)

	Unit	Material	Labor	Equip.	Total	Specification
Single						
Porcelain						
Demolish	Ea.		34		34	Includes material and labor to install a
Install	Ea.	1150	176		1326	single bowl enamel finished cast iron
Demolish and Install	Ea.	1150	210		1360	kitchen sink with faucet and drain.
Reinstall	Ea.		140.57		140.57	
Clean	Ea.	.01	1.45		1.46	
Minimum Charge	Job		273		273	
Stainless Steel						
Demolish	Ea.		34		34	Includes material and labor to install a
Install	Ea.	790	176		966	single bowl stainless steel kitchen sink
Demolish and Install	Ea.	790	210		1000	with faucet and drain.
Reinstall	Ea.		140.57		140.57	
Clean	Ea.	.01	1.45		1.46	
Minimum Charge	Job		273		273	
Double						
Porcelain						
Demolish	Ea.		39		39	Includes material and labor to install a
Install	Ea.	440	205		645	double bowl enamel finished cast iron
Demolish and Install	Ea.	440	244		684	kitchen sink with faucet and drain.
Reinstall	Ea.		164		164	
Clean	Ea.	.15	18.70		18.85	
Minimum Charge	Job		273		273	
Stainless Steel						
Demolish	Ea.		39		39	Includes material and labor to install a
Install	Ea.	510	205		715	double bowl stainless steel kitchen sink
Demolish and Install	Ea.	510	244		754	with faucet and drain.
Reinstall	Ea.		164		164	
Clean	Ea.	.15	18.70		18.85	
Minimum Charge	Job		273		273	

For customer support on your Contractor's Pricing Guide: Residential Repair & Remodeling, call 800.448.8182.

Finish Mechanical

Sink (assembly)

Sink (assembly)	Unit	Material	Labor	Equip.	Total	Specification
Bar						
Demolish	Ea.		34		34	Includes material and labor to install a
Install	Ea.	230	137		367	small single bowl stainless steel bar /
Demolish and Install	Ea.	230	171		401	vegetable / kitchen sink with faucet
Reinstall	Ea.		109.36		109.36	and drain.
Clean	Ea.	.01	1.45		1.46	
Minimum Charge	Job		273		273	
Floor						
Demolish	Ea.		109		109	Includes material and labor to install
Install	Ea.	1650	224		1874	an enameled cast iron floor mounted
Demolish and Install	Ea.	1650	333		1983	corner service sink with faucet and
Reinstall	Ea.		178.91		178.91	drain.
Clean	Ea.	.15	9.35		9.50	
Minimum Charge	Job		273		273	
Pedestal						
Demolish	Ea.		34		34	Includes material and labor to install a
Install	Ea.	425	154		579	vitreous china pedestal lavatory with
Demolish and Install	Ea.	425	188		613	faucet set and pop-up drain.
Reinstall	Ea.		123		123	
Clean	Ea.	.01	1.45		1.46	
Minimum Charge	Job		273		273	
Vanity Lavatory						
Demolish	Ea.		34		34	Includes material and labor to install a
Install	Ea.	278	182		460	vitreous china lavatory with faucet set
Demolish and Install	Ea.	278	216		494	and pop-up drain.
Reinstall	Ea.		145.78		145.78	
Clean	Ea.	.01	1.45		1.46	
Minimum Charge	Job		273		273	
Rough-in						
Install	Ea.	570	460		1030	Includes labor and materials to install copper supply pipe with valves and drain, waste and vent pipe with all couplings, hangers and fasteners necessary.
Laundry						
Demolish	Ea.		37.50		37.50	Includes material and labor to install a
Install	Ea.	161	151		312	laundry sink (wall mounted or with
Demolish and Install	Ea.	161	188.50		349.50	legs) including faucet and pop up
Clean	Ea.	.01	1.45		1.46	drain.
Minimum Charge	Job		273		273	

Sink Only

Sink Only	Unit	Material	Labor	Equip.	Total	Specification
Single						
Porcelain						
Demolish	Ea.		34		34	Includes material and labor to install a
Install	Ea.	1050	195		1245	single bowl enamel finished cast iron
Demolish and Install	Ea.	1050	229		1279	kitchen sink.
Reinstall	Ea.		156.23		156.23	
Clean	Ea.	.01	1.45		1.46	
Minimum Charge	Job		273		273	

For customer support on your Contractor's Pricing Guide: Residential Repair & Remodeling, call 800.448.8182.

Finish Mechanical

Sink Only		Unit	Material	Labor	Equip.	Total	Specification
Stainless Steel							
	Demolish	Ea.		34		34	Includes material and labor to install a
	Install	Ea.	680	137		817	single bowl stainless steel kitchen sink.
	Demolish and Install	Ea.	680	171		851	
	Reinstall	Ea.		109.36		109.36	
	Clean	Ea.	.01	1.45		1.46	
	Minimum Charge	Job		273		273	
Double							
Porcelain							
	Demolish	Ea.		39		39	Includes material and labor to install a
	Install	Ea.	315	228		543	double bowl enamel finished cast iron
	Demolish and Install	Ea.	315	267		582	kitchen sink.
	Reinstall	Ea.		182.27		182.27	
	Clean	Ea.	.15	18.70		18.85	
	Minimum Charge	Job		273		273	
Stainless Steel							
	Demolish	Ea.		39		39	Includes material and labor to install a
	Install	Ea.	1025	176		1201	double bowl stainless steel kitchen
	Demolish and Install	Ea.	1025	215		1240	sink.
	Reinstall	Ea.		141.11		141.11	
	Clean	Ea.	.15	18.70		18.85	
	Minimum Charge	Job		273		273	
Bar							
	Demolish	Ea.		34		34	Includes material and labor to install a
	Install	Ea.	92	137		229	small single bowl stainless steel bar /
	Demolish and Install	Ea.	92	171		263	vegetable / kitchen sink.
	Reinstall	Ea.		109.36		109.36	
	Clean	Ea.	.01	1.45		1.46	
	Minimum Charge	Job		273		273	
Floor							
	Demolish	Ea.		109		109	Includes material and labor to install
	Install	Ea.	1525	249		1774	an enameled cast iron floor mounted
	Demolish and Install	Ea.	1525	358		1883	corner service sink.
	Reinstall	Ea.		198.84		198.84	
	Clean	Ea.	.15	9.35		9.50	
	Minimum Charge	Job		273		273	
Pedestal							
	Demolish	Ea.		34		34	Includes material and labor to install a
	Install	Ea.	330	171		501	vitreous china pedestal lavatory.
	Demolish and Install	Ea.	330	205		535	
	Reinstall	Ea.		136.70		136.70	
	Clean	Ea.	.01	1.45		1.46	
	Minimum Charge	Job		273		273	
Vanity Lavatory							
	Demolish	Ea.		34		34	Includes material and labor to install a
	Install	Ea.	178	171		349	vitreous china lavatory.
	Demolish and Install	Ea.	178	205		383	
	Reinstall	Ea.		136.70		136.70	
	Clean	Ea.	.01	1.45		1.46	
	Minimum Charge	Job		273		273	
Rough-in							
	Install	Ea.	650	430		1080	Includes labor and materials to install copper supply pipe with valves and drain, waste and vent pipe with all couplings, hangers and fasteners necessary.

254

For customer support on your Contractor's Pricing Guide: Residential Repair & Remodeling, call 800.448.8182.

Finish Mechanical

Sink Only

	Unit	Material	Labor	Equip.	Total	Specification
Faucet Sink						
Good Grade						
Demolish	Ea.		27.50		27.50	Includes material and labor to install a
Install	Ea.	92	37.50		129.50	sink faucet and fittings.
Demolish and Install	Ea.	92	65		157	
Reinstall	Ea.		30.19		30.19	
Clean	Ea.	.07	11.70		11.77	
Minimum Charge	Job		273		273	
Better Grade						
Demolish	Ea.		27.50		27.50	Includes material and labor to install a
Install	Ea.	154	37.50		191.50	sink faucet and fittings.
Demolish and Install	Ea.	154	65		219	
Reinstall	Ea.		30.19		30.19	
Clean	Ea.	.07	11.70		11.77	
Minimum Charge	Job		273		273	
Premium Grade						
Demolish	Ea.		27.50		27.50	Includes material and labor to install a
Install	Ea.	300	47		347	sink faucet and fittings.
Demolish and Install	Ea.	300	74.50		374.50	
Reinstall	Ea.		37.65		37.65	
Clean	Ea.	.07	11.70		11.77	
Minimum Charge	Job		273		273	
Drain & Basket						
Demolish	Ea.		17.10		17.10	Includes material and labor to install a
Install	Ea.	17.75	34		51.75	sink drain and basket.
Demolish and Install	Ea.	17.75	51.10		68.85	
Reinstall	Ea.		27.34		27.34	
Clean	Ea.	.07	11.70		11.77	
Minimum Charge	Job		273		273	

Bathtub

	Unit	Material	Labor	Equip.	Total	Specification
Enameled						
Steel						
Demolish	Ea.		98.50		98.50	Includes labor and material to install a
Install	Ea.	635	179		814	formed steel bathtub with spout,
Demolish and Install	Ea.	635	277.50		912.50	mixing valve, shower head and
Reinstall	Ea.		143.13		143.13	pop-up drain.
Clean	Ea.	.31	31		31.31	
Minimum Charge	Job		273		273	
Cast Iron						
Demolish	Ea.		98.50		98.50	Includes material and labor to install a
Install	Ea.	1600	224		1824	cast iron bathtub with spout, mixing
Demolish and Install	Ea.	1600	322.50		1922.50	valve, shower head and pop-up drain.
Reinstall	Ea.		178.91		178.91	
Clean	Ea.	.31	31		31.31	
Minimum Charge	Job		273		273	
Fiberglass						
Demolish	Ea.		98.50		98.50	Includes material and labor to install
Install	Ea.	2975	179		3154	an acrylic soaking tub with spout,
Demolish and Install	Ea.	2975	277.50		3252.50	mixing valve, shower head and
Reinstall	Ea.		143.13		143.13	pop-up drain.
Clean	Ea.	.31	31		31.31	
Minimum Charge	Job		273		273	

For customer support on your Contractor's Pricing Guide: Residential Repair & Remodeling, call 800.448.8182.

Finish Mechanical

Bathtub

	Unit	Material	Labor	Equip.	Total	Specification
Institutional						
Demolish	Ea.		98.50		98.50	Includes material and labor to install a
Install	Ea.	1475	330		1805	hospital / institutional type bathtub
Demolish and Install	Ea.	1475	428.50		1903.50	with spout, mixing valve, shower head
Reinstall	Ea.		262.40		262.40	and pop-up drain.
Clean	Ea.	.31	31		31.31	
Minimum Charge	Job		273		273	
Whirlpool (Acrylic)						
Demolish	Ea.		305		305	Includes material and labor to install a
Install	Ea.	1375	985		2360	molded fiberglass whirlpool tub.
Demolish and Install	Ea.	1375	1290		2665	
Reinstall	Ea.		787.20		787.20	
Clean	Ea.	.31	31		31.31	
Minimum Charge	Job		273		273	
Pump / Motor						
Demolish	Ea.		76		76	Includes material and labor to install a
Install	Ea.	390	164		554	whirlpool tub pump / motor.
Demolish and Install	Ea.	390	240		630	
Reinstall	Ea.		131.36		131.36	
Clean	Ea.	.08	23.50		23.58	
Minimum Charge	Job		273		273	
Heater / Motor						
Demolish	Ea.		76		76	Includes material and labor to install a
Install	Ea.	208	243		451	whirlpool tub heater.
Demolish and Install	Ea.	208	319		527	
Reinstall	Ea.		194.42		194.42	
Clean	Ea.	.08	23.50		23.58	
Minimum Charge	Job		273		273	
Thermal Cover						
Demolish	Ea.		12.15		12.15	Includes material and labor to install a
Install	Ea.	505	34		539	whirlpool tub thermal cover.
Demolish and Install	Ea.	505	46.15		551.15	
Reinstall	Ea.		27.34		27.34	
Clean	Ea.	.76	23.50		24.26	
Minimum Charge	Job		273		273	
Tub / Shower Combination						
Demolish	Ea.		87		87	Includes material and labor to install a
Install	Ea.	945	246		1191	fiberglass module tub with shower
Demolish and Install	Ea.	945	333		1278	surround with spout, mixing valve,
Reinstall	Ea.		196.80		196.80	shower head and pop-up drain.
Clean	Ea.	.39	47		47.39	
Minimum Charge	Job		273		273	
Accessories						
Sliding Door						
Demolish	Ea.		26		26	Includes material and labor to install a
Install	Ea.	700	112		812	48" aluminum framed tempered glass
Demolish and Install	Ea.	700	138		838	shower door.
Reinstall	Ea.		89.87		89.87	
Clean	Ea.	.64	15.60		16.24	
Minimum Charge	Job		273		273	
Shower Head						
Demolish	Ea.		6.85		6.85	Includes material and labor to install a
Install	Ea.	61	23		84	water saving shower head.
Demolish and Install	Ea.	61	29.85		90.85	
Reinstall	Ea.		18.23		18.23	
Clean	Ea.	.01	4.68		4.69	
Minimum Charge	Job		273		273	

For customer support on your Contractor's Pricing Guide: Residential Repair & Remodeling, call 800.448.8182.

Finish Mechanical

Bathtub

	Unit	Material	Labor	Equip.	Total	Specification
Faucet Set						Includes material and labor to install a combination spout / diverter for a bathtub.
Demolish	Ea.		34		34	
Install	Ea.	94.50	68.50		163	
Demolish and Install	Ea.	94.50	102.50		197	
Reinstall	Ea.		54.68		54.68	
Clean	Ea.	.07	11.70		11.77	
Minimum Charge	Job		273		273	
Shower Rod						Includes material and labor to install a chrome plated shower rod.
Demolish	Ea.		4.33		4.33	
Install	Ea.	11.15	37		48.15	
Demolish and Install	Ea.	11.15	41.33		52.48	
Reinstall	Ea.		29.78		29.78	
Clean	Ea.		4.68		4.68	
Minimum Charge	Job		273		273	
Rough-in						Includes labor and materials to install copper supply pipe with valves and drain, waste and vent pipe with all couplings, hangers and fasteners necessary.
Install	Ea.	650	475		1125	

Shower

	Unit	Material	Labor	Equip.	Total	Specification
Fiberglass						
32" x 32"						Includes material and labor to install a fiberglass shower stall with door, mixing valve and shower head and drain fitting.
Demolish	Ea.		205		205	
Install	Ea.	415	179		594	
Demolish and Install	Ea.	415	384		799	
Reinstall	Ea.		143.13		143.13	
Clean	Ea.	.39	47		47.39	
Minimum Charge	Job		273		273	
36" x 36"						Includes material and labor to install a fiberglass shower stall with door, mixing valve and shower head and drain fitting.
Demolish	Ea.		205		205	
Install	Ea.	460	179		639	
Demolish and Install	Ea.	460	384		844	
Reinstall	Ea.		143.13		143.13	
Clean	Ea.	.39	47		47.39	
Minimum Charge	Job		273		273	
35" x 60"						Includes material and labor to install a handicap fiberglass shower stall with door & seat, mixing valve and shower head and drain fitting.
Demolish	Ea.		205		205	
Install	Ea.	875	246		1121	
Demolish and Install	Ea.	875	451		1326	
Reinstall	Ea.		196.80		196.80	
Clean	Ea.	.39	47		47.39	
Minimum Charge	Job		273		273	
Accessories						
Single Door						Includes material and labor to install a shower door.
Demolish	Ea.		26		26	
Install	Ea.	144	59.50		203.50	
Demolish and Install	Ea.	144	85.50		229.50	
Reinstall	Ea.		47.43		47.43	
Clean	Ea.	.64	15.60		16.24	
Minimum Charge	Job		273		273	

For customer support on your Contractor's Pricing Guide: Residential Repair & Remodeling, call 800.448.8182.

Finish Mechanical

Shower

Shower	Unit	Material	Labor	Equip.	Total	Specification
Shower Head						
Demolish	Ea.		6.85		6.85	Includes material and labor to install a
Install	Ea.	61	23		84	water saving shower head.
Demolish and Install	Ea.	61	29.85		90.85	
Reinstall	Ea.		18.23		18.23	
Clean	Ea.	.01	4.68		4.69	
Minimum Charge	Job		273		273	
Faucet Set						
Demolish	Ea.		34		34	Includes material and labor to install a
Install	Ea.	94.50	68.50		163	combination spout / diverter for a
Demolish and Install	Ea.	94.50	102.50		197	bathtub.
Reinstall	Ea.		54.68		54.68	
Clean	Ea.	.07	11.70		11.77	
Minimum Charge	Job		273		273	
Shower Pan						
Demolish	Ea.		30.50		30.50	Includes material and labor to install
Install	Ea.	123	68.50		191.50	the base portion of a fiberglass
Demolish and Install	Ea.	123	99		222	shower unit.
Reinstall	Ea.		54.68		54.68	
Clean	Ea.	.31	31		31.31	
Minimum Charge	Job		273		273	
Rough-in						
Install	Ea.	690	480		1170	Includes labor and materials to install copper supply pipe with valves and drain, waste and vent pipe with all couplings, hangers and fasteners necessary.

Ductwork

Ductwork	Unit	Material	Labor	Equip.	Total	Specification
Return Grill						
Under 10" Wide						
Demolish	Ea.		11.85		11.85	Includes material and labor to install a
Install	Ea.	46	29.50		75.50	10" x 10" air return register.
Demolish and Install	Ea.	46	41.35		87.35	
Reinstall	Ea.		23.72		23.72	
Clean	Ea.	.08	10.40		10.48	
Minimum Charge	Job		267		267	
12" to 20" Wide						
Demolish	Ea.		11.85		11.85	Includes material and labor to install a
Install	Ea.	89	31.50		120.50	16" x 16" air return register.
Demolish and Install	Ea.	89	43.35		132.35	
Reinstall	Ea.		25.11		25.11	
Clean	Ea.	.08	10.40		10.48	
Minimum Charge	Job		267		267	
Over 20" Wide						
Demolish	Ea.		17.95		17.95	Includes material and labor to install a
Install	Ea.	182	48.50		230.50	24" x 24" air return register.
Demolish and Install	Ea.	182	66.45		248.45	
Reinstall	Ea.		38.81		38.81	
Clean	Ea.	.08	11.70		11.78	
Minimum Charge	Job		267		267	

Finish Mechanical

Ductwork	Unit	Material	Labor	Equip.	Total	Specification
Supply Grill						
Baseboard Type						
Demolish	Ea.		8.45		8.45	Includes material and labor to install a
Install	Ea.	15.25	26.50		41.75	14" x 6" baseboard air supply
Demolish and Install	Ea.	15.25	34.95		50.20	register.
Reinstall	Ea.		21.34		21.34	
Clean	Ea.	.08	10.40		10.48	
Minimum Charge	Job		267		267	
10" Wide						
Demolish	Ea.		9.90		9.90	Includes material and labor to install a
Install	Ea.	37	26.50		63.50	10" x 6" air supply register.
Demolish and Install	Ea.	37	36.40		73.40	
Reinstall	Ea.		21.34		21.34	
Clean	Ea.	.08	10.40		10.48	
Minimum Charge	Job		267		267	
12" to 15" Wide						
Demolish	Ea.		9.90		9.90	Includes material and labor to install a
Install	Ea.	31.50	31.50		63	14" x 8" air supply register.
Demolish and Install	Ea.	31.50	41.40		72.90	
Reinstall	Ea.		25.11		25.11	
Clean	Ea.	.08	10.40		10.48	
Minimum Charge	Job		267		267	
18" to 24" Wide						
Demolish	Ea.		11.85		11.85	Includes material and labor to install a
Install	Ea.	71	41		112	24" x 8" air supply register.
Demolish and Install	Ea.	71	52.85		123.85	
Reinstall	Ea.		32.84		32.84	
Clean	Ea.	.08	11.70		11.78	
Minimum Charge	Job		267		267	
30" to 36" Wide						
Demolish	Ea.		14.80		14.80	Includes material and labor to install a
Install	Ea.	93.50	38		131.50	30" x 8" air supply register.
Demolish and Install	Ea.	93.50	52.80		146.30	
Reinstall	Ea.		30.49		30.49	
Clean	Ea.	.08	23.50		23.58	
Minimum Charge	Job		267		267	
Diffuser						
12" x 12"						
Demolish	Ea.		11.85		11.85	Includes material and labor to install a
Install	Ea.	84	44.50		128.50	12" x 12" aluminum diffuser.
Demolish and Install	Ea.	84	56.35		140.35	
Reinstall	Ea.		35.57		35.57	
Clean	Ea.	.08	10.40		10.48	
Minimum Charge	Job		267		267	
14" x 14"						
Demolish	Ea.		11.85		11.85	Includes material and labor to install a
Install	Ea.	150	44.50		194.50	14" x 14" aluminum diffuser.
Demolish and Install	Ea.	150	56.35		206.35	
Reinstall	Ea.		35.57		35.57	
Clean	Ea.	.08	11.70		11.78	
Minimum Charge	Job		267		267	
18" x 18"						
Demolish	Ea.		17.95		17.95	Includes material and labor to install a
Install	Ea.	144	59.50		203.50	18" x 18" aluminum diffuser.
Demolish and Install	Ea.	144	77.45		221.45	
Reinstall	Ea.		47.43		47.43	
Clean	Ea.	.08	15.60		15.68	
Minimum Charge	Job		267		267	

Finish Mechanical

Ductwork	Unit	Material	Labor	Equip.	Total	Specification
12″ Round						
Demolish	Ea.		16.45		16.45	Includes material and labor to install a
Install	Ea.	29.50	44.50		74	12″ diameter aluminum diffuser with a
Demolish and Install	Ea.	29.50	60.95		90.45	butterfly damper.
Reinstall	Ea.		35.57		35.57	
Clean	Ea.	.08	10.40		10.48	
Minimum Charge	Job		267		267	
14″ to 20″ Round						
Demolish	Ea.		22		22	Includes material and labor to install a
Install	Ea.	164	59.50		223.50	20″ diameter aluminum diffuser with a
Demolish and Install	Ea.	164	81.50		245.50	butterfly damper.
Reinstall	Ea.		47.43		47.43	
Clean	Ea.	.08	11.70		11.78	
Minimum Charge	Job		267		267	
Flush-out / Sanitize						
Install	L.F.	.31	4.68		4.99	Includes labor and material costs to
Minimum Charge	Job		267		267	clean ductwork or flue.

For customer support on your Contractor's Pricing Guide: Residential Repair & Remodeling, call 800.448.8182.

Lighting Fixture   **Duplex Receptacle**

Electrical Per S.F.	Unit	Material	Labor	Equip.	Total	Specification
Residential						
Light Fixtures						
Install	S.F.	.23	.28		.51	Includes labor and material to replace house light fixtures. Apply S.F. cost to floor area including garages but not basements.

Circuits	Unit	Material	Labor	Equip.	Total	Specification
120V						
Outlet with Wiring						
Demolish	Ea.		31.50		31.50	Includes material and labor to install
Install	Ea.	9.30	39		48.30	general purpose outlet with cover, up
Demolish and Install	Ea.	9.30	70.50		79.80	to 20 feet of #12/2 wire.
Reinstall	Ea.		31.25		31.25	
Clean	Ea.		2.34		2.34	
Minimum Charge	Job		284		284	
GFCI Outlet w / Wiring						
Demolish	Ea.		31.50		31.50	Cost includes material and labor to
Install	Ea.	28	46		74	install 120V GFCI receptacle and wall
Demolish and Install	Ea.	28	77.50		105.50	plate, with up to 20 feet of #12/2
Reinstall	Ea.		36.94		36.94	wire.
Clean	Ea.		2.34		2.34	
Minimum Charge	Job		284		284	
Exterior Outlet w / Wiring						
Demolish	Ea.		31.50		31.50	Cost includes material and labor to
Install	Ea.	45.50	47.50		93	install 110V waterproof external
Demolish and Install	Ea.	45.50	79		124.50	receptacle and cover plate with up to
Reinstall	Ea.		37.89		37.89	20' of #12/2 wire.
Clean	Ea.		2.34		2.34	
Minimum Charge	Job		284		284	
240V Outlet w / Wiring						
Demolish	Ea.		31.50		31.50	Cost includes material and labor to
Install	Ea.	65	88.50		153.50	install 240V receptacle and cover
Demolish and Install	Ea.	65	120		185	plate with up to 20' of #12/2 wire.
Reinstall	Ea.		70.94		70.94	
Clean	Ea.		2.34		2.34	
Minimum Charge	Job		284		284	

Finish Electrical

Circuits

	Unit	Material	Labor	Equip.	Total	Specification
TV / Cable Jack w / Wiring						
Demolish	Ea.		31.50		31.50	Includes material and labor to install
Install	Ea.	17.40	35.50		52.90	TV/cable outlet box and cover plate
Demolish and Install	Ea.	17.40	67		84.40	including 20' of wiring.
Reinstall	Ea.		28.42		28.42	
Clean	Ea.		2.34		2.34	
Minimum Charge	Job		284		284	
Phone Jack with Wiring						
Demolish	Ea.		31.50		31.50	Includes material and labor to install
Install	Ea.	9.10	22		31.10	phone jack and cover plate including
Demolish and Install	Ea.	9.10	53.50		62.60	up to 20' of telephone cable.
Reinstall	Ea.		17.49		17.49	
Clean	Ea.		1.46		1.46	
Minimum Charge	Job		284		284	

Light Switch

	Unit	Material	Labor	Equip.	Total	Specification
Single						
Switch						
Demolish	Ea.		6.30		6.30	Includes material and labor to install
Install	Ea.	3.67	27		30.67	single switch and wall plate. Wiring is
Demolish and Install	Ea.	3.67	33.30		36.97	not included.
Reinstall	Ea.		21.65		21.65	
Clean	Ea.		2.34		2.34	
Minimum Charge	Job		284		284	
Plate						
Demolish	Ea.		3.16		3.16	Includes material and labor to install a
Install	Ea.	.39	7.10		7.49	one gang brown wall switch plate.
Demolish and Install	Ea.	.39	10.26		10.65	
Reinstall	Ea.		5.68		5.68	
Clean	Ea.	.03	.47		.50	
Minimum Charge	Job		284		284	
Double						
Switch						
Demolish	Ea.		12.65		12.65	Includes material and labor to install
Install	Ea.	5.95	40.50		46.45	double switch and wall plate. Wiring
Demolish and Install	Ea.	5.95	53.15		59.10	is not included.
Reinstall	Ea.		32.48		32.48	
Clean	Ea.	.03	7.50		7.53	
Minimum Charge	Job		284		284	
Plate						
Demolish	Ea.		3.16		3.16	Includes material and labor to install a
Install	Ea.	1.12	10.70		11.82	two gang brown wall switch plate.
Demolish and Install	Ea.	1.12	13.86		14.98	
Reinstall	Ea.		8.58		8.58	
Clean	Ea.	.03	.47		.50	
Minimum Charge	Job		284		284	
Triple						
Switch						
Demolish	Ea.		19.15		19.15	Includes material and labor to install
Install	Ea.	15.40	47.50		62.90	triple switch and wall plate. Wiring is
Demolish and Install	Ea.	15.40	66.65		82.05	not included.
Reinstall	Ea.		37.89		37.89	
Clean	Ea.	.03	11.20		11.23	
Minimum Charge	Job		284		284	

Finish Electrical

Light Switch		Unit	Material	Labor	Equip.	Total	Specification
Plate							
	Demolish	Ea.		3.16		3.16	Includes material and labor to install a
	Install	Ea.	1.82	17.75		19.57	three gang brown wall switch plate.
	Demolish and Install	Ea.	1.82	20.91		22.73	
	Reinstall	Ea.		14.21		14.21	
	Clean	Ea.	.03	.47		.50	
	Minimum Charge	Job		284		284	
Dimmer							
Switch							
	Demolish	Ea.		6.30		6.30	Includes material and labor to install
	Install	Ea.	37	33.50		70.50	dimmer switch. Wiring is not included.
	Demolish and Install	Ea.	37	39.80		76.80	
	Reinstall	Ea.		26.75		26.75	
	Clean	Ea.		2.34		2.34	
	Minimum Charge	Job		284		284	
Plate							
	Demolish	Ea.		3.16		3.16	Includes material and labor to install a
	Install	Ea.	.39	7.10		7.49	one gang brown wall switch plate.
	Demolish and Install	Ea.	.39	10.26		10.65	
	Reinstall	Ea.		5.68		5.68	
	Clean	Ea.	.03	.47		.50	
	Minimum Charge	Job		284		284	
With Wiring							
Single							
	Demolish	Ea.		31.50		31.50	Includes material and labor to install
	Install	Ea.	8.65	33		41.65	single switch and plate including up to
	Demolish and Install	Ea.	8.65	64.50		73.15	20' of romex cable.
	Reinstall	Ea.		26.59		26.59	
	Clean	Ea.		2.34		2.34	
	Minimum Charge	Job		284		284	
Double							
	Demolish	Ea.		31.50		31.50	Includes material and labor to install
	Install	Ea.	26	57		83	double switch and plate including up
	Demolish and Install	Ea.	26	88.50		114.50	to 20' of romex cable.
	Reinstall	Ea.		45.47		45.47	
	Clean	Ea.	.03	7.50		7.53	
	Minimum Charge	Job		284		284	
Triple							
	Demolish	Ea.		47.50		47.50	Includes material and labor to install
	Install	Ea.	22.50	64		86.50	triple switch and plate including up to
	Demolish and Install	Ea.	22.50	111.50		134	20' of romex cable.
	Reinstall	Ea.		51.15		51.15	
	Clean	Ea.	.03	11.20		11.23	
	Minimum Charge	Job		284		284	
Dimmer							
	Demolish	Ea.		31.50		31.50	Includes material and labor to install
	Install	Ea.	42	39		81	dimmer switch and cover plate
	Demolish and Install	Ea.	42	70.50		112.50	including up to 20' of romex cable.
	Reinstall	Ea.		31.25		31.25	
	Clean	Ea.		2.34		2.34	
	Minimum Charge	Job		284		284	

Finish Electrical

Cover for Outlet / Switch

	Unit	Material	Labor	Equip.	Total	Specification
High Grade						
Demolish	Ea.		3.16		3.16	Includes material and labor to install
Install	Ea.	1.64	7.10		8.74	wall plate, stainless steel, 1 gang.
Demolish and Install	Ea.	1.64	10.26		11.90	
Reinstall	Ea.		5.68		5.68	
Clean	Ea.	.03	.47		.50	
Minimum Charge	Job		284		284	
Deluxe Grade						
Demolish	Ea.		3.16		3.16	Includes material and labor to install
Install	Ea.	6.50	7.10		13.60	wall plate, brushed brass, 1 gang.
Demolish and Install	Ea.	6.50	10.26		16.76	
Reinstall	Ea.		5.68		5.68	
Clean	Ea.	.03	.47		.50	
Minimum Charge	Job		284		284	

Residential Light Fixture

	Unit	Material	Labor	Equip.	Total	Specification
Ceiling						
Good Quality						
Demolish	Ea.		13.15		13.15	Includes material and labor to install
Install	Ea.	24	14.20		38.20	standard quality incandescent ceiling
Demolish and Install	Ea.	24	27.35		51.35	light fixture. Wiring is not included.
Reinstall	Ea.		11.37		11.37	
Clean	Ea.	.03	31		31.03	
Minimum Charge	Job		284		284	
Better Quality						
Demolish	Ea.		13.15		13.15	Includes material and labor to install
Install	Ea.	95.50	30		125.50	custom quality incandescent ceiling
Demolish and Install	Ea.	95.50	43.15		138.65	light fixture. Wiring is not included.
Reinstall	Ea.		23.93		23.93	
Clean	Ea.	.03	31		31.03	
Minimum Charge	Job		284		284	
Premium Quality						
Demolish	Ea.		13.15		13.15	Includes material and labor to install
Install	Ea.	380	30		410	deluxe quality incandescent ceiling
Demolish and Install	Ea.	380	43.15		423.15	light fixture. Wiring is not included.
Reinstall	Ea.		23.93		23.93	
Clean	Ea.	.03	31		31.03	
Minimum Charge	Job		284		284	
Recessed Ceiling						
Demolish	Ea.		26.50		26.50	Includes material and labor to install
Install	Ea.	53	18.95		71.95	standard quality recessed eyeball
Demolish and Install	Ea.	53	45.45		98.45	spotlight with housing. Wiring is not
Reinstall	Ea.		15.16		15.16	included.
Clean	Ea.	.01	3.90		3.91	
Minimum Charge	Job		284		284	
Recessed Eyeball						
Demolish	Ea.		26.50		26.50	Includes material and labor to install
Install	Ea.	32.50	20.50		53	custom quality recessed eyeball
Demolish and Install	Ea.	32.50	47		79.50	spotlight with housing. Wiring is not
Reinstall	Ea.		16.24		16.24	included.
Clean	Ea.	.01	3.90		3.91	
Minimum Charge	Job		284		284	

For customer support on your Contractor's Pricing Guide: Residential Repair & Remodeling, call 800.448.8182.

Finish Electrical

Residential Light Fixture		Unit	Material	Labor	Equip.	Total	Specification
Recessed Shower Type							
	Demolish	Ea.		26.50		26.50	Includes material and labor to install
	Install	Ea.	82	18.95		100.95	deluxe quality recessed eyeball
	Demolish and Install	Ea.	82	45.45		127.45	spotlight with housing. Wiring is not
	Reinstall	Ea.		15.16		15.16	included.
	Clean	Ea.	.01	3.90		3.91	
	Minimum Charge	Job		284		284	
Run Wiring							
	Install	Ea.	18.05	22.50		40.55	Includes labor and material to install
	Minimum Charge	Job		284		284	an electric fixture utility box with
							non-metallic cable.
Fluorescent							
Good Quality							
	Demolish	Ea.		28.50		28.50	Includes material and labor to install
	Install	Ea.	47	81		128	standard quality 12" wide by 48" long
	Demolish and Install	Ea.	47	109.50		156.50	decorative surface mounted fluorescent
	Reinstall	Ea.		64.96		64.96	fixture w/acrylic diffuser.
	Clean	Ea.	.01	3.90		3.91	
	Minimum Charge	Job		284		284	
Better Quality							
	Demolish	Ea.		28.50		28.50	Includes material and labor to install
	Install	Ea.	96.50	91.50		188	an interior fluorescent lighting fixture.
	Demolish and Install	Ea.	96.50	120		216.50	
	Reinstall	Ea.		73.34		73.34	
	Clean	Ea.	.01	3.90		3.91	
	Minimum Charge	Job		284		284	
Premium Quality							
	Demolish	Ea.		28.50		28.50	Includes material and labor to install
	Install	Ea.	98	107		205	deluxe quality 24" wide by 48" long
	Demolish and Install	Ea.	98	135.50		233.50	decorative surface mounted fluorescent
	Reinstall	Ea.		85.80		85.80	fixture w/acrylic diffuser.
	Clean	Ea.	.01	3.90		3.91	
	Minimum Charge	Job		284		284	
Run Wiring							
	Install	Ea.	18.05	22.50		40.55	Includes labor and material to install
	Minimum Charge	Job		284		284	an electric fixture utility box with
							non-metallic cable.
Exterior							
Good Quality							
	Demolish	Ea.		25.50		25.50	Includes material and labor to install
	Install	Ea.	67	35.50		102.50	standard quality incandescent outdoor
	Demolish and Install	Ea.	67	61		128	light fixture. Wiring is not included.
	Reinstall	Ea.		28.42		28.42	
	Clean	Ea.	.08	15.60		15.68	
	Minimum Charge	Job		284		284	
Better Quality							
	Demolish	Ea.		25.50		25.50	Includes material and labor to install
	Install	Ea.	122	35.50		157.50	custom quality incandescent outdoor
	Demolish and Install	Ea.	122	61		183	light fixture. Wiring is not included.
	Reinstall	Ea.		28.42		28.42	
	Clean	Ea.	.08	15.60		15.68	
	Minimum Charge	Job		284		284	
Premium Quality							
	Demolish	Ea.		25.50		25.50	Includes material and labor to install
	Install	Ea.	345	142		487	deluxe quality incandescent outdoor
	Demolish and Install	Ea.	345	167.50		512.50	light fixture. Wiring is not included.
	Reinstall	Ea.		113.68		113.68	
	Clean	Ea.	.08	15.60		15.68	
	Minimum Charge	Job		284		284	

Finish Electrical

Residential Light Fixture

	Unit	Material	Labor	Equip.	Total	Specification
Outdoor w / Pole						Includes material and labor to install
Demolish	Ea.		25.50		25.50	post lantern type incandescent outdoor
Install	Ea.	365	142		507	light fixture. Wiring is not included.
Demolish and Install	Ea.	365	167.50		532.50	
Reinstall	Ea.		113.68		113.68	
Clean	Ea.	.08	15.60		15.68	
Minimum Charge	Job		284		284	
Run Wiring						Includes labor and material to install
Install	Ea.	18.05	22.50		40.55	an electric fixture utility box with
Minimum Charge	Job		284		284	non-metallic cable.
Track Lighting						
Demolish	L.F.		8.75		8.75	Includes material and labor to install
Install	L.F.	29	23.50		52.50	track lighting.
Demolish and Install	L.F.	29	32.25		61.25	
Reinstall	L.F.		18.95		18.95	
Clean	L.F.	.08	7.50		7.58	
Minimum Charge	Job		284		284	

Commercial Light Fixture

	Unit	Material	Labor	Equip.	Total	Specification
Fluorescent						
2 Bulb, 2'						
Demolish	Ea.		25.50		25.50	Includes material and labor to install a
Install	Ea.	38	71		109	2' long surface mounted fluorescent
Demolish and Install	Ea.	38	96.50		134.50	light fixture with two 40 watt bulbs.
Reinstall	Ea.		56.84		56.84	
Clean	Ea.	.39	26.50		26.89	
Minimum Charge	Job		284		284	
2 Bulb, 4'						
Demolish	Ea.		25.50		25.50	Includes material and labor to install a
Install	Ea.	157	81		238	4' long surface mounted fluorescent
Demolish and Install	Ea.	157	106.50		263.50	light fixture with two 40 watt bulbs.
Reinstall	Ea.		64.96		64.96	
Clean	Ea.	.39	26.50		26.89	
Minimum Charge	Job		284		284	
3 Bulb, 4'						
Demolish	Ea.		25.50		25.50	Includes material and labor to install a
Install	Ea.	108	81		189	4' long pendant mounted fluorescent
Demolish and Install	Ea.	108	106.50		214.50	light fixture with three 34 watt bulbs.
Reinstall	Ea.		64.96		64.96	
Clean	Ea.	.39	26.50		26.89	
Minimum Charge	Job		284		284	
4 Bulb, 4'						
Demolish	Ea.		25.50		25.50	Includes material and labor to install a
Install	Ea.	97.50	87.50		185	4' long pendant mounted fluorescent
Demolish and Install	Ea.	97.50	113		210.50	light fixture with four 34 watt bulbs.
Reinstall	Ea.		69.96		69.96	
Clean	Ea.	.39	26.50		26.89	
Minimum Charge	Job		284		284	
2 Bulb, 8'						
Demolish	Ea.		31.50		31.50	Includes material and labor to install
Install	Ea.	122	107		229	an 8' long surface mounted
Demolish and Install	Ea.	122	138.50		260.50	fluorescent light fixture with two 110
Reinstall	Ea.		85.80		85.80	watt bulbs.
Clean	Ea.	.39	26.50		26.89	
Minimum Charge	Job		284		284	

Finish Electrical

Commercial Light Fixture		Unit	Material	Labor	Equip.	Total	Specification
Run Wiring							
	Install	Ea.	18.05	22.50		40.55	Includes labor and material to install
	Minimum Charge	Job		284		284	an electric fixture utility box with
							non-metallic cable.
Ceiling, Recessed							
Eyeball Type							
	Demolish	Ea.		26.50		26.50	Includes material and labor to install
	Install	Ea.	32.50	20.50		53	custom quality recessed eyeball
	Demolish and Install	Ea.	32.50	47		79.50	spotlight with housing. Wiring is not
	Reinstall	Ea.		16.24		16.24	included.
	Clean	Ea.	.01	3.90		3.91	
	Minimum Charge	Job		284		284	
Shower Type							
	Demolish	Ea.		26.50		26.50	Includes material and labor to install
	Install	Ea.	82	18.95		100.95	deluxe quality recessed eyeball
	Demolish and Install	Ea.	82	45.45		127.45	spotlight with housing. Wiring is not
	Reinstall	Ea.		15.16		15.16	included.
	Clean	Ea.	.01	3.90		3.91	
	Minimum Charge	Job		284		284	
Run Wiring							
	Install	Ea.	18.05	22.50		40.55	Includes labor and material to install
	Minimum Charge	Job		284		284	an electric fixture utility box with
							non-metallic cable.
Ceiling Grid Type							
2 Tube							
	Demolish	Ea.		37		37	Cost includes material and labor to
	Install	Ea.	63.50	107		170.50	install 2 lamp fluorescent light fixture
	Demolish and Install	Ea.	63.50	144		207.50	in 2' x 4' ceiling grid system including
	Reinstall	Ea.		85.80		85.80	lens and tubes.
	Clean	Ea.	.39	26.50		26.89	
	Minimum Charge	Job		284		284	
4 Tube							
	Demolish	Ea.		42		42	Cost includes material and labor to
	Install	Ea.	73	121		194	install 4 lamp fluorescent light fixture
	Demolish and Install	Ea.	73	163		236	in 2' x 4' ceiling grid system including
	Reinstall	Ea.		96.75		96.75	lens and tubes.
	Clean	Ea.	.39	26.50		26.89	
	Minimum Charge	Job		284		284	
Acrylic Diffuser							
	Demolish	Ea.		6.60		6.60	Includes material and labor to install
	Install	Ea.	93	17.20		110.20	fluorescent light fixture w/acrylic
	Demolish and Install	Ea.	93	23.80		116.80	diffuser.
	Reinstall	Ea.		13.78		13.78	
	Clean	Ea.	.24	4.68		4.92	
	Minimum Charge	Job		284		284	
Louver Diffuser							
	Demolish	Ea.		7.90		7.90	Includes material and labor to install
	Install	Ea.	105	17.20		122.20	fluorescent light fixture w/louver
	Demolish and Install	Ea.	105	25.10		130.10	diffuser.
	Reinstall	Ea.		13.78		13.78	
	Clean	Ea.	.24	4.68		4.92	
	Minimum Charge	Job		284		284	
Run Wiring							
	Install	Ea.	18.05	22.50		40.55	Includes labor and material to install
	Minimum Charge	Job		284		284	an electric fixture utility box with
							non-metallic cable.

For customer support on your Contractor's Pricing Guide: Residential Repair & Remodeling, call 800.448.8182.

Commercial Light Fixture	Unit	Material	Labor	Equip.	Total	Specification
Track Lighting						
Demolish	L.F.		8.75		8.75	Includes material and labor to install
Install	L.F.	29	23.50		52.50	track lighting.
Demolish and Install	L.F.	29	32.25		61.25	
Reinstall	L.F.		18.95		18.95	
Clean	L.F.	.08	7.50		7.58	
Minimum Charge	Job		284		284	
Emergency						
Demolish	Ea.		45		45	Includes material and labor to install
Install	Ea.	157	142		299	battery-powered emergency lighting
Demolish and Install	Ea.	157	187		344	unit with two (2) lights, including
Reinstall	Ea.		113.68		113.68	wiring.
Clean	Ea.	.19	26.50		26.69	
Minimum Charge	Job		284		284	
Run Wiring						
Install	Ea..	18.05	22.50		40.55	Includes labor and material to install
Minimum Charge	Job		284		284	an electric fixture utility box with non-metallic cable.

Fan	Unit	Material	Labor	Equip.	Total	Specification
Ceiling Paddle						
Good Quality						
Demolish	Ea.		35		35	Includes material and labor to install
Install	Ea.	116	57		173	standard ceiling fan.
Demolish and Install	Ea.	116	92		208	
Reinstall	Ea.		45.47		45.47	
Clean	Ea.	.15	23.50		23.65	
Minimum Charge	Job		284		284	
Better Quality						
Demolish	Ea.		35		35	Includes material and labor to install
Install	Ea.	273	57		330	better ceiling fan.
Demolish and Install	Ea.	273	92		365	
Reinstall	Ea.		45.47		45.47	
Clean	Ea.	.15	23.50		23.65	
Minimum Charge	Job		284		284	
Premium Quality						
Demolish	Ea.		35		35	Includes material and labor to install
Install	Ea.	262	71		333	deluxe ceiling fan.
Demolish and Install	Ea.	262	106		368	
Reinstall	Ea.		56.84		56.84	
Clean	Ea.	.15	23.50		23.65	
Minimum Charge	Job		284		284	
Light Kit						
Install	Ea.	45	71		116	Includes labor and material to install a
Minimum Charge	Job		284		284	ceiling fan light kit.
Bathroom Exhaust						
Ceiling Mounted						
Demolish	Ea.		27		27	Cost includes material and labor to
Install	Ea.	49.50	56.50		106	install 60 CFM ceiling mounted
Demolish and Install	Ea.	49.50	83.50		133	bathroom exhaust fan, up to 10' of
Reinstall	Ea.		45.34		45.34	wiring with one wall switch.
Clean	Ea.	.08	9.35		9.43	
Minimum Charge	Job		284		284	

Fan

Fan	Unit	Material	Labor	Equip.	Total	Specification
Wall Mounted						
Demolish	Ea.		27		27	Cost includes material and labor to
Install	Ea.	66.50	56.50		123	install 110 CFM wall mounted
Demolish and Install	Ea.	66.50	83.50		150	bathroom exhaust fan, up to 10' of
Reinstall	Ea.		45.34		45.34	wiring with one wall switch.
Clean	Ea.	.08	9.35		9.43	
Minimum Charge	Job		284		284	
Lighted						
Demolish	Ea.		27		27	Cost includes material and labor to
Install	Ea.	74.50	56.50		131	install 50 CFM ceiling mounted lighted
Demolish and Install	Ea.	74.50	83.50		158	exhaust fan, up to 10' of wiring with
Reinstall	Ea.		45.34		45.34	one wall switch..
Clean	Ea.	.08	9.35		9.43	
Minimum Charge	Job		284		284	
Blower Heater w / Light						
Demolish	Ea.		27		27	Includes material and labor to install
Install	Ea.	87.50	94.50		182	ceiling blower type ventilator with
Demolish and Install	Ea.	87.50	121.50		209	switch, light and snap-on grill. Includes
Reinstall	Ea.		75.56		75.56	1300 watt/120 volt heater.
Clean	Ea.	.08	9.35		9.43	
Minimum Charge	Job		284		284	
Fan Heater w / Light						
Demolish	Ea.		27		27	Includes material and labor to install
Install	Ea.	120	94.50		214.50	circular ceiling fan heater unit, 1500
Demolish and Install	Ea.	120	121.50		241.50	watt, surface mounted.
Reinstall	Ea.		75.56		75.56	
Clean	Ea.	.08	9.35		9.43	
Minimum Charge	Job		284		284	
Whole House Exhaust						
36" Ceiling Mounted						
Install	Ea.	1275	142		1417	Includes material and labor to install a
Clean	Ea.	.02	7.45		7.47	whole house exhaust fan.
Minimum Charge	Job		284		284	
Kitchen Exhaust						
Standard Model						
Install	Ea.	74.50	38		112.50	Includes labor and materials to install
Clean	Ea.	.02	7.45		7.47	a bathroom or kitchen vent fan.
Minimum Charge	Job		284		284	
Low Noise Model						
Install	Ea.	119	38		157	Includes labor and materials to install
Clean	Ea.	.02	7.45		7.47	a bathroom or kitchen vent fan, low
Minimum Charge	Job		284		284	noise model.

Low Voltage Systems

Low Voltage Systems	Unit	Material	Labor	Equip.	Total	Specification
Smoke Detector						
Pre-wired						
Demolish	Ea.		13.15		13.15	Includes material and labor to install
Install	Ea.	23.50	26.50		50	AC type smoke detector including
Demolish and Install	Ea.	23.50	39.65		63.15	wiring.
Reinstall	Ea.		21.32		21.32	
Clean	Ea.	.03	11.70		11.73	
Minimum Charge	Job		284		284	

Low Voltage Systems	Unit	Material	Labor	Equip.	Total	Specification
Run Wiring						
Install	Ea.	18.05	22.50		40.55	Includes labor and material to install
Minimum Charge	Job		284		284	an electric fixture utility box with
						non-metallic cable.
Door Bell / Chimes						
Good Quality						
Demolish	Ea.		39.50		39.50	Includes material and labor to install
Install	Ea.	52.50	142		194.50	surface mounted 2 note door bell.
Demolish and Install	Ea.	52.50	181.50		234	Wiring is not included.
Reinstall	Ea.		113.68		113.68	
Clean	Ea.	.08	11.70		11.78	
Minimum Charge	Job		284		284	
Better Quality						
Demolish	Ea.		39.50		39.50	Includes material and labor to install
Install	Ea.	94	142		236	standard quality surface mounted 2
Demolish and Install	Ea.	94	181.50		275.50	note door bell. Wiring is not included.
Reinstall	Ea.		113.68		113.68	
Clean	Ea.	.08	11.70		11.78	
Minimum Charge	Job		284		284	
Premium Quality						
Demolish	Ea.		39.50		39.50	Includes material and labor to install
Install	Ea.	103	142		245	deluxe quality surface mounted 2 note
Demolish and Install	Ea.	103	181.50		284.50	door bell. Wiring is not included.
Reinstall	Ea.		113.68		113.68	
Clean	Ea.	.08	11.70		11.78	
Minimum Charge	Job		284		284	
Run Wiring						
Install	Ea.	18.05	22.50		40.55	Includes labor and material to install
Minimum Charge	Job		284		284	an electric fixture utility box with
						non-metallic cable.
Intercom System						
Intercom (master)						
Demolish	Ea.		105		105	Includes material and labor to install
Install	Ea.	775	284		1059	master intercom station with AM/FM
Demolish and Install	Ea.	775	389		1164	music and room monitoring. Wiring is
Reinstall	Ea.		227.36		227.36	not included.
Clean	Ea.	.08	31		31.08	
Minimum Charge	Job		284		284	
Intercom (remote)						
Demolish	Ea.		63		63	Includes material and labor to install
Install	Ea.	70.50	71		141.50	remote station for master intercom
Demolish and Install	Ea.	70.50	134		204.50	system. Wiring is not included.
Reinstall	Ea.		56.84		56.84	
Clean	Ea.	.08	11.70		11.78	
Minimum Charge	Job		284		284	
Security System						
Alarm Panel						
Demolish	Ea.		126		126	Includes material and labor to install a
Install	Ea.	86.50	284		370.50	6 zone burglar alarm panel.
Demolish and Install	Ea.	86.50	410		496.50	
Reinstall	Ea.		227.36		227.36	
Clean	Ea.	.08	11.70		11.78	
Minimum Charge	Job		284		284	

Low Voltage Systems

	Unit	Material	Labor	Equip.	Total	Specification
Motion Detector						
Demolish	Ea.		90		90	Includes material and labor to install
Install	Ea.	180	247		427	an ultrasonic motion detector for a
Demolish and Install	Ea.	180	337		517	security system.
Reinstall	Ea.		197.70		197.70	
Clean	Ea.	.08	11.70		11.78	
Minimum Charge	Job		284		284	
Door Switch						
Demolish	Ea.		39.50		39.50	Includes material and labor to install a
Install	Ea.	63.50	107		170.50	door / window contact for a security
Demolish and Install	Ea.	63.50	146.50		210	system.
Reinstall	Ea.		85.80		85.80	
Clean	Ea.	.08	3.74		3.82	
Minimum Charge	Job		284		284	
Window Switch						
Demolish	Ea.		39.50		39.50	Includes material and labor to install a
Install	Ea.	63.50	107		170.50	door / window contact for a security
Demolish and Install	Ea.	63.50	146.50		210	system.
Reinstall	Ea.		85.80		85.80	
Clean	Ea.	.08	3.74		3.82	
Minimum Charge	Job		284		284	
Satellite System						
Dish						
Demolish	Ea.		175		175	Includes material, labor and
Install	Ea.	1750	475		2225	equipment to install 10' mesh dish.
Demolish and Install	Ea.	1750	650		2400	
Reinstall	Ea.		378.93		378.93	
Clean	Ea.	1.52	47		48.52	
Minimum Charge	Job		284		284	
Motor						
Demolish	Ea.		87.50		87.50	Includes material and labor to install a
Install	Ea.	255	237		492	television satellite dish motor unit.
Demolish and Install	Ea.	255	324.50		579.50	
Reinstall	Ea.		189.47		189.47	
Clean	Ea.	.08	11.70		11.78	
Minimum Charge	Job		284		284	
Television Antenna						
40' Pole						
Demolish	Ea.		98.50		98.50	Includes material and labor to install
Install	Ea.	23.50	178		201.50	stand-alone 1"-2" metal pole, 10' long.
Demolish and Install	Ea.	23.50	276.50		300	Cost does not include antenna.
Reinstall	Ea.		142.10		142.10	
Clean	Ea.	.31	47		47.31	
Paint	Ea.	14.60	50.50		65.10	
Minimum Charge	Job		284		284	
Rotor Unit						
Demolish	Ea.		79		79	Includes material and labor to install a
Install	Ea.	62.50	71		133.50	television antenna rotor.
Demolish and Install	Ea.	62.50	150		212.50	
Reinstall	Ea.		56.84		56.84	
Clean	Ea.	.31	47		47.31	
Minimum Charge	Job		284		284	
Single Booster						
Demolish	Ea.		79		79	Includes material and labor to install
Install	Ea.	30.50	71		101.50	standard grade VHF television signal
Demolish and Install	Ea.	30.50	150		180.50	booster unit and all hardware required
Reinstall	Ea.		56.84		56.84	for connection to TV cable.
Clean	Ea.	.01	5.85		5.86	
Minimum Charge	Job		284		284	

Gazebo

Gutter

Shutter

Window Treatment	Unit	Material	Labor	Equip.	Total	Specification
Drapery						
Good Grade						
Install	L.F.	48.50	7.50		56	Includes material and labor to install
Reinstall	L.F.		5.98		5.98	good quality drapery.
Clean	L.F.	1.66	.19		1.85	
Minimum Charge	Job		187		187	
Premium Grade						
Install	L.F.	37	7.50		44.50	Includes material and labor to install
Reinstall	L.F.		5.98		5.98	premium quality drapery.
Clean	L.F.	1.66	.21		1.87	
Minimum Charge	Job		187		187	
Lining						
Install	L.F.	28	7.50		35.50	Includes material and labor to install
Reinstall	L.F.		5.98		5.98	drapery lining.
Clean	L.F.	1.09	.11		1.20	
Minimum Charge	Job		187		187	
Valance Board						
Demolish	L.F.		2.08		2.08	Includes material and labor to install a
Install	L.F.	7.55	14.95		22.50	drapery valance board.
Demolish and Install	L.F.	7.55	17.03		24.58	
Reinstall	L.F.		11.97		11.97	
Clean	L.F.	1.09	.37		1.46	
Minimum Charge	Job		187		187	
Rod						
Demolish	L.F.		5.20		5.20	Includes material and labor to install
Install	L.F.	9.65	8.20		17.85	standard traverse drapery hardware.
Demolish and Install	L.F.	9.65	13.40		23.05	
Reinstall	L.F.		6.56		6.56	
Clean	L.F.	.03	.37		.40	
Paint	L.F.	.15	.78		.93	
Minimum Charge	Job		187		187	
Blinds						
Mini-type						
Demolish	S.F.		1.05		1.05	Includes material and labor to install
Install	S.F.	5.80	.82		6.62	horizontal 1" slat blinds.
Demolish and Install	S.F.	5.80	1.87		7.67	
Reinstall	S.F.		.66		.66	
Clean	S.F.	1.34	.04		1.38	
Minimum Charge	Job		187		187	

273

Window Treatment	Unit	Material	Labor	Equip.	Total	Specification
Vertical Type						
Demolish	S.F.		1.05		1.05	Includes material and labor to install
Install	S.F.	.17	1.05		1.22	vertical 3" - 5" cloth or PVC blinds.
Demolish and Install	S.F.	.17	2.10		2.27	
Reinstall	S.F.		.84		.84	
Clean	S.F.	1.40	.04		1.44	
Minimum Charge	Job		187		187	
Vertical Type per L.F.						
Demolish	L.F.		1.73		1.73	Includes material and labor to install
Install	L.F.	.99	4.99		5.98	vertical 3" - 5" cloth or PVC blinds x
Demolish and Install	L.F.	.99	6.72		7.71	72" long.
Reinstall	L.F.		3.99		3.99	
Clean	L.F.	8.75	.26		9.01	
Minimum Charge	Job		187		187	
Venetian Type						
Demolish	S.F.		1.05		1.05	Includes material and labor to install
Install	S.F.	7.35	.82		8.17	horizontal 2" slat blinds.
Demolish and Install	S.F.	7.35	1.87		9.22	
Reinstall	S.F.		.66		.66	
Clean	S.F.	1.40	.04		1.44	
Minimum Charge	Job		187		187	
Wood Shutter						
27" x 36", 4 Panel						
Demolish	Ea.		33.50		33.50	Includes material and labor to install
Install	Pr.	119	28.50		147.50	movable louver, up to 33" x 36", 4
Demolish and Install	Pr.	119	62		181	panel shutter.
Reinstall	Pr.		22.78		22.78	
Clean	Ea.	.03	12.45		12.48	
Paint	Ea.	2.33	29		31.33	
Minimum Charge	Job		242		242	
33" x 36", 4 Panel						
Demolish	Ea.		33.50		33.50	Includes material and labor to install
Install	Pr.	119	28.50		147.50	movable louver, up to 33" x 36", 4
Demolish and Install	Pr.	119	62		181	panel shutter.
Reinstall	Pr.		22.78		22.78	
Clean	Ea.	.03	12.45		12.48	
Paint	Ea.	2.43	31		33.43	
Minimum Charge	Job		242		242	
39" x 36", 4 Panel						
Demolish	Ea.		33.50		33.50	Includes material and labor to install
Install	Pr.	129	28.50		157.50	movable louver, up to 47" x 36", 4
Demolish and Install	Pr.	129	62		191	panel shutter.
Reinstall	Pr.		22.78		22.78	
Clean	Ea.	.03	12.45		12.48	
Paint	Ea.	2.43	31		33.43	
Minimum Charge	Job		242		242	
47" x 36", 4 Panel						
Demolish	Ea.		33.50		33.50	Includes material and labor to install
Install	Pr.	129	28.50		157.50	movable louver, up to 47" x 36", 4
Demolish and Install	Pr.	129	62		191	panel shutter.
Reinstall	Pr.		22.78		22.78	
Clean	Ea.	.03	12.45		12.48	
Paint	Ea.	2.53	33.50		36.03	
Minimum Charge	Job		242		242	

For customer support on your Contractor's Pricing Guide: Residential Repair & Remodeling, call 800.448.8182.

Stairs	Unit	Material	Labor	Equip.	Total	Specification
Concrete Steps						
Precast						Includes material and labor to install
Demolish	Ea.		71	49.50	120.50	precast concrete front entrance stairs,
Install	Flight	660	168	128	956	4' wide, with 48" platform and 2
Demolish and Install	Flight	660	239	177.50	1076.50	risers.
Clean	Ea.		5.20		5.20	
Paint	Ea.	3.97	50.50		54.47	
Minimum Charge	Job		445		445	
Metal						
Concrete Filled						Includes material and labor to install
Demolish	Ea.		21		21	steel, cement filled metal pan stairs
Install	Riser	620	75	5.40	700.40	with picket rail.
Demolish and Install	Riser	620	96	5.40	721.40	
Clean	Ea.	.04	3.15		3.19	
Paint	Ea.	1.14	2.99		4.13	
Minimum Charge	Job		279		279	
Landing						
Demolish	S.F.		5.20		5.20	Includes material and labor to install
Install	S.F.	122	8.85	.64	131.49	pre-erected conventional steel pan
Demolish and Install	S.F.	122	14.05	.64	136.69	landing.
Clean	S.F.	.04	.67		.71	
Paint	S.F.	.40	.86		1.26	
Minimum Charge	Job		279		279	
Railing						
Demolish	L.F.		2.87		2.87	Includes material and labor to install
Install	L.F.	20.50	12.80	.92	34.22	primed steel pipe.
Demolish and Install	L.F.	20.50	15.67	.92	37.09	
Clean	L.F.	.31	2.34		2.65	
Paint	L.F.	1.40	6.75		8.15	
Minimum Charge	Job		279		279	

Elevator	Unit	Material	Labor	Equip.	Total	Specification
2 Stop Residential						
Install	Ea.	14100	7525		21625	Includes labor, material and
Clean	Ea.	.55	23.50		24.05	equipment to install 2 stop elevator for
Minimum Charge	Job		750		750	residential use.
2 Stop Commercial						
Install	Ea.	52500	15000		67500	Includes labor, material and
Clean	Ea.	.55	23.50		24.05	equipment to install 2 stop elevator for
Minimum Charge	Job		750		750	apartment or commercial building with
						2,000 lb/13 passenger capacity.
3 Stop Commercial						
Install	Ea.	64000	23100		87100	Includes labor, material and
Clean	Ea.	.55	23.50		24.05	equipment to install 3 stop elevator for
Minimum Charge	Job		750		750	apartment or commercial building with
						2,000 lb/13 passenger capacity.
4 Stop Commercial						
Install	Ea.	75500	31300		106800	Includes labor, material and
Clean	Ea.	.55	23.50		24.05	equipment to install 4 stop elevator for
Minimum Charge	Job		750		750	apartment or commercial building with
						2,000 lb/13 passenger capacity.

Improvements / Appliances / Treatments

Elevator		Unit	Material	Labor	Equip.	Total	Specification
5 Stop Commercial							Includes labor, material and
	Install	Ea.	87000	39600		126600	equipment to install 5 stop elevator for
	Clean	Ea.	.55	23.50		24.05	apartment or commercial building with
	Minimum Charge	Job		750		750	2,000 lb/13 passenger capacity.

Rangetop		Unit	Material	Labor	Equip.	Total	Specification
4-burner							
	Demolish	Ea.		38		38	Cost includes material and labor to
	Install	Ea.	325	94.50		419.50	install 4-burner electric surface unit.
	Demolish and Install	Ea.	325	132.50		457.50	
	Reinstall	Ea.		75.79		75.79	
	Clean	Ea.	.02	37.50		37.52	
	Minimum Charge	Job		93.50		93.50	
6-burner							
	Demolish	Ea.		38		38	Includes material and labor to install a
	Install	Ea.	2150	114		2264	6-burner cooktop unit.
	Demolish and Install	Ea.	2150	152		2302	
	Reinstall	Ea.		90.94		90.94	
	Clean	Ea.	.02	37.50		37.52	
	Minimum Charge	Job		93.50		93.50	
Premium Brand w / Grill							
	Demolish	Ea.		38		38	Includes material and labor to install
	Install	Ea.	1950	114		2064	electric downdraft cooktop with grill.
	Demolish and Install	Ea.	1950	152		2102	
	Reinstall	Ea.		90.94		90.94	
	Clean	Ea.	.02	37.50		37.52	
	Minimum Charge	Job		93.50		93.50	

Rangehood		Unit	Material	Labor	Equip.	Total	Specification
Vented							
30"							
	Demolish	Ea.		34.50		34.50	Includes material and labor to install a
	Install	Ea.	128	114		242	vented range hood.
	Demolish and Install	Ea.	128	148.50		276.50	
	Reinstall	Ea.		90.94		90.94	
	Clean	Ea.	.02	10.65		10.67	
	Minimum Charge	Job		93.50		93.50	
36"							
	Demolish	Ea.		34.50		34.50	Includes material and labor to install a
	Install	Ea.	162	114		276	vented range hood.
	Demolish and Install	Ea.	162	148.50		310.50	
	Reinstall	Ea.		90.94		90.94	
	Clean	Ea.	.02	10.65		10.67	
	Minimum Charge	Job		93.50		93.50	
42"							
	Demolish	Ea.		34.50		34.50	Includes material and labor to install a
	Install	Ea.	173	125		298	vented range hood.
	Demolish and Install	Ea.	173	159.50		332.50	
	Reinstall	Ea.		100.18		100.18	
	Clean	Ea.	.02	10.65		10.67	
	Minimum Charge	Job		93.50		93.50	

For customer support on your Contractor's Pricing Guide: Residential Repair & Remodeling, call 800.448.8182.

Improvements / Appliances / Treatments

Rangehood		Unit	Material	Labor	Equip.	Total	Specification
Ventless							
30"							
	Demolish	Ea.		34.50		34.50	Includes material and labor to install a
	Install	Ea.	985	71		1056	ventless range hood.
	Demolish and Install	Ea.	985	105.50		1090.50	
	Reinstall	Ea.		56.84		56.84	
	Clean	Ea.	.02	10.65		10.67	
	Minimum Charge	Job		93.50		93.50	
36"							
	Demolish	Ea.		34.50		34.50	Includes material and labor to install a
	Install	Ea.	221	114		335	ventless range hood.
	Demolish and Install	Ea.	221	148.50		369.50	
	Reinstall	Ea.		90.94		90.94	
	Clean	Ea.	.02	10.65		10.67	
	Minimum Charge	Job		93.50		93.50	
42"							
	Demolish	Ea.		34.50		34.50	Includes material and labor to install a
	Install	Ea.	119	94.50		213.50	ventless range hood.
	Demolish and Install	Ea.	119	129		248	
	Reinstall	Ea.		75.79		75.79	
	Clean	Ea.	.02	10.65		10.67	
	Minimum Charge	Job		93.50		93.50	

Oven		Unit	Material	Labor	Equip.	Total	Specification
Free Standing							
	Demolish	Ea.		38		38	Cost includes material and labor to
	Install	Ea.	730	75		805	install 30" free standing electric range
	Demolish and Install	Ea.	730	113		843	with self-cleaning oven.
	Reinstall	Ea.		59.84		59.84	
	Clean	Ea.	.02	37.50		37.52	
	Minimum Charge	Job		93.50		93.50	
Single Wall Oven							
	Demolish	Ea.		38		38	Includes material and labor to install a
	Install	Ea.	675	142		817	built-in oven.
	Demolish and Install	Ea.	675	180		855	
	Reinstall	Ea.		113.68		113.68	
	Clean	Ea.	.02	37.50		37.52	
	Minimum Charge	Job		93.50		93.50	
Double Wall Oven							
	Demolish	Ea.		38		38	Includes material and labor to install a
	Install	Ea.	2650	242		2892	conventional double oven.
	Demolish and Install	Ea.	2650	280		2930	
	Reinstall	Ea.		193.60		193.60	
	Clean	Ea.	.02	37.50		37.52	
	Minimum Charge	Job		93.50		93.50	
Drop-in Type w / Range							
	Demolish	Ea.		38		38	Includes material and labor to install a
	Install	Ea.	1325	94.50		1419.50	built-in cooking range with oven.
	Demolish and Install	Ea.	1325	132.50		1457.50	
	Reinstall	Ea.		75.79		75.79	
	Clean	Ea.	.02	37.50		37.52	
	Minimum Charge	Job		93.50		93.50	

For customer support on your Contractor's Pricing Guide: Residential Repair & Remodeling, call 800.448.8182.

Improvements / Appliances / Treatments

Oven

Oven	Unit	Material	Labor	Equip.	Total	Specification
Hi/Lo w / Range, Microwave						
Demolish	Ea.		38		38	Includes material and labor to install a
Install	Ea.	3050	485		3535	double oven, one conventional, one
Demolish and Install	Ea.	3050	523		3573	microwave.
Reinstall	Ea.		387.20		387.20	
Clean	Ea.	.02	37.50		37.52	
Minimum Charge	Job		93.50		93.50	
Hi/Lo w/ Range / Double Oven						
Demolish	Ea.		38		38	Includes material and labor to install a
Install	Ea.	2650	242		2892	conventional double oven.
Demolish and Install	Ea.	2650	280		2930	
Reinstall	Ea.		193.60		193.60	
Clean	Ea.	.02	37.50		37.52	
Minimum Charge	Job		93.50		93.50	
Countertop Microwave						
Demolish	Ea.		13		13	Includes material and labor to install a
Install	Ea.	152	47		199	1.0 cubic foot microwave oven.
Demolish and Install	Ea.	152	60		212	
Reinstall	Ea.		37.40		37.40	
Clean	Ea.	.02	18.65		18.67	
Minimum Charge	Job		93.50		93.50	
Cabinet / Wall Mounted Microwave						
Demolish	Ea.		26		26	Includes material and labor to install a
Install	Ea.	885	142		1027	space saver microwave oven.
Demolish and Install	Ea.	885	168		1053	
Reinstall	Ea.		113.68		113.68	
Clean	Ea.	.02	18.65		18.67	
Minimum Charge	Job		93.50		93.50	

Dryer

Dryer	Unit	Material	Labor	Equip.	Total	Specification
Basic Grade						
Demolish	Ea.		17.30		17.30	Includes material and labor to install
Install	Ea.	435	288		723	electric or gas dryer. Vent kit not
Demolish and Install	Ea.	435	305.30		740.30	included.
Reinstall	Ea.		230.53		230.53	
Clean	Ea.	.02	37.50		37.52	
Minimum Charge	Job		93.50		93.50	
Good Grade						
Demolish	Ea.		17.30		17.30	Includes material and labor to install
Install	Ea.	545	288		833	electric or gas dryer. Vent kit not
Demolish and Install	Ea.	545	305.30		850.30	included.
Reinstall	Ea.		230.53		230.53	
Clean	Ea.	.02	37.50		37.52	
Minimum Charge	Job		93.50		93.50	
Premium Grade						
Demolish	Ea.		17.30		17.30	Includes material and labor to install
Install	Ea.	2100	430		2530	electric or gas dryer. Vent kit not
Demolish and Install	Ea.	2100	447.30		2547.30	included.
Reinstall	Ea.		345.79		345.79	
Clean	Ea.	.02	37.50		37.52	
Minimum Charge	Job		93.50		93.50	

Improvements / Appliances / Treatments

Dryer

Dryer	Unit	Material	Labor	Equip.	Total	Specification
Washer / Dryer Combination Unit						
Demolish	Ea.		17.30		17.30	Includes material and labor to install
Install	Ea.	2025	75		2100	electric washer/dryer combination
Demolish and Install	Ea.	2025	92.30		2117.30	unit, 27" wide, stackable or unitized.
Reinstall	Ea.		59.84		59.84	
Clean	Ea.	.02	37.50		37.52	
Minimum Charge	Job		93.50		93.50	

Washer

Washer	Unit	Material	Labor	Equip.	Total	Specification
Basic Grade						
Demolish	Ea.		17.30		17.30	Includes material and labor to install
Install	Ea.	770	182		952	washing machine. New hoses not
Demolish and Install	Ea.	770	199.30		969.30	included.
Reinstall	Ea.		145.81		145.81	
Clean	Ea.	.02	37.50		37.52	
Minimum Charge	Job		93.50		93.50	
Good Grade						
Demolish	Ea.		17.30		17.30	Includes material and labor to install
Install	Ea.	1275	273		1548	washing machine. New hoses not
Demolish and Install	Ea.	1275	290.30		1565.30	included.
Reinstall	Ea.		218.72		218.72	
Clean	Ea.	.02	37.50		37.52	
Minimum Charge	Job		93.50		93.50	
Premium Grade						
Demolish	Ea.		17.30		17.30	Includes material and labor to install
Install	Ea.	1050	545		1595	washing machine with digital
Demolish and Install	Ea.	1050	562.30		1612.30	computer readouts and large capacity.
Reinstall	Ea.		437.44		437.44	New hoses not included.
Clean	Ea.	.02	37.50		37.52	
Minimum Charge	Job		93.50		93.50	
Washer / Dryer Combination Unit						
Demolish	Ea.		17.30		17.30	Includes material and labor to install
Install	Ea.	2025	75		2100	electric washer/dryer combination
Demolish and Install	Ea.	2025	92.30		2117.30	unit, 27" wide, stackable or unitized.
Reinstall	Ea.		59.84		59.84	
Clean	Ea.	.02	37.50		37.52	
Minimum Charge	Job		93.50		93.50	

Refrigerator

Refrigerator	Unit	Material	Labor	Equip.	Total	Specification
12 Cubic Foot						
Demolish	Ea.		52		52	Includes material and labor to install a
Install	Ea.	615	75		690	refrigerator.
Demolish and Install	Ea.	615	127		742	
Reinstall	Ea.		59.84		59.84	
Clean	Ea.	.02	37.50		37.52	
Minimum Charge	Job		93.50		93.50	
16 Cubic Foot						
Demolish	Ea.		52		52	Includes material and labor to install a
Install	Ea.	750	83		833	refrigerator.
Demolish and Install	Ea.	750	135		885	
Reinstall	Ea.		66.49		66.49	
Clean	Ea.	.02	37.50		37.52	
Minimum Charge	Job		93.50		93.50	

279

Improvements / Appliances / Treatments

Refrigerator		Unit	Material	Labor	Equip.	Total	Specification
18 Cubic Foot							
	Demolish	Ea.		52		52	Includes material and labor to install a
	Install	Ea.	1000	93.50		1093.50	refrigerator.
	Demolish and Install	Ea.	1000	145.50		1145.50	
	Reinstall	Ea.		74.80		74.80	
	Clean	Ea.	.02	37.50		37.52	
	Minimum Charge	Job		93.50		93.50	
21 Cubic Foot							
	Demolish	Ea.		52		52	Includes material and labor to install a
	Install	Ea.	1750	107		1857	refrigerator.
	Demolish and Install	Ea.	1750	159		1909	
	Reinstall	Ea.		85.49		85.49	
	Clean	Ea.	.02	37.50		37.52	
	Minimum Charge	Job		93.50		93.50	
24 Cubic Foot							
	Demolish	Ea.		52		52	Includes material and labor to install a
	Install	Ea.	2350	107		2457	refrigerator.
	Demolish and Install	Ea.	2350	159		2509	
	Reinstall	Ea.		85.49		85.49	
	Clean	Ea.	.02	37.50		37.52	
	Minimum Charge	Job		93.50		93.50	
27 Cubic Foot							
	Demolish	Ea.		52		52	Includes material and labor to install a
	Install	Ea.	3350	75		3425	refrigerator.
	Demolish and Install	Ea.	3350	127		3477	
	Reinstall	Ea.		59.84		59.84	
	Clean	Ea.	.02	37.50		37.52	
	Minimum Charge	Job		93.50		93.50	
Ice Maker							
	Install	Ea.	95	68.50		163.50	Includes labor and material to install
	Minimum Charge	Job		93.50		93.50	an automatic ice maker in a refrigerator.

Freezer		Unit	Material	Labor	Equip.	Total	Specification
15 Cubic Foot							
	Demolish	Ea.		52		52	Cost includes material and labor to
	Install	Ea.	715	75		790	install 15 cubic foot freezer.
	Demolish and Install	Ea.	715	127		842	
	Reinstall	Ea.		59.84		59.84	
	Clean	Ea.	.02	37.50		37.52	
	Minimum Charge	Job		93.50		93.50	
18 Cubic Foot							
	Demolish	Ea.		52		52	Cost includes material and labor to
	Install	Ea.	945	75		1020	install 18 cubic foot freezer.
	Demolish and Install	Ea.	945	127		1072	
	Reinstall	Ea.		59.84		59.84	
	Clean	Ea.	.02	37.50		37.52	
	Minimum Charge	Job		93.50		93.50	
21 Cubic Foot							
	Demolish	Ea.		52		52	Includes material and labor to install a
	Install	Ea.	715	75		790	15 to 23 cubic foot deep freezer.
	Demolish and Install	Ea.	715	127		842	
	Reinstall	Ea.		59.84		59.84	
	Clean	Ea.	.02	37.50		37.52	
	Minimum Charge	Job		93.50		93.50	

For customer support on your Contractor's Pricing Guide: Residential Repair & Remodeling, call 800.448.8182.

Improvements / Appliances / Treatments

Freezer

	Unit	Material	Labor	Equip.	Total	Specification
24 Cubic Foot						
Demolish	Ea.		52		52	Includes material and labor to install a
Install	Ea.	790	150		940	24 cubic foot deep freezer.
Demolish and Install	Ea.	790	202		992	
Reinstall	Ea.		119.68		119.68	
Clean	Ea.	.02	37.50		37.52	
Minimum Charge	Job		93.50		93.50	
Chest-type						
Demolish	Ea.		52		52	Cost includes material and labor to
Install	Ea.	715	75		790	install 15 cubic foot freezer.
Demolish and Install	Ea.	715	127		842	
Clean	Ea.	.02	37.50		37.52	
Minimum Charge	Job		93.50		93.50	

Dishwasher

	Unit	Material	Labor	Equip.	Total	Specification
4 Cycle						
Demolish	Ea.		41.50		41.50	Includes material and labor to install
Install	Ea.	695	172		867	good grade automatic dishwasher.
Demolish and Install	Ea.	695	213.50		908.50	
Reinstall	Ea.		137.78		137.78	
Clean	Ea.	.02	37.50		37.52	
Minimum Charge	Job		93.50		93.50	

Garbage Disposal

	Unit	Material	Labor	Equip.	Total	Specification
In Sink Unit						
Demolish	Ea.		38		38	Cost includes material and labor to
Install	Ea.	250	69		319	install 1/2 HP custom disposal unit.
Demolish and Install	Ea.	250	107		357	
Reinstall	Ea.		55.11		55.11	
Clean	Ea.	.01	10.65		10.66	
Minimum Charge	Job		93.50		93.50	

Trash Compactor

	Unit	Material	Labor	Equip.	Total	Specification
4 to 1 Compaction						
Demolish	Ea.		39.50		39.50	Includes material and labor to install a
Install	Ea.	1600	97		1697	trash compactor.
Demolish and Install	Ea.	1600	136.50		1736.50	
Reinstall	Ea.		77.44		77.44	
Clean	Ea.	.02	23.50		23.52	
Minimum Charge	Job		93.50		93.50	

Kitchenette Unit

	Unit	Material	Labor	Equip.	Total	Specification
Range, Refrigerator and Sink						
Demolish	Ea.		158		158	Includes material and labor to install
Install	Ea.	1900	575		2475	range/refrigerator/sink unit, 72"
Demolish and Install	Ea.	1900	733		2633	wide, base unit only.
Reinstall	Ea.		459.27		459.27	
Clean	Ea.	.39	18.65		19.04	
Minimum Charge	Job		93.50		93.50	

For customer support on your Contractor's Pricing Guide: Residential Repair & Remodeling, call 800.448.8182.

Bath Accessories		Unit	Material	Labor	Equip.	Total	Specification
Towel Bar							
	Demolish	Ea.		8.65		8.65	Includes material and labor to install
	Install	Ea.	59	20		79	towel bar up to 24" long.
	Demolish and Install	Ea.	59	28.65		87.65	
	Reinstall	Ea.		16.13		16.13	
	Clean	Ea.		4.68		4.68	
	Minimum Charge	Job		242		242	
Toothbrush Holder							
	Demolish	Ea.		8.65		8.65	Includes material and labor to install a
	Install	Ea.	13.35	24		37.35	surface mounted tumbler and
	Demolish and Install	Ea.	13.35	32.65		46	toothbrush holder.
	Reinstall	Ea.		19.36		19.36	
	Clean	Ea.		4.68		4.68	
	Minimum Charge	Job		242		242	
Grab Rail							
	Demolish	Ea.		13		13	Includes material and labor to install a
	Install	Ea.	24	21		45	24" long grab bar. Blocking is not
	Demolish and Install	Ea.	24	34		58	included.
	Reinstall	Ea.		16.83		16.83	
	Clean	Ea.	.15	7.80		7.95	
	Minimum Charge	Job		242		242	
Soap Dish							
	Demolish	Ea.		8.65		8.65	Includes material and labor to install a
	Install	Ea.	5.65	24		29.65	surface mounted soap dish.
	Demolish and Install	Ea.	5.65	32.65		38.30	
	Reinstall	Ea.		19.36		19.36	
	Clean	Ea.		4.68		4.68	
	Minimum Charge	Job		242		242	
Toilet Paper Roller							
	Demolish	Ea.		8.65		8.65	Includes material and labor to install a
	Install	Ea.	16.05	16.15		32.20	surface mounted toilet tissue dispenser.
	Demolish and Install	Ea.	16.05	24.80		40.85	
	Reinstall	Ea.		12.91		12.91	
	Clean	Ea.	.15	11.70		11.85	
	Minimum Charge	Job		242		242	
Dispenser							
Soap							
	Demolish	Ea.		11.55		11.55	Includes material and labor to install a
	Install	Ea.	58	24		82	surface mounted soap dispenser.
	Demolish and Install	Ea.	58	35.55		93.55	
	Reinstall	Ea.		19.36		19.36	
	Clean	Ea.	.15	11.70		11.85	
	Minimum Charge	Job		242		242	
Towel							
	Demolish	Ea.		13		13	Includes material and labor to install a
	Install	Ea.	73.50	30.50		104	surface mounted towel dispenser.
	Demolish and Install	Ea.	73.50	43.50		117	
	Reinstall	Ea.		24.20		24.20	
	Clean	Ea.	.15	15.60		15.75	
	Minimum Charge	Job		242		242	
Seat Cover							
	Demolish	Ea.		8.65		8.65	Includes material and labor to install a
	Install	Ea.	40.50	32.50		73	surface mounted toilet seat cover
	Demolish and Install	Ea.	40.50	41.15		81.65	dispenser.
	Reinstall	Ea.		25.81		25.81	
	Clean	Ea.	.15	11.70		11.85	
	Minimum Charge	Job		242		242	

For customer support on your Contractor's Pricing Guide: Residential Repair & Remodeling, call 800.448.8182.

Improvements / Appliances / Treatments

Bath Accessories		Unit	Material	Labor	Equip.	Total	Specification
Sanitary Napkin							
	Demolish	Ea.		22		22	Includes material and labor to install a
	Install	Ea.	48	51		99	surface mounted sanitary napkin
	Demolish and Install	Ea.	48	73		121	dispenser.
	Reinstall	Ea.		40.76		40.76	
	Clean	Ea.	.08	15.60		15.68	
	Minimum Charge	Job		242		242	
Partition							
Urinal							
	Demolish	Ea.		26		26	Includes material and labor to install a
	Install	Ea.	268	121		389	urinal screen.
	Demolish and Install	Ea.	268	147		415	
	Reinstall	Ea.		96.80		96.80	
	Clean	Ea.	.23	10.40		10.63	
	Minimum Charge	Job		242		242	
Toilet							
	Demolish	Ea.		77		77	Includes material and labor to install
	Install	Ea.	860	194		1054	baked enamel standard size partition,
	Demolish and Install	Ea.	860	271		1131	floor or ceiling mounted with one wall
	Reinstall	Ea.		154.88		154.88	and one door.
	Clean	Ea.	1.52	23.50		25.02	
	Minimum Charge	Job		242		242	
Handicap							
	Demolish	Ea.		77		77	Includes material and labor to install
	Install	Ea.	1300	194		1494	baked enamel handicap partition,
	Demolish and Install	Ea.	1300	271		1571	floor or ceiling mounted with one wall
	Reinstall	Ea.		154.88		154.88	and one door.
	Clean	Ea.	1.52	23.50		25.02	
	Minimum Charge	Job		242		242	

Fire Prevention / Protection		Unit	Material	Labor	Equip.	Total	Specification
Fire Extinguisher							
5# Carbon Dioxide							
	Install	Ea.	244			244	Includes labor and material to install one wall mounted 5 lb. carbon dioxide factory charged unit, complete with hose, horn and wall mounting bracket.
10# Carbon Dioxide							
	Install	Ea.	365			365	Includes labor and material to install one wall mounted 10 lb. carbon dioxide factory charged unit, complete with hose, horn and wall mounting bracket.
15# Carbon Dioxide							
	Install	Ea.	445			445	Includes labor and material to install one wall mounted 15 lb. carbon dioxide factory charged unit, complete with hose, horn and wall mounting bracket.

For customer support on your Contractor's Pricing Guide: Residential Repair & Remodeling, call 800.448.8182.

Improvements / Appliances / Treatments

Fire Prevention / Protection		Unit	Material	Labor	Equip.	Total	Specification
5# Dry Chemical							
	Install	Ea.	41.50			41.50	Includes labor and material to install one wall mounted 5 lb. dry chemical factory charged unit, complete with hose, horn and wall mounting bracket.
20# Dry Chemical							
	Install	Ea.	167			167	Includes labor and material to install one wall mounted 20 lb. dry chemical factory charged unit, complete with hose, horn and wall mounting bracket.

Gutters		Unit	Material	Labor	Equip.	Total	Specification
Galvanized							
	Demolish	L.F.		1.73		1.73	Cost includes material and labor to install 5" galvanized steel gutter with enamel finish, fittings, hangers, and corners.
	Install	L.F.	3.96	4.27		8.23	
	Demolish and Install	L.F.	3.96	6		9.96	
	Reinstall	L.F.		3.42		3.42	
	Clean	L.F.		.50		.50	
	Paint	L.F.	.39	1.24		1.63	
	Minimum Charge	Job		267		267	
Aluminum							
	Demolish	L.F.		1.73		1.73	Cost includes material and labor to install 5" aluminum gutter with enamel finish, fittings, hangers, and corners.
	Install	L.F.	2.64	4.27		6.91	
	Demolish and Install	L.F.	2.64	6		8.64	
	Reinstall	L.F.		3.42		3.42	
	Clean	L.F.		.50		.50	
	Paint	L.F.	.39	1.24		1.63	
	Minimum Charge	Job		267		267	
Copper							
	Demolish	L.F.		1.73		1.73	Cost includes material and labor to install 5" half-round copper gutter with fittings, hangers, and corners.
	Install	L.F.	9.15	4.27		13.42	
	Demolish and Install	L.F.	9.15	6		15.15	
	Reinstall	L.F.		3.42		3.42	
	Clean	L.F.		.50		.50	
	Minimum Charge	Job		267		267	
Plastic							
	Demolish	L.F.		1.73		1.73	Cost includes material and labor to install 4" white vinyl gutter fittings, hangers, and corners.
	Install	L.F.	2.34	4.21		6.55	
	Demolish and Install	L.F.	2.34	5.94		8.28	
	Reinstall	L.F.		3.37		3.37	
	Clean	L.F.		.50		.50	
	Paint	L.F.	.39	1.24		1.63	
	Minimum Charge	Job		267		267	

Downspouts		Unit	Material	Labor	Equip.	Total	Specification
Galvanized							
	Demolish	L.F.		1.19		1.19	Cost includes material and labor to install 4" diameter downspout with enamel finish.
	Install	L.F.	5.15	3.68		8.83	
	Demolish and Install	L.F.	5.15	4.87		10.02	
	Reinstall	L.F.		2.94		2.94	
	Clean	L.F.		.50		.50	
	Paint	L.F.	.39	1.24		1.63	
	Minimum Charge	Job		267		267	

Downspouts	Unit	Material	Labor	Equip.	Total	Specification
Aluminum						
Demolish	L.F.		1.19		1.19	Includes material and labor to install
Install	L.F.	3.19	3.81		7	aluminum enameled downspout.
Demolish and Install	L.F.	3.19	5		8.19	
Reinstall	L.F.		3.05		3.05	
Clean	L.F.		.50		.50	
Paint	L.F.	.39	1.24		1.63	
Minimum Charge	Job		267		267	
Copper						
Demolish	L.F.		1.19		1.19	Includes material and labor to install
Install	L.F.	10.60	2.81		13.41	round copper downspout.
Demolish and Install	L.F.	10.60	4		14.60	
Reinstall	L.F.		2.25		2.25	
Clean	L.F.		.50		.50	
Paint	L.F.	.39	1.24		1.63	
Minimum Charge	Job		267		267	
Plastic						
Demolish	L.F.		1.19		1.19	Cost includes material and labor to
Install	L.F.	2.70	2.54		5.24	install 2" x 3" vinyl rectangular
Demolish and Install	L.F.	2.70	3.73		6.43	downspouts.
Reinstall	L.F.		2.03		2.03	
Clean	L.F.		.50		.50	
Paint	L.F.	.39	1.24		1.63	
Minimum Charge	Job		267		267	

Shutters	Unit	Material	Labor	Equip.	Total	Specification
Exterior						
25" High						
Demolish	Ea.		13		13	Includes material and labor to install
Install	Pr.	226	48.50		274.50	pair of exterior shutters 25" x 16",
Demolish and Install	Pr.	226	61.50		287.50	unpainted, installed as fixed shutters
Reinstall	Pr.		48.40		48.40	on concrete or wood frame.
Clean	Ea.	.03	12.45		12.48	
Paint	Ea.	3.09	21.50		24.59	
Minimum Charge	Job		242		242	
39" High						
Demolish	Ea.		13		13	Includes material and labor to install
Install	Pr.	305	48.50		353.50	pair of shutters 39" x 16", unpainted,
Demolish and Install	Pr.	305	61.50		366.50	installed as fixed shutters on concrete
Reinstall	Pr.		48.40		48.40	or wood frame.
Clean	Ea.	.03	12.45		12.48	
Paint	Ea.	4.80	29.50		34.30	
Minimum Charge	Job		242		242	
51" High						
Demolish	Ea.		13		13	Includes material and labor to install
Install	Pr.	385	48.50		433.50	pair of shutters 51" x 16", unpainted,
Demolish and Install	Pr.	385	61.50		446.50	installed as fixed shutters on concrete
Reinstall	Pr.		48.40		48.40	or wood frame.
Clean	Ea.	.03	12.45		12.48	
Paint	Ea.	6.20	30.50		36.70	
Minimum Charge	Job		242		242	

Shutters

	Unit	Material	Labor	Equip.	Total	Specification
59" High						
Demolish	Ea.		13		13	Includes material and labor to install
Install	Pr.	510	48.50		558.50	pair of shutters 59" x 16", unpainted,
Demolish and Install	Pr.	510	61.50		571.50	installed as fixed shutters on concrete
Reinstall	Pr.		48.40		48.40	or wood frame.
Clean	Ea.	.03	12.45		12.48	
Paint	Ea.	7.20	33.50		40.70	
Minimum Charge	Job		242		242	
67" High						
Demolish	Ea.		13		13	Includes material and labor to install
Install	Pr.	675	54		729	pair of shutters 67" x 16", unpainted,
Demolish and Install	Pr.	675	67		742	installed as fixed shutters on concrete
Reinstall	Pr.		53.78		53.78	or wood frame.
Clean	Ea.	.03	12.45		12.48	
Paint	Ea.	8.25	37.50		45.75	
Minimum Charge	Job		242		242	
Interior Wood						
23" x 24", 2 Panel						
Demolish	Ea.		13		13	Cost includes material and labor to
Install	Ea.	179	28.50		207.50	install 23" x 24", 4 panel interior
Demolish and Install	Ea.	179	41.50		220.50	movable pine shutter, 1-1/4" louver,
Reinstall	Ea.		22.78		22.78	knobs and hooks and hinges.
Clean	Ea.	.03	12.45		12.48	
Paint	Ea.	2.33	29		31.33	
Minimum Charge	Job		242		242	
27" x 24", 4 Panel						
Demolish	Ea.		13		13	Cost includes material and labor to
Install	Ea.	193	28.50		221.50	install 27" x 24", 4 panel interior
Demolish and Install	Ea.	193	41.50		234.50	movable pine shutter, 1-1/4" louver,
Reinstall	Ea.		22.78		22.78	knobs and hooks and hinges.
Clean	Ea.	.03	12.45		12.48	
Paint	Ea.	2.33	29		31.33	
Minimum Charge	Job		242		242	
31" x 24", 4 Panel						
Demolish	Ea.		13		13	Cost includes material and labor to
Install	Ea.	212	28.50		240.50	install 31" x 24", 4 panel interior
Demolish and Install	Ea.	212	41.50		253.50	movable pine shutter, 1-1/4" louver,
Reinstall	Ea.		22.78		22.78	knobs and hooks and hinges.
Clean	Ea.	.03	12.45		12.48	
Paint	Ea.	2.33	29		31.33	
Minimum Charge	Job		242		242	
35" x 24", 4 Panel						
Demolish	Ea.		13		13	Cost includes material and labor to
Install	Ea.	228	28.50		256.50	install 35" x 24", 4 panel interior
Demolish and Install	Ea.	228	41.50		269.50	movable pine shutter, 1-1/4" louver,
Reinstall	Ea.		22.78		22.78	knobs and hooks and hinges.
Clean	Ea.	.03	12.45		12.48	
Paint	Ea.	2.33	29		31.33	
Minimum Charge	Job		242		242	
39" x 24", 4 Panel						
Demolish	Ea.		13		13	Cost includes material and labor to
Install	Ea.	253	28.50		281.50	install 39" x 24", 4 panel interior
Demolish and Install	Ea.	253	41.50		294.50	movable pine shutter, 1-1/4" louver,
Reinstall	Ea.		22.78		22.78	knobs and hooks and hinges.
Clean	Ea.	.03	12.45		12.48	
Paint	Ea.	2.33	29		31.33	
Minimum Charge	Job		242		242	

Improvements / Appliances / Treatments

Shutters

	Unit	Material	Labor	Equip.	Total	Specification
47" x 24", 4 Panel						
Demolish	Ea.		13		13	Cost includes material and labor to
Install	Ea.	287	28.50		315.50	install 47" x 24", 4 panel interior
Demolish and Install	Ea.	287	41.50		328.50	movable pine shutter, 1-1/4" louver,
Reinstall	Ea.		22.78		22.78	knobs and hooks and hinges.
Clean	Ea.	.03	12.45		12.48	
Paint	Ea.	2.43	31		33.43	
Minimum Charge	Job		242		242	

Awning

	Unit	Material	Labor	Equip.	Total	Specification
Standard Aluminum						
36" Wide x 30" Deep						
Demolish	Ea.		13		13	Includes material and labor to install
Install	Ea.	197	40.50		237.50	one aluminum awning with
Demolish and Install	Ea.	197	53.50		250.50	weather-resistant finish.
Reinstall	Ea.		32.27		32.27	
Clean	Ea.		15.60		15.60	
Minimum Charge	Job		242		242	
48" Wide x 30" Deep						
Demolish	Ea.		13		13	Includes material and labor to install
Install	Ea.	244	48.50		292.50	one aluminum awning with
Demolish and Install	Ea.	244	61.50		305.50	weather-resistant finish.
Reinstall	Ea.		38.72		38.72	
Clean	Ea.		15.60		15.60	
Minimum Charge	Job		242		242	
60" Wide x 30" Deep						
Demolish	Ea.		26		26	Includes material and labor to install
Install	Ea.	274	54		328	one aluminum awning with
Demolish and Install	Ea.	274	80		354	weather-resistant finish.
Reinstall	Ea.		43.02		43.02	
Clean	Ea.		15.60		15.60	
Minimum Charge	Job		242		242	
72" Wide x 30" Deep						
Demolish	Ea.		26		26	Includes material and labor to install
Install	Ea.	305	60.50		365.50	one aluminum awning with
Demolish and Install	Ea.	305	86.50		391.50	weather-resistant finish.
Reinstall	Ea.		48.40		48.40	
Clean	Ea.		15.60		15.60	
Minimum Charge	Job		242		242	
Roll-up Type						
Demolish	S.F.		.94		.94	Includes material and labor to install
Install	S.F.	18.90	4.84		23.74	an aluminum roll-up awning.
Demolish and Install	S.F.	18.90	5.78		24.68	
Reinstall	S.F.		3.87		3.87	
Clean	S.F.	.01	.78		.79	
Minimum Charge	Job		242		242	
Canvas						
30" Wide						
Demolish	Ea.		7.80		7.80	Includes material and labor to install a
Install	Ea.	220	40.50		260.50	canvas window awning.
Demolish and Install	Ea.	220	48.30		268.30	
Reinstall	Ea.		32.27		32.27	
Clean	Ea.		4.68		4.68	
Minimum Charge	Job		242		242	

For customer support on your Contractor's Pricing Guide: Residential Repair & Remodeling, call 800.448.8182.

Improvements / Appliances / Treatments

Awning

	Unit	Material	Labor	Equip.	Total	Specification
36″ Wide						
Demolish	Ea.		7.80		7.80	Includes material and labor to install a
Install	Ea.	255	48.50		303.50	canvas window awning.
Demolish and Install	Ea.	255	56.30		311.30	
Reinstall	Ea.		38.72		38.72	
Clean	Ea.		5.10		5.10	
Minimum Charge	Job		242		242	
42″ Wide						
Demolish	Ea.		7.80		7.80	Includes material and labor to install a
Install	Ea.	268	60.50		328.50	canvas window awning.
Demolish and Install	Ea.	268	68.30		336.30	
Reinstall	Ea.		48.40		48.40	
Clean	Ea.		6.10		6.10	
Minimum Charge	Job		242		242	
48″ Wide						
Demolish	Ea.		7.80		7.80	Includes material and labor to install a
Install	Ea.	283	69		352	canvas window awning.
Demolish and Install	Ea.	283	76.80		359.80	
Reinstall	Ea.		55.31		55.31	
Clean	Ea.		7.05		7.05	
Minimum Charge	Job		242		242	

Cloth Canopy Cover

	Unit	Material	Labor	Equip.	Total	Specification
8′ x 10′						
Demolish	Ea.		420		420	Includes material and labor to install
Install	Ea.	240	242		482	one 8′ x 10′ cloth canopy patio cover
Demolish and Install	Ea.	240	662		902	with front bar and tension support
Reinstall	Ea.		193.60		193.60	rafters and 9″ valance.
Clean	Ea.	1.22	62.50		63.72	
Minimum Charge	Job		485		485	
8′ x 15′						
Demolish	Ea.		420		420	Includes material and labor to install
Install	Ea.	250	242		492	one 8′ x 15′ cloth canopy patio cover
Demolish and Install	Ea.	250	662		912	with front bar and tension support
Reinstall	Ea.		193.60		193.60	rafters and 9″ valance.
Clean	Ea.	1.22	93.50		94.72	
Minimum Charge	Job		485		485	

Vent

	Unit	Material	Labor	Equip.	Total	Specification
Small Roof Turbine						
Demolish	Ea.		13		13	Includes material and labor to install
Install	Ea.	76	48		124	spinner ventilator, wind driven,
Demolish and Install	Ea.	76	61		137	galvanized, 4″ neck diam, 180 CFM.
Reinstall	Ea.		38.42		38.42	
Clean	Ea.		8.30		8.30	
Minimum Charge	Job		273		273	
Large Roof Turbine						
Demolish	Ea.		10.40		10.40	Includes material and labor to install
Install	Ea.	89	68.50		157.50	spinner ventilator, wind driven,
Demolish and Install	Ea.	89	78.90		167.90	galvanized, 8″ neck diam, 360 CFM.
Reinstall	Ea.		54.88		54.88	
Clean	Ea.		9.35		9.35	
Minimum Charge	Job		273		273	

For customer support on your Contractor's Pricing Guide: Residential Repair & Remodeling, call 800.448.8182.

Vent		Unit	Material	Labor	Equip.	Total	Specification
Gable Louvered							
	Demolish	Ea.		26		26	Includes material and labor to install a
	Install	Ea.	69	16.15		85.15	vinyl gable end vent.
	Demolish and Install	Ea.	69	42.15		111.15	
	Reinstall	Ea.		12.91		12.91	
	Clean	Ea.		4.99		4.99	
	Paint	Ea.	1.95	6.20		8.15	
	Minimum Charge	Job		242		242	
Ridge Vent							
	Demolish	L.F.		1.34		1.34	Includes material and labor to install a
	Install	L.F.	3.03	3.44		6.47	mill finish aluminum ridge vent strip.
	Demolish and Install	L.F.	3.03	4.78		7.81	
	Minimum Charge	Job		242		242	
Foundation							
	Demolish	Ea.		9.45		9.45	Includes material and labor to install
	Install	Ea.	22	15.95		37.95	galvanized foundation block vent.
	Demolish and Install	Ea.	22	25.40		47.40	
	Reinstall	Ea.		12.77		12.77	
	Clean	Ea.	.08	9.35		9.43	
	Minimum Charge	Job		242		242	
Frieze							
	Demolish	L.F.		.92		.92	Includes material and labor to install
	Install	L.F.	1.55	1.94		3.49	pine frieze board.
	Demolish and Install	L.F.	1.55	2.86		4.41	
	Reinstall	L.F.		1.55		1.55	
	Clean	L.F.	.02	.31		.33	
	Minimum Charge	Job		242		242	
Water Heater Cap							
	Demolish	Ea.		19		19	Includes material and labor to install a
	Install	Ea.	48.50	66.50		115	galvanized steel water heater cap.
	Demolish and Install	Ea.	48.50	85.50		134	
	Reinstall	Ea.		53.36		53.36	
	Clean	Ea.	.53	23.50		24.03	

Cupolas		Unit	Material	Labor	Equip.	Total	Specification
22" x 22"							
	Demolish	Ea.		240		240	Includes material and labor to install a
	Install	Ea.	440	121		561	cedar cupola with aluminum roof
	Demolish and Install	Ea.	440	361		801	covering.
	Reinstall	Ea.		96.80		96.80	
	Minimum Charge	Job		242		242	
29" x 29"							
	Demolish	Ea.		240		240	Includes material and labor to install a
	Install	Ea.	830	121		951	cedar cupola with aluminum roof
	Demolish and Install	Ea.	830	361		1191	covering.
	Reinstall	Ea.		96.80		96.80	
	Minimum Charge	Job		242		242	
35" x 35"							
	Demolish	Ea.		240		240	Includes material and labor to install a
	Install	Ea.	1000	161		1161	cedar cupola with aluminum roof
	Demolish and Install	Ea.	1000	401		1401	covering.
	Reinstall	Ea.		129.07		129.07	
	Minimum Charge	Job		242		242	

Cupolas

	Unit	Material	Labor	Equip.	Total	Specification
47" x 47"						
Demolish	Ea.		240		240	Includes material and labor to install a
Install	Ea.	1625	242		1867	cedar cupola with aluminum roof
Demolish and Install	Ea.	1625	482		2107	covering.
Reinstall	Ea.		193.60		193.60	
Minimum Charge	Job		242		242	
Weather Vane						
Install	Ea.	204	60.50		264.50	Includes minimum labor and
Minimum Charge	Job		242		242	equipment to install residential type weathervane.

Aluminum Carport

	Unit	Material	Labor	Equip.	Total	Specification
20' x 20' Complete						
Demolish	Ea.		260		260	Includes material, labor and
Install	Ea.	4475	795	207	5477	equipment to install an aluminum
Demolish and Install	Ea.	4475	1055	207	5737	carport. Foundations are not included.
Reinstall	Ea.		635.52	165.98	801.50	
Clean	Ea.	.08	93.50		93.58	
Minimum Charge	Job		485		485	
24' x 24' Complete						
Demolish	Ea.		260		260	Includes material, labor and
Install	Ea.	5200	1050	277	6527	equipment to install an aluminum
Demolish and Install	Ea.	5200	1310	277	6787	carport. Foundations are not included.
Reinstall	Ea.		847.36	221.31	1068.67	
Clean	Ea.	.08	187		187.08	
Minimum Charge	Job		485		485	
Metal Support Posts						
Demolish	Ea.		3.46		3.46	Includes material and labor to install
Install	Ea.	460	60.50		520.50	metal support posts for a carport.
Demolish and Install	Ea.	460	63.96		523.96	
Reinstall	Ea.		48.40		48.40	
Clean	Ea.	.08	3.90		3.98	
Paint	Ea.	.30	8.05		8.35	
Minimum Charge	Job		485		485	

Storage Shed

	Unit	Material	Labor	Equip.	Total	Specification
Aluminum						
Pre-fab (8' x 10')						
Demolish	Ea.		75.50		75.50	Includes material and labor to install
Install	Ea.	2075	720		2795	an aluminum storage shed.
Demolish and Install	Ea.	2075	795.50		2870.50	
Clean	Ea.	.51	47		47.51	
Minimum Charge	Ea.		555		555	
Pre-fab (10' x 12')						
Demolish	Ea.		140		140	Includes material and labor to install
Install	Ea.	1875	720		2595	an aluminum storage shed.
Demolish and Install	Ea.	1875	860		2735	
Clean	Ea.	.51	52		52.51	
Minimum Charge	Ea.		555		555	

For customer support on your Contractor's Pricing Guide: Residential Repair & Remodeling, call 800.448.8182.

Storage Shed

Storage Shed	Unit	Material	Labor	Equip.	Total	Specification
Wood						
Pre-fab (8' x 10')						
Demolish	Ea.		75.50		75.50	Includes material and labor to install a
Install	Ea.	2025	720		2745	pre-fabricated wood storage shed.
Demolish and Install	Ea.	2025	795.50		2820.50	
Clean	Ea.	.51	47		47.51	
Minimum Charge	Ea.		555		555	
Pre-fab (10' x 12')						
Demolish	Ea.		140		140	Includes material and labor to install a
Install	Ea.	2625	720		3345	pre-fabricated wood storage shed.
Demolish and Install	Ea.	2625	860		3485	
Clean	Ea.	.51	52		52.51	
Minimum Charge	Ea.		555		555	

Gazebo

Gazebo	Unit	Material	Labor	Equip.	Total	Specification
Good Grade						
Demolish	S.F.		4.16		4.16	Includes material and labor to install a
Install	S.F.	63.50	36.50		100	gazebo.
Demolish and Install	S.F.	63.50	40.66		104.16	
Reinstall	S.F.		29.04		29.04	
Clean	S.F.	.07	2.49		2.56	
Minimum Charge	Job		242		242	
Custom Grade						
Demolish	S.F.		4.16		4.16	Includes material and labor to install a
Install	S.F.	114	36.50		150.50	gazebo.
Demolish and Install	S.F.	114	40.66		154.66	
Reinstall	S.F.		29.04		29.04	
Clean	S.F.	.07	2.49		2.56	
Minimum Charge	Job		242		242	
Bench Seating						
Demolish	L.F.		6.95		6.95	Includes material and labor to install
Install	L.F.	3.58	24		27.58	pressure treated bench seating.
Demolish and Install	L.F.	3.58	30.95		34.53	
Reinstall	L.F.		19.36		19.36	
Clean	L.F.	.07	1.87		1.94	
Paint	L.F.	1.02	3.23		4.25	
Minimum Charge	Job		242		242	
Re-screen						
Install	S.F.	2.45	1.94		4.39	Includes labor and material to
Minimum Charge	Job		242		242	re-screen wood frame.

Pool

Pool	Unit	Material	Labor	Equip.	Total	Specification
Screen Enclosure						
Re-screen						
Install	SF Wall	.03	5.55		5.58	Includes material and labor to install a
Minimum Charge	Job		242		242	screen enclosure for a pool.
Screen Door						
Demolish	Ea.		9.60		9.60	Includes material and labor to install
Install	Ea.	320	80.50		400.50	wood screen door with aluminum cloth
Demolish and Install	Ea.	320	90.10		410.10	screen. Frame and hardware not
Reinstall	Ea.		64.53		64.53	included.
Clean	Ea.	.06	14.90		14.96	
Paint	Ea.	7.15	16.15		23.30	
Minimum Charge	Job		242		242	

Pool

	Unit	Material	Labor	Equip.	Total	Specification
Heater / Motor						
Demolish	Ea.		49.50		49.50	Includes material and labor to install
Install	Ea.	2725	635		3360	swimming pool heater, not incl. base
Demolish and Install	Ea.	2725	684.50		3409.50	or pad, elec., 12 kW, 4,800 gal
Reinstall	Ea.		509.72		509.72	pool, incl. pump.
Clean	Ea.	1.52	47		48.52	
Minimum Charge	Job		284		284	
Filter System						
Demolish	Ea.		83		83	Includes material, labor and
Install	Total	1400	435		1835	equipment to install sand filter system
Demolish and Install	Total	1400	518		1918	tank including hook-up to existing
Reinstall	Total		349.95		349.95	equipment. Pump and motor not
Minimum Charge	Job		273		273	included.
Pump / Motor						
Demolish	Ea.		49.50		49.50	Includes material and labor to install a
Install	Ea.	825	109		934	22 GPM pump.
Demolish and Install	Ea.	825	158.50		983.50	
Reinstall	Ea.		87.49		87.49	
Clean	Ea.	1.52	47		48.52	
Minimum Charge	Job		284		284	
Pump Water						
Install	Day		510	106	616	Includes labor and equipment to pump
Minimum Charge	Job		238		238	water from a swimming pool.
Replaster						
Install	SF Surf	.88	12.50		13.38	Includes labor and material to
Minimum Charge	Job		238		238	replaster gunite pool with a surface
						area of 500 to 600 S.F.
Solar Panels						
Demolish	S.F.		2.03		2.03	Includes material and labor to install
Install	S.F.	36.50	3.60		40.10	solar panel.
Demolish and Install	S.F.	36.50	5.63		42.13	
Reinstall	S.F.		2.88		2.88	
Minimum Charge	Ea.		555		555	

Construction Clean-up	Unit	Material	Labor	Equip.	Total	Specification
Final						
Clean	S.F.		.33		.33	Includes labor for final site/job
Minimum Charge	Job		187		187	clean-up including detail cleaning.

Costs shown in *Residential Costs with RSMeans data* are based on national averages for materials and installation. To adjust these costs to a specific location, simply multiply the base cost by the factor for that city. The data is arranged alphabetically by state and postal zip code numbers. For a city not listed, use the factor for a nearby city with similar economic characteristics.

STATE	CITY	Residential
ALABAMA		
350-352	Birmingham	.87
354	Tuscaloosa	.86
355	Jasper	.84
356	Decatur	.85
357-358	Huntsville	.86
359	Gadsden	.81
360-361	Montgomery	.85
362	Anniston	.80
363	Dothan	.86
364	Evergreen	.77
365-366	Mobile	.83
367	Selma	.80
368	Phenix City	.82
369	Butler	.79
ALASKA		
995-996	Anchorage	1.14
997	Fairbanks	1.19
998	Juneau	1.16
999	Ketchikan	1.19
ARIZONA		
850,853	Phoenix	.87
851,852	Mesa/Tempe	.84
855	Globe	.82
856-857	Tucson	.86
859	Show Low	.85
860	Flagstaff	.92
863	Prescott	.90
864	Kingman	.85
865	Chambers	.89
ARKANSAS		
716	Pine Bluff	.76
717	Camden	.77
718	Texarkana	.80
719	Hot Springs	.75
720-722	Little Rock	.84
723	West Memphis	.83
724	Jonesboro	.81
725	Batesville	.77
726	Harrison	.80
727	Fayetteville	.76
728	Russellville	.78
729	Fort Smith	.85
CALIFORNIA		
900-902	Los Angeles	1.14
903-905	Inglewood	1.13
906-908	Long Beach	1.11
910-912	Pasadena	1.09
913-916	Van Nuys	1.13
917-918	Alhambra	1.14
919-921	San Diego	1.09
922	Palm Springs	1.09
923-924	San Bernardino	1.12
925	Riverside	1.12
926-927	Santa Ana	1.13
928	Anaheim	1.11
930	Oxnard	1.12
931	Santa Barbara	1.12
932-933	Bakersfield	1.11
934	San Luis Obispo	1.15
935	Mojave	1.13
936-938	Fresno	1.17
939	Salinas	1.21
940-941	San Francisco	1.27
942,956-958	Sacramento	1.16
943	Palo Alto	1.23
944	San Mateo	1.27
945	Vallejo	1.20
946	Oakland	1.25
947	Berkeley	1.30
948	Richmond	1.29
949	San Rafael	1.28
950	Santa Cruz	1.22
951	San Jose	1.27
952	Stockton	1.18
953	Modesto	1.16

STATE	CITY	Residential
CALIFORNIA (CONT'D)		
954	Santa Rosa	1.23
955	Eureka	1.23
959	Marysville	1.18
960	Redding	1.17
961	Susanville	1.17
COLORADO		
800-802	Denver	.91
803	Boulder	.92
804	Golden	.87
805	Fort Collins	.89
806	Greeley	.87
807	Fort Morgan	.91
808-809	Colorado Springs	.85
810	Pueblo	.88
811	Alamosa	.86
812	Salida	.88
813	Durango	.91
814	Montrose	.84
815	Grand Junction	.94
816	Glenwood Springs	.85
CONNECTICUT		
060	New Britain	1.07
061	Hartford	1.06
062	Willimantic	1.07
063	New London	1.06
064	Meriden	1.07
065	New Haven	1.07
066	Bridgeport	1.08
067	Waterbury	1.07
068	Norwalk	1.07
069	Stamford	1.08
D.C.		
200-205	Washington	.94
DELAWARE		
197	Newark	1.02
198	Wilmington	1.02
199	Dover	1.02
FLORIDA		
320,322	Jacksonville	.83
321	Daytona Beach	.83
323	Tallahassee	.84
324	Panama City	.83
325	Pensacola	.83
326,344	Gainesville	.79
327-328,347	Orlando	.84
329	Melbourne	.82
330-332,340	Miami	.85
333	Fort Lauderdale	.81
334,349	West Palm Beach	.83
335-336,346	Tampa	.83
337	St. Petersburg	.82
338	Lakeland	.78
339,341	Fort Myers	.79
342	Sarasota	.84
GEORGIA		
300-303,399	Atlanta	.92
304	Statesboro	.81
305	Gainesville	.82
306	Athens	.80
307	Dalton	.82
308-309	Augusta	.88
310-312	Macon	.83
313-314	Savannah	.87
315	Waycross	.79
316	Valdosta	.80
317,398	Albany	.84
318-319	Columbus	.84
HAWAII		
967	Hilo	1.13
968	Honolulu	1.16

Location Factors - Residential

STATE	CITY	Residential	STATE	CITY	Residential
STATES & POSS.			**KENTUCKY (CONT'D)**		
969	Guam	.99	406	Frankfort	.89
			407-409	Corbin	.79
IDAHO			410	Covington	.81
832	Pocatello	.89	411-412	Ashland	.83
833	Twin Falls	.90	413-414	Campton	.83
834	Idaho Falls	.87	415-416	Pikeville	.86
835	Lewiston	.97	417-418	Hazard	.82
836-837	Boise	.89	420	Paducah	.85
838	Coeur d'Alene	.97	421-422	Bowling Green	.85
			423	Owensboro	.86
ILLINOIS			424	Henderson	.85
600-603	North Suburban	1.14	425-426	Somerset	.80
604	Joliet	1.18	427	Elizabethtown	.79
605	South Suburban	1.15			
606-608	Chicago	1.21	**LOUISIANA**		
609	Kankakee	1.06	700-701	New Orleans	.89
610-611	Rockford	1.10	703	Thibodaux	.81
612	Rock Island	.91	704	Hammond	.77
613	La Salle	1.07	705	Lafayette	.87
614	Galesburg	.97	706	Lake Charles	.88
615-616	Peoria	1.04	707-708	Baton Rouge	.86
617	Bloomington	.95	710-711	Shreveport	.85
618-619	Champaign	.98	712	Monroe	.78
620-622	East St. Louis	.99	713-714	Alexandria	.79
623	Quincy	1.00			
624	Effingham	1.00	**MAINE**		
625	Decatur	1.02	039	Kittery	.88
626-627	Springfield	1.02	040-041	Portland	.94
628	Centralia	.99	042	Lewiston	.93
629	Carbondale	.96	043	Augusta	.92
			044	Bangor	.90
INDIANA			045	Bath	.87
460	Anderson	.87	046	Machias	.90
461-462	Indianapolis	.91	047	Houlton	.91
463-464	Gary	.97	048	Rockland	.90
465-466	South Bend	.89	049	Waterville	.87
467-468	Fort Wayne	.88			
469	Kokomo	.89	**MARYLAND**		
470	Lawrenceburg	.86	206	Waldorf	.91
471	New Albany	.86	207-208	College Park	.85
472	Columbus	.91	209	Silver Spring	.88
473	Muncie	.90	210-212	Baltimore	.93
474	Bloomington	.95	214	Annapolis	.91
475	Washington	.91	215	Cumberland	.87
476-477	Evansville	.91	216	Easton	.84
478	Terre Haute	.89	217	Hagerstown	.89
479	Lafayette	.91	218	Salisbury	.86
			219	Elkton	.90
IOWA					
500-503,509	Des Moines	.92	**MASSACHUSETTS**		
504	Mason City	.80	010-011	Springfield	1.05
505	Fort Dodge	.81	012	Pittsfield	1.01
506-507	Waterloo	.87	013	Greenfield	1.01
508	Creston	.84	014	Fitchburg	1.02
510-511	Sioux City	.90	015-016	Worcester	1.08
512	Sibley	.74	017	Framingham	1.08
513	Spencer	.76	018	Lowell	1.10
514	Carroll	.85	019	Lawrence	1.10
515	Council Bluffs	.84	020-022, 024	Boston	1.15
516	Shenandoah	.83	023	Brockton	1.08
520	Dubuque	.85	025	Buzzards Bay	1.04
521	Decorah	.84	026	Hyannis	1.03
522-524	Cedar Rapids	.94	027	New Bedford	1.07
525	Ottumwa	.87			
526	Burlington	.91	**MICHIGAN**		
527-528	Davenport	.98	480,483	Royal Oak	.95
			481	Ann Arbor	.98
KANSAS			482	Detroit	1.01
660-662	Kansas City	.98	484-485	Flint	.94
664-666	Topeka	.87	486	Saginaw	.90
667	Fort Scott	.89	487	Bay City	.90
668	Emporia	.81	488-489	Lansing	.92
669	Belleville	.81	490	Battle Creek	.87
670-672	Wichita	.83	491	Kalamazoo	.88
673	Independence	.92	492	Jackson	.87
674	Salina	.81	493,495	Grand Rapids	.89
675	Hutchinson	.78	494	Muskegon	.87
676	Hays	.83	496	Traverse City	.81
677	Colby	.85	497	Gaylord	.85
678	Dodge City	.82	498-499	Iron Mountain	.84
679	Liberal	.80			
			MINNESOTA		
KENTUCKY			550-551	Saint Paul	1.07
400-402	Louisville	.89	553-555	Minneapolis	1.08
403-405	Lexington	.88	556-558	Duluth	1.02

For customer support on your Contractor's Pricing Guide: Residential Repair & Remodeling, call 800.448.8182.

Location Factors - Residential

STATE	CITY	Residential
MINNESOTA (CONT'D)		
559	Rochester	1.03
560	Mankato	.97
561	Windom	.87
562	Willmar	.89
563	St. Cloud	.99
564	Brainerd	.90
565	Detroit Lakes	.86
566	Bemidji	.88
567	Thief River Falls	.87
MISSISSIPPI		
386	Clarksdale	.73
387	Greenville	.79
388	Tupelo	.73
389	Greenwood	.79
390-392	Jackson	.86
393	Meridian	.79
394	Laurel	.77
395	Biloxi	.86
396	McComb	.73
397	Columbus	.74
MISSOURI		
630-631	St. Louis	1.01
633	Bowling Green	.94
634	Hannibal	.90
635	Kirksville	.85
636	Flat River	.94
637	Cape Girardeau	.88
638	Sikeston	.84
639	Poplar Bluff	.83
640-641	Kansas City	1.01
644-645	St. Joseph	.96
646	Chillicothe	.91
647	Harrisonville	.92
648	Joplin	.90
650-651	Jefferson City	.91
652	Columbia	.88
653	Sedalia	.85
654-655	Rolla	.92
656-658	Springfield	.87
MONTANA		
590-591	Billings	.92
592	Wolf Point	.86
593	Miles City	.89
594	Great Falls	.92
595	Havre	.81
596	Helena	.90
597	Butte	.84
598	Missoula	.85
599	Kalispell	.85
NEBRASKA		
680-681	Omaha	.90
683-685	Lincoln	.89
686	Columbus	.86
687	Norfolk	.85
688	Grand Island	.88
689	Hastings	.86
690	McCook	.80
691	North Platte	.85
692	Valentine	.79
693	Alliance	.80
NEVADA		
889-891	Las Vegas	1.03
893	Ely	.99
894-895	Reno	.91
897	Carson City	.92
898	Elko	.99
NEW HAMPSHIRE		
030	Nashua	.99
031	Manchester	.98
032-033	Concord	.98
034	Keene	.91
035	Littleton	.94
036	Charleston	.89
037	Claremont	.90
038	Portsmouth	.93

STATE	CITY	Residential
NEW JERSEY		
070-071	Newark	1.16
072	Elizabeth	1.20
073	Jersey City	1.14
074-075	Paterson	1.16
076	Hackensack	1.14
077	Long Branch	1.10
078	Dover	1.14
079	Summit	1.15
080,083	Vineland	1.10
081	Camden	1.17
082,084	Atlantic City	1.21
085-086	Trenton	1.17
087	Point Pleasant	1.12
088-089	New Brunswick	1.20
NEW MEXICO		
870-872	Albuquerque	.87
873	Gallup	.86
874	Farmington	.86
875	Santa Fe	.87
877	Las Vegas	.86
878	Socorro	.86
879	Truth/Consequences	.85
880	Las Cruces	.82
881	Clovis	.84
882	Roswell	.84
883	Carrizozo	.84
884	Tucumcari	.84
NEW YORK		
100-102	New York	1.30
103	Staten Island	1.22
104	Bronx	1.25
105	Mount Vernon	1.09
106	White Plains	1.16
107	Yonkers	1.17
108	New Rochelle	1.14
109	Suffern	1.08
110	Queens	1.25
111	Long Island City	1.31
112	Brooklyn	1.33
113	Flushing	1.28
114	Jamaica	1.28
115,117,118	Hicksville	1.17
116	Far Rockaway	1.27
119	Riverhead	1.19
120-122	Albany	1.02
123	Schenectady	1.03
124	Kingston	1.07
125-126	Poughkeepsie	1.13
127	Monticello	1.09
128	Glens Falls	.93
129	Plattsburgh	.97
130-132	Syracuse	.99
133-135	Utica	.98
136	Watertown	.92
137-139	Binghamton	.99
140-142	Buffalo	1.06
143	Niagara Falls	.96
144-146	Rochester	1.00
147	Jamestown	.89
148-149	Elmira	.92
NORTH CAROLINA		
270,272-274	Greensboro	.99
271	Winston-Salem	.99
275-276	Raleigh	.98
277	Durham	.99
278	Rocky Mount	.94
279	Elizabeth City	.92
280	Gastonia	1.00
281-282	Charlotte	1.00
283	Fayetteville	.95
284	Wilmington	.95
285	Kinston	.92
286	Hickory	.95
287-288	Asheville	.96
289	Murphy	.97
NORTH DAKOTA		
580-581	Fargo	.88
582	Grand Forks	.88
583	Devils Lake	.90
584	Jamestown	.86
585	Bismarck	.90

STATE	CITY	Residential
NORTH DAKOTA (CONT'D)		
586	Dickinson	.85
587	Minot	.84
588	Williston	.88
OHIO		
430-432	Columbus	.93
433	Marion	.89
434-436	Toledo	.96
437-438	Zanesville	.86
439	Steubenville	.89
440	Lorain	.94
441	Cleveland	.97
442-443	Akron	.96
444-445	Youngstown	.94
446-447	Canton	.94
448-449	Mansfield	.88
450	Hamilton	.89
451-452	Cincinnati	.90
453-454	Dayton	.91
455	Springfield	.91
456	Chillicothe	.91
457	Athens	.86
458	Lima	.86
OKLAHOMA		
730-731	Oklahoma City	.85
734	Ardmore	.81
735	Lawton	.84
736	Clinton	.81
737	Enid	.83
738	Woodward	.82
739	Guymon	.82
740-741	Tulsa	.84
743	Miami	.81
744	Muskogee	.84
745	McAlester	.77
746	Ponca City	.79
747	Durant	.77
748	Shawnee	.77
749	Poteau	.78
OREGON		
970-972	Portland	1.03
973	Salem	1.01
974	Eugene	1.03
975	Medford	.99
976	Klamath Falls	1.00
977	Bend	1.05
978	Pendleton	1.00
979	Vale	1.01
PENNSYLVANIA		
150-152	Pittsburgh	1.03
153	Washington	.94
154	Uniontown	.89
155	Bedford	.88
156	Greensburg	.92
157	Indiana	.91
158	Dubois	.88
159	Johnstown	.90
160	Butler	.89
161	New Castle	.90
162	Kittanning	.87
163	Oil City	.88
164-165	Erie	.95
166	Altoona	.88
167	Bradford	.90
168	State College	.91
169	Wellsboro	.87
170-171	Harrisburg	.98
172	Chambersburg	.85
173-174	York	.90
175-176	Lancaster	.93
177	Williamsport	.88
178	Sunbury	.87
179	Pottsville	.87
180	Lehigh Valley	.97
181	Allentown	1.03
182	Hazleton	.87
183	Stroudsburg	.90
184-185	Scranton	.97
186-187	Wilkes-Barre	.91
188	Montrose	.87
189	Doylestown	1.00

STATE	CITY	Residential
PENNSYLVANIA (CONT'D)		
190-191	Philadelphia	1.18
193	Westchester	1.03
194	Norristown	1.01
195-196	Reading	.98
PUERTO RICO		
009	San Juan	.71
RHODE ISLAND		
028	Newport	1.05
029	Providence	1.07
SOUTH CAROLINA		
290-292	Columbia	.98
293	Spartanburg	.96
294	Charleston	.98
295	Florence	.89
296	Greenville	.95
297	Rock Hill	.94
298	Aiken	.95
299	Beaufort	.97
SOUTH DAKOTA		
570-571	Sioux Falls	.92
572	Watertown	.82
573	Mitchell	.81
574	Aberdeen	.88
575	Pierre	.93
576	Mobridge	.77
577	Rapid City	.87
TENNESSEE		
370-372	Nashville	.90
373-374	Chattanooga	.86
375,380-381	Memphis	.87
376	Johnson City	.75
377-379	Knoxville	.83
382	McKenzie	.72
383	Jackson	.83
384	Columbia	.75
385	Cookeville	.69
TEXAS		
750	McKinney	.78
751	Waxahachie	.79
752-753	Dallas	.85
754	Greenville	.78
755	Texarkana	.81
756	Longview	.77
757	Tyler	.79
758	Palestine	.75
759	Lufkin	.77
760-761	Fort Worth	.84
762	Denton	.87
763	Wichita Falls	.86
764	Eastland	.84
765	Temple	.82
766-767	Waco	.86
768	Brownwood	.82
769	San Angelo	.80
770-772	Houston	.85
773	Huntsville	.78
774	Wharton	.77
775	Galveston	.80
776-777	Beaumont	.88
778	Bryan	.75
779	Victoria	.79
780	Laredo	.82
781-782	San Antonio	.83
783-784	Corpus Christi	.87
785	McAllen	.85
786-787	Austin	.82
788	Del Rio	.81
789	Giddings	.80
790-791	Amarillo	.83
792	Childress	.82
793-794	Lubbock	.83
795-796	Abilene	.84
797	Midland	.87
798-799,885	El Paso	.83
UTAH		
840-841	Salt Lake City	.85
842,844	Ogden	.84
843	Logan	.84

For customer support on your Contractor's Pricing Guide: Residential Repair & Remodeling, call 800.448.8182.

Location Factors - Residential

STATE	CITY	Residential
UTAH (CONT'D)		
845	Price	.83
846-847	Provo	.85
VERMONT		
050	White River Jct.	.91
051	Bellows Falls	.96
052	Bennington	.95
053	Brattleboro	.96
054	Burlington	.95
056	Montpelier	.97
057	Rutland	.93
058	St. Johnsbury	.93
059	Guildhall	.94
VIRGINIA		
220-221	Fairfax	.98
222	Arlington	1.01
223	Alexandria	1.06
224-225	Fredericksburg	.98
226	Winchester	.97
227	Culpeper	.99
228	Harrisonburg	.95
229	Charlottesville	.94
230-232	Richmond	.99
233-235	Norfolk	.97
236	Newport News	.97
237	Portsmouth	.92
238	Petersburg	.96
239	Farmville	.96
240-241	Roanoke	.99
242	Bristol	.90
243	Pulaski	.92
244	Staunton	.92
245	Lynchburg	.95
246	Grundy	.91
WASHINGTON		
980-981,987	Seattle	1.05
982	Everett	1.10
983-984	Tacoma	1.05
985	Olympia	1.02
986	Vancouver	1.02
988	Wenatchee	.97
989	Yakima	.99
990-992	Spokane	.96
993	Richland	.93
994	Clarkston	.88
WEST VIRGINIA		
247-248	Bluefield	.89
249	Lewisburg	.88
250-253	Charleston	.96
254	Martinsburg	.86
255-257	Huntington	.96
258-259	Beckley	.89
260	Wheeling	.90
261	Parkersburg	.91
262	Buckhannon	.90
263-264	Clarksburg	.89
265	Morgantown	.89
266	Gassaway	.89
267	Romney	.88
268	Petersburg	.90
WISCONSIN		
530,532	Milwaukee	1.04
531	Kenosha	1.05
534	Racine	1.02
535	Beloit	.95
537	Madison	.99
538	Lancaster	.94
539	Portage	.90
540	New Richmond	.92
541-543	Green Bay	1.07
544	Wausau	.93
545	Rhinelander	.91
546	La Crosse	.93
547	Eau Claire	.98
548	Superior	.91
549	Oshkosh	.93
WYOMING		
820	Cheyenne	.88
821	Yellowstone Nat. Pk.	.84
822	Wheatland	.84

STATE	CITY	Residential
WYOMING (CONT'D)		
823	Rawlins	.85
824	Worland	.85
825	Riverton	.84
826	Casper	.87
827	Newcastle	.84
828	Sheridan	.89
829-831	Rock Springs	.92
CANADIAN FACTORS (reflect Canadian currency)		
ALBERTA		
	Calgary	1.07
	Edmonton	1.07
	Fort McMurray	1.10
	Lethbridge	1.08
	Lloydminster	1.03
	Medicine Hat	1.05
	Red Deer	1.05
BRITISH COLUMBIA		
	Kamloops	.98
	Prince George	.98
	Vancouver	1.06
	Victoria	1.01
MANITOBA		
	Brandon	1.11
	Portage la Prairie	1.00
	Winnipeg	.98
NEW BRUNSWICK		
	Bathurst	.89
	Dalhousie	.90
	Fredericton	.97
	Moncton	.94
	Newcastle	.89
	St. John	1.03
NEWFOUNDLAND		
	Corner Brook	1.06
	St. Johns	1.05
NORTHWEST TERRITORIES		
	Yellowknife	1.14
NOVA SCOTIA		
	Bridgewater	.91
	Dartmouth	1.00
	Halifax	1.04
	New Glasgow	1.00
	Sydney	.99
	Truro	.91
	Yarmouth	1.00
ONTARIO		
	Barrie	1.13
	Brantford	1.11
	Cornwall	1.10
	Hamilton	1.10
	Kingston	1.10
	Kitchener	1.06
	London	1.11
	North Bay	1.20
	Oshawa	1.12
	Ottawa	1.09
	Owen Sound	1.12
	Peterborough	1.09
	Sarnia	1.11
	Sault Ste. Marie	1.04
	St. Catharines	1.06
	Sudbury	1.03
	Thunder Bay	1.10
	Timmins	1.08
	Toronto	1.13
	Windsor	1.10
PRINCE EDWARD ISLAND		
	Charlottetown	.92
	Summerside	.95
QUEBEC		
	Cap-de-la-Madeleine	1.11
	Charlesbourg	1.11
	Chicoutimi	1.15
	Gatineau	1.11
	Granby	1.11

Location Factors - Residential

STATE	CITY	Residential
QUEBEC (CONT'D)		
	Hull	1.11
	Joliette	1.11
	Laval	1.11
	Montreal	1.08
	Quebec City	1.09
	Rimouski	1.16
	Rouyn-Noranda	1.11
	Saint-Hyacinthe	1.11
	Sherbrooke	1.11
	Sorel	1.11
	Saint-Jerome	1.11
	Trois-Rivieres	1.24
SASKATCHEWAN		
	Moose Jaw	.90
	Prince Albert	.89
	Regina	1.08
	Saskatoon	1.01
YUKON		
	Whitehorse	1.05

For customer support on your Contractor's Pricing Guide: Residential Repair & Remodeling, call 800.448.8182.

Abbreviations

	Area
STM	American Society for Testing and Materials
F.	Board feet
arp.	Carpenter
F.	Cubic feet
WJ	Composite wood joist
Y.	Cubic yard
a.	Each
quip.	Equipment
xp.	Exposure
t.	Exterior
	Fahrenheit
	Foot, feet
al.	Gallon
.	Hour
	Inch, inches
st.	Installation
t.	Interior
.	Pound
F.	Linear feet
L	Laminated veneer lumber
at.	Material
ax.	Maximum
BF	Thousand board feet
BM	Thousand feet board measure
SF	Thousand square feet
n.	Minimum
C.	On center
&P	Overhead and profit
WJ	Open web wood joist
.	Ounce
	Pair
an.	Quantity
F.	Square foot
.	Square, 100 square feet
.	Square yard
L.F.	Vertical linear feet
	Foot, feet
	Inch, inches
	Degrees

Index

RSMeans data Training

Virtual classes offered monthly. Call for the schedule.   📞 877.620.6245

RSMeans offers training classes in a variety of formats—eLearning training modules, instructor-led classes, on-site training, and virtual training classes, as well as custom solutions. Due to our current COVID-19 situation, RSMeans has adjusted our current training offerings and instructor-led classes for 2022 have not been determined. Please visit our website at rsmeans.com/products/training/seminars for our current schedule of class offerings.

Training classes that are offered in one or more of our formats are: Building Systems and the Construction Process; Construction Cost Estimating: Concepts & Practice; Introduction to Estimating; Scope of Work for Facilities Estimating; Facilities Construction Estimating; Fundamentals to Construction Cost Estimating; Maintenance & Repair Estimating for Facilities; Mechanical & Electrical Estimating; RSMeans CostWorks CD; and RSMeans Online Training.

Registration Information

How to register
By Phone
Register by phone at 877.620.6245

Online
Register online at RSMeans.com/products/services/training

Note: Purchase orders or credits cards are required to register.

Instructor-led Government Pricing
All federal government employees save off the regular seminar price. Other promotional discounts cannot be combined with the government discount. Call 781.422.5115 for government pricing.

If you have any questions or you would like to register for any class or purchase a training module, call us at 877.620.6245.

Construction Cost Estimating: Concepts and Practice

This one- or two-day introductory course to improve estimating skills and effectiveness starts with the details of interpreting bid documents and ends with the summary of the estimate and bid submission.

Some of what you'll learn:
- Using the plans and specifications to create estimates
- The takeoff process—deriving all tasks with correct quantities
- Summarizing the estimate to arrive at the final number
- Formulas for area and cubic measure, adding waste and adjusting productivity to specific projects
- Evaluating subcontractors' proposals and prices
- Adding insurance and bonds
- Understanding how labor costs are calculated
- Submitting bids and proposals

Who should attend: project managers, architects, engineers, owners' representatives, contractors, and anyone who's responsible for budgeting or estimating construction projects.

Fundamentals to Construction Cost Estimating

This self-paced course focuses on unit price estimating for facilities renovation, repair, and remodeling. Combining hand on skill building with a discussion of best estimating practice and real-life problems, this informative online course will bo your estimating skills, bring you up to date with key concept and provide tips and guidelines that can save you time and h you avoid costly estimating oversights and errors.

Some of what you'll learn:
- The three phases of estimating
- Bid document review
- How to perform quantity takeoffs
- How to price the work

Who should attend: facilities professionals and anyone involv in facilities planning, design, and construction. For novice estimators, this course can serve as a primer to Facilities Construction Estimating Self-Paced Training or the following live training courses: Facilities Construction Estimating, Mechanical & Electrical Estimating, Assessing Scope of Work Facilities Estimating, or Maintenance & Repair Estimating.

sit

echanical & Electrical Estimating

 two-day course teaches attendees how to prepare more
urate and complete mechanical/electrical estimates, avoid
pitfalls of omission and double-counting, and understand
composition and rationale within the RSMeans mechanical/
ctrical database.

e of what you'll learn:
he unique way mechanical and electrical systems are
terrelated
&E estimates—conceptual, planning, budgeting, and
dding stages
rder of magnitude, square foot, assemblies, and unit
rice estimating
omparative cost analysis of equipment and design
ternatives

 should attend: architects, engineers, facilities
nagers, mechanical and electrical contractors, and
ers who need a highly reliable method for developing,
erstanding, and evaluating mechanical and electrical
tracts.

Facilities Construction Estimating

In this two-day course, professionals working in facilities
management can get help with their daily challenges
to establish budgets for all phases of a project.

Some of what you'll learn:
- Determining the full scope of a project
- Understanding of RSMeans data and what is included
  in prices
- Identifying appropriate factors to be included in
  your estimate
- Creative solutions to estimating issues
- Organizing estimates for presentation and discussion
- Special estimating techniques for repair/remodel and
  maintenance projects
- Appropriate use of contingency, city cost indexes, and
  reference notes
- Techniques to get to the correct estimate quickly

Who should attend: facility managers, engineers, contractors,
facility tradespeople, planners, and project managers.

essing Scope of Work for Facilities
nstruction Estimating

 two-day practical training program addresses the vital
ortance of understanding the scope of projects in order
roduce accurate cost estimates for facilities repair and
odeling.

e of what you'll learn:
iscussions of site visits, plans/specs, record drawings
 facilities, and site-specific lists
eview of CSI divisions, including means, methods,
aterials, and the challenges of scoping each topic
xercises in scope identification and scope writing for
curate estimating of projects
ands-on exercises that require scope, takeoff,
d pricing

 should attend: corporate and government estimators,
nners, facility managers, and others who need to produce
urate project estimates.

CostWorks CD Self-Paced Training
Course

This self-paced course helps users become more familiar with the
functionality of the CD. Each menu, icon, screen, and function
found in the program is explained in depth. Time is devoted to
hands-on estimating exercises.

Some of what you'll learn:
- Searching the database using all navigation methods
- Exporting RSMeans data to your preferred spreadsheet
  format
- Viewing crews, assembly components, and much more
- Automatically regionalizing the database

This training session requires users to bring a laptop
computer to class and to follow along using a purchased
copy of RSMeans CostWorks CD.

When you register for this course you will receive an outline for
your laptop requirements.

aintenance & Repair Estimating
r Facilities

 two-day course teaches attendees how to plan, budget,
 estimate the cost of ongoing and preventive maintenance
 repair for existing buildings and grounds.

e of what you'll learn:
he most financially favorable maintenance, repair,
d replacement scheduling and estimating
reventive planning and facilities upgrading
etermining both in-house and contract-out service costs
nnual, asset-protecting M&R plan

 should attend: facility managers, maintenance
ervisors, buildings and grounds superintendents, plant
nagers, planners, estimators, and others involved in
lities planning and budgeting.

RSMeans Online Training

Construction estimating is vital to the decision-making process
at each state of every project. Our online solution works the
way you do. It's systematic, flexible, and intuitive. In this self-
paced course you will see how you can estimate any phase of
any project faster and better.

Some of what you'll learn:
- Customizing our online estimating solution
- Making the most of RSMeans "Circle Reference" numbers
- How to integrate your cost data
- Generating reports, exporting estimates to MS Excel,
  sharing, collaborating, and more

Also offered as a self-paced or on-site training program!

RSMeans data Training

319

RSMeans Training Cancellation Policy

Virtual Training (Individual)

If you are unable to attend a training class, a colleague may substitute at any time before the session starts by notifying the training registrar at 1.781.422.5115 or your sales representative. Gordian is not responsible for sending duplicate materials to the substitute colleague.

If you are unable to find a substitute and cancel fewer than ten (10) business days prior to the training class, you will either forfeit the registration fee or be allow to apply your registration fee to the same class only once within the same calendar year. No-shows will forfeit their registration fees and no rescheduling is permi In the unfortunate circumstance that Gordian must cancel your event, Gordian will work with you to reschedule your attendance in the same training class at a lat date or will fully refund your registration fee.

Any on-demand (self-paced) training modules are not eligible for cancellation, substitution, transfer, return, or refund.

Virtual Training (Private Group)

If you provide notice more than ten (10) business days prior to the training class, Gordian will work to reschedule your event. If you cancel your event fewer than (10) business days prior to the training class, you are responsible for paying for the trainers' time for the cancelled training, as well as the full cost of materials an shipping. In the unfortunate circumstance that Gordian must cancel your event, Gordian will work with you to reschedule your attendance in the same training cla a later date or will fully refund your registration fee.

Public Training Events (Individual)

If you are unable to attend a training conference, a colleague may substitute no fewer than ten (10) business days before the session starts by notifying the traini registrar at 1.781.422.5115 or your sales representative. Gordian is not responsible for sending duplicate materials to the substitute colleague.

If you are unable to find a substitute and cancel fewer than ten (10) business days prior to the conference, you will forfeit the registration fee. No-shows will forfeit their registration fees and no rescheduling is permitted. In the unfortunate circumstance that Gordian must cancel your event, Gordian will fully refund you registration fee.

On-Site Training (Private Group)

If you provide notice more than ten (10) business days prior to the training class, Gordian will work with you to reschedule your event, but you are responsible for rescheduling fees associated with the trainer's time, airfare, and lodging. If you cancel fewer than ten (10) business days before your event, you are responsible fo the full cost of the event. In the unfortunate event that Gordian must cancel your event, Gordian will work with you to reschedule your attendance at a later date will fully refund your registration fee.

AACE Approved Courses

The AACE recognizes .1 CEUs per hour of training attended for all of Gordian's virtual training courses.

AIA Continuing Education

We are registered with the AIA Continuing Education System (AIA/CES) and are committed to developing quality learning activities in accordance with the CES cri Many seminars meet the AIA/CES criteria for Quality Level 2. AIA members may receive 14 learning units (LUs) for each two-day RSMeans course.

Daily Course Schedule

The first day of each seminar session begins at 8:30 a.m. and ends at 4:30 p.m. The second day begins at 8:00 a.m. and ends at 4:00 p.m. Participants are urged bring a hand-held calculator since many actual problems will be worked out in each session.

Continental Breakfast

Your registration includes the cost of a continental breakfast and a morning and afternoon refreshment break. These informal segments allow you to discuss topi mutual interest with other seminar attendees. You are free to make your own lunch and dinner arrangements.

Hotel/Transportation Arrangements

We arrange to hold a block of rooms at most host hotels. To take advantage of special group rates when making your reservation, be sure to mention that you are attending the RSMeans Institute data seminar. You are, of course, free to stay at the lodging place of your choice. Hotel reservations and transportation arrangem should be made directly by seminar attendees.

Important

Class sizes are limited, so please register as soon as possible.